Chiral Separations

Chiral Separations
Applications and Technology

Satinder Ahuja, Editor
Ahuja Consulting

American Chemical Society, Washington, DC

Library of Congress Cataloging-in-Publication Data

Chiral separations: applications and technology / Satinder Ahuja, editor

    p. Cm.— (ACS professional reference books)

    Includes bibliographical references and index.

    ISBN 0–8412–3407–8 (alk. paper)

    1. Separation (Technology). 2. Chirality—Industrial applications.

    I. Ahuja, Satinder, 1933– . II. Series: ACS professional reference book.

TP156.S45C495
660′.2842—dc20   96–34589
                                       CIP

The paper used in this publication meets the minimum requirements of American National Standard for Information Sciences—Permanence of Paper for Printed Library Materials, ANSI Z39.48–1984.   ∞

PRINTED IN THE UNITED STATES OF AMERICA

Advisory Board

About the Editor

SATINDER AHUJA is President and Chief Consultant of Ahuja & Co., Monsey, New York. Dr. Ahuja concluded a very successful career in the pharmaceutical industry, including development of new drugs and delivery systems. He is a recognized expert in the area of chromatography, problem solving, and good manufacturing practices/good laboratory practices related issues. His assistance in these areas is being sought nationally and internationally. He is a member of the American Chemical Society and the American Association of Pharmaceutical Scientists and is the editor or author of nine books, including *Chromatography of Pharmaceuticals: Natural, Synthetic and Recombinant Products; Trace and Ultratrace Analysis by HPLC; Selectivity and Detectability Optimization in HPLC;* and *Chiral Separations by Liquid Chromatography.* Dr. Ahuja earned his Ph.D. in pharmaceutical analytical chemistry at the Philadelphia College of Pharmacy and Science and helps develop budding scientists by teaching at Pace University. He has won a number of awards and has served on distinguished panels for the National Science Foundation and the United Nations.

Table of Contents

Contributors

Satinder Ahuja   1 and 139
Ahuja Consulting
27 Monsey Heights
Monsey, NY 10952

Daniel W. Armstrong   201
Department of Chemistry
University of Missouri
Rolla, MO 65401

L. Jonathan Brice   309
School of Chemical Sciences
University of Illinois
Urbana, IL 61801

Charlotte A. Brunner   155
Food and Drug Administration
Center for Drug Evaluation and
    Research
HFD–473
200 C Street, S.W.
Washington, DC 20204

Hans-Peter Buser   93
R&D Crop Protection Division and
    Pharmaceuticals Division,
    R-1047.2.12
Ciba-Geigy, Ltd.
CH–4002 Basel, Switzerland

David L. Coffen   59
ArQule Inc.
200 Boston Avenue
Medford, MA 02155

E. J. Corey   37
Department of Chemistry
Harvard University
Cambridge, MA 02138

Thomas D. Doyle   155
Food and Drug Administration
Center for Drug Evaluation
    and Research
HFD–473
200 C Street, S.W.
Washington, DC 20204

K. Helen Ekberg-Ott   201
Department of Chemistry
University of Missouri
Rolla, MO 65401

Eric R. Francotte   93 and 271
Pharmaceuticals Division
    Research and Development,
    K-122.P.25
Ciba-Geigy Ltd.
CH–4002 Basel, Switzerland

Kathleen D. Giacomini   173
Department of Pharmacy,
    926–S
University of California
San Francisco, CA 94143

Robert L. Hunt   155
Food and Drug Administration
Center for Drug Evaluation
    and Research
HFD–473
200 C Street, S.W.
Washington, DC 20204

Ronda J. Ott   173
Division of Pharmacokinetics
    and Drug Metabolism
Burroughs Wellcome Company
3030 Cornwallis
Research Triangle Park,
NC 27709

William H. Pirkle   309
School of Chemical Sciences
University of Illinois
Urbana, IL 61801

Paul van Eikeren   9
Vice President, Chemistry
Argonaut Technologies, Inc.
887 Industrial Road, Suite G
San Carlos, CA 94070

Preface

The importance of chiral separations in a number of fields cannot be overemphasized. Chirality can confer entirely different biological properties to molecules that are alike in every way except that their mirror image is not superimposable. To better understand chirality, it is a good idea to take a look at your hands: you will notice that the right hand and the left hand are not superimposable. As a matter of fact, the word chiral is derived from the Greek word *cheiro*, meaning hand. Chiral molecules are also called enantiomers.

A new field, chiral technology, is emerging to deal with the technological issues of chirality. It is likely to revolutionize our thinking, because new viable chiral molecules will have to be developed by means other than traditional synthetic routes. This revolution will be further energized by the facts that the world market for enantiomeric drugs exceeds 45 billion dollars a year and yet the differences in the physiologic properties of the racemic drugs (mixtures of enantiomers) have not been fully examined in many cases. This is primarily due to the difficulty of obtaining both enantiomers in optically pure forms. To ensure the safety and effectiveness of currently used and newly developing drugs, it is important to prepare or isolate these enantiomers so that they can be examined separately. A number of preparation and isolation methods are discussed in this book.

The symmetric factor classifies molecules with different spatial arrangements as either enantiomers or diastereomers. A molecule can have only one enantiomer; however, it can have several diastereomers or none. Two stereoisomers cannot be both enantiomers and diastereomers of each other simultaneously. Diastereomers differ in energy content and therefore exhibit different physical and chemical properties. Enantiomers, on the other hand, have identical physical properties and consequently are very difficult to separate and quantitate. Chromatographic methods offer distinct advantages over other methods and are therefore more commonly used. However, it is always necessary to use some kind of chiral discriminator or selector. Two different types of selectors can be used: a chiral additive in the mobile phase or a chiral stationary phase. Chiral separation methods are extensively discussed in this book, and numerous examples from fields such as pharmaceuticals, agriculture, and food have been included (Chapters 5–9).

Chemical synthesis still remains a favorite route of preparation of enantiomers, especially when a large amount of a given enantiomer is required. E. J. Corey, a Nobel laureate, discusses catalytic enantioselective Diels–Alder reactions and enantioselective additions of carbon to carbonyl groups in Chapter 3. Many other examples are discussed in Chapter 2.

There are two basic approaches to obtaining chiral compounds: asymmetric synthesis and resolution. Asymmetric synthesis may require auxiliary chiral synthesis or asymmetric catalysis, and resolution involves separation methods (discussed extensively in this book) or enantioselective catalysis, which may be direct or subtractive. Asymmetric synthesis should be , in principle, the most cost-effective method of producing single-enantiomer products, because all the precursors are converted to the desired enantiomer. However, the decision to implement an asymmetric synthesis approach should be based on an assessment of efficiency and cost. The factors to be considered are discussed in Chapter 2. The manufacture of a number of enantiomeric compounds and case studies of diastereomeric resolution are discussed in great detail.

Enzyme-catalyzed reactions are ideally suited to production of single enantiomers because the interaction of proteins with small chiral molecules, in general, is highly stereospecific and shows clear preference for single enantiomers (Chapter 4). Various classes of enzymes such as oxidoreductases, transferases, hydrolases, lases, isomerases, and ligases can be used. Hydrolases are by far the most important class of enzymes suitable for preparation of chiral compounds, and their applications have been developed extensively during the past two decades. A number of examples are described. Additional examples include preparation of chiral alcohols, chiral acids and esters, amino acids, prochiral diols, and prochiral diesters.

It can be argued that separation methods can compete with synthetic methods for production of individual enantiomers of interest. This is possible because a number of preparative separation methods are available (Chapter 10). These methods should complement the synthetic methods and enzyme catalysis methods described earlier.

The book details the general strategy, available chiral stationary phases, criteria for selecting the chiral stationary phase, choice of chromatograpic methods, and improvement of separations. Many applications are described covering subjects such as productivity in simulated moving bed mode, separations by supercritical fluid chromatography, and resolution of a large number of racemic compounds with a variety of chiral stationary phases.

The usefulness of membrane systems for enantioselective transport of chiral molecules is covered in Chapter 11. Basic treatments of membrane theory, morphology, and transport considerations are included to better acquaint the reader with this mode of separation. Discussed in some detail are the types of enantioselective transport systems, grouped by structural class of the organic carrier molecule. Examples of compounds separable by this technique, including details of separation conditions, are provided for norephedrine, ephedrine, metazapine, phenylglycine, salbutamol, terbutaline, propranolol, and ibuprofen. Membrane separations may well provide the basis of chiral technology, which will be hard to match in terms of cost and convenience.

Satinder Ahuja
Ahuja Consulting
Monsey, NY 10952

1

Chiral Separations and Technology: An Overview

Satinder Ahuja

The word *chiral* was derived from the Greek word *cheir,* which means hand. Molecules that relate to each other like a pair of hands are called chiral molecules. Molecules that are mirror images of each other are also called *enantiomers.* These molecules generally have a tetrahedral carbon atom with different substituents. A molecule need not have a tetrahedral carbon atom in order to exist in enantiomeric forms, but it is necessary that the molecule as a whole be chiral. The symmetric factor classifies molecules with different spatial arrangement of atoms as either enantiomers or diastereoisomers. Diastereoisomers or diastereomers are basically stereoisomers with two or more centers of asymmetry that are not enantiomers of each other. A molecule can have only one enantiomer; however, it can have several diastereomers or none. Two stereoisomers cannot be both enantiomers and diastereomers of each other simultaneously. Stereoisomers can occur when a molecule has one or more centers of chirality, helicity, planar chirality, axial chirality, torsional chirality, or topological asymmetry (*1*).

Stereochemistry and Biological Activity

The need to determine the stereochemical composition of chemical compounds, especially those of pharmaceutical importance, is now well recognized (*2*). Dextromethorphan provides a dramatic example in that it is an over-the-counter cough suppressant, whereas levomethorphan, its stereoisomer, is a controlled

3407–8/96/0001$15.00/0

narcotic. Similarly, it has been reported that the teratogenic activity of thalidomide may reside exclusively in the (S)-enantiomer (3), though this conclusion remains controversial. Other examples relating to a variety of compounds with different pharmacologic or toxicologic activities can be found in later chapters. All of this background emphasizes the need for careful investigation because the structures of 12 of the 20 most prescribed drugs in the United States and 114 of the top 200 contain one or more asymmetric centers (4). About half of over two thousand drugs listed in the *USP Dictionary of USAN and International Drug Names* contain at least one asymmetric center, and 25% have been used in racemic or diastereomeric forms (5).

Chirality Selection

The drugs that are the best candidates for selection of appropriate enantiomers are listed in Table I (6). This list represents a tremendous marketing potential for

Table I. Candidates for Racemic Switches

Cardiovascular		*Anti-inflammatory and analgesics*	
Acebutolol	Alprenolol	Cicloprofen	Corticosteroids
Atenolol	Betaxolol	Dihydroxythebaine	Fenbufen
Bisoprolol	Bopindolol	Fenoprofen	Flurbiprofen
Bucumolol	Bufetolol	Ibuprofen	Indoprofen
Bufuralol	Bunitrolol	Ketoprofen	Minoxiprofen
Bupranolol	Butofilolol	Pirprofen	Suprofen
Carazolol	Carvedilol	Triamcinolone	
Disopyramide	Dobutamine		
Indenolol	Mepindolol	*Anticancer*	
Metipranolol	Metoprolol	Cytarabine	
Nadolol	Nicardipine		
Oxprenolol	Pindolol	*Antibiotics, anti-infectives,*	
Propranolol	Sotalol	*and antivirals*	
Toliprolol	Verapamil	Ciprofloxacin	Norfloxacin
Xibenolol		Ofloxacin	
Central nervous system		*Hormones and genitourinary*	
Dobutamine	Fluoxetine	Benzyl glutamate	Butoconazole
Ketamine	Lorazepam	Calcitonin	Estradiol
Meclizine	Oxaprotiline	Fluorogestone	Gonadorelin
Phenylpropanolamine	Thioridazine	Norgestrel	Testosterone
Polychloramphetamine	Tomoxetine		
Toloxatone	Viloxazine		
		Antihistamine and cough–cold	
Respiratory		Astemizole	Terfenadine
Albuterol	Metaproterenol		
Terbutaline			

SOURCE: Technology Catalysts International.

the desirable enantiomers. The world market for enantiomeric drugs exceeds $45 billion a year (Table II) and is expected to keep growing at a good pace well into the next century.

Table II. Sales of Enantiopure Drugs

Name	Sale (billion $)
Cardiovascular	12.5
Antibiotics	12.5
Hormones	6.5
Central nervous system	3.0
Anti-inflammatory	1.5
Anticancer	1.5
Other	7.5
Total	45.5

NOTE: 1994 figures in billions of dollars.
SOURCE: Data taken from Reference 6.

The differences in the physiologic properties between enantiomers of racemic drugs have not been fully examined in many cases, mainly because of difficulties in obtaining both enantiomers in optically pure forms. It is entirely possible that a patient may be taking a useless, or even undesirable, drug when ingesting a racemic mixture, since some enantiomers exhibit different pharmacologic activities. To ensure the safety and effectiveness of currently used and newly developing drugs, it is important to isolate and examine the enantiomers separately.

Accurate assessment of the isomeric purity of substances is critical, since isomeric impurities may have unwanted toxicologic, pharmacologic, or other effects. Such impurities may be carried through synthesis, preferentially react at one or more steps, and yield an undesirable level of another impurity. Some examples of activity differences between enantiomers are given in Table III.

Chiral Synthesis

Chemical synthesis still remains a favorite route of preparation of enantiomers, especially when a large amount of a given enantiomer is required. Professor E. J. Corey, a Nobel laureate, discusses catalytic enantioselective Diels–Alder reactions and enantioselective additions of carbon to carbonyl groups (Chapter 3). Many other examples are also discussed.

There are two basic approaches to obtaining chiral compounds: asymmetric synthesis and resolution (Chapter 2). Asymmetric synthesis may require auxiliary chiral synthesis or asymmetric catalysis, and resolution involves separation methods (discussed extensively in this book) or enantioselective catalysis, which may

Table III. Activities of Some Isomeric Compounds

Name	Activity
Drugs	
Amphetamine	*d*-Isomer is potent central nervous system stimulant, while *l*-isomer has little, if any, effect.
Epinephrine	*l*-Isomer is 10 times more active as a vasoconstrictor than *d*-isomer.
Propranolol	Racemic compound is used as drug; however, only (*S*)-(−)-isomer has the desired adrenergic blocking activity.
Food	
Asparagine	*D*-Asparagine tastes sweet, while *L*-enantiomer tastes bitter.
Limonene	(*S*)-Limonene smells like lemons, while (*R*)-limonene smells like oranges.
Carvone	(*S*)-(+)-Carvone smells like caraway, while (*R*)-(−)-carvone smells like spearmint.
Vitamin	
Ascorbic acid	(+)-Isomer is good antiscorbutic, while (−)-isomer has no such properties.
Insecticide	
Bermethrine	*d*-Isomer is much more toxic than *l*-isomer.

be direct or subtractive. Asymmetric synthesis should be, in principle, the most cost-effective method for producing single-enantiomer products, because all the precursors are converted to the desired enantiomer. However, the decision to implement an asymmetric synthesis approach should be based on an assessment of efficiency and cost. The factors to consider include (1) the catalyst efficiency (that is, the number of product molecules produced per molecule of the catalyst); (2) the availability of the metal, the ligand, and the starting materials (especially critical for low-value products); (3) reaction conditions, such as very low temperatures or high pressures, and reaction kinetics. Two examples, manufacture of (*S*)-naproxen and (*S*)-ibuprofen, are detailed in this book.

 Case studies of the diastereomeric resolution of lactam antibiotics and the enzymatic resolution of diltiazem are also discussed in detail.

Chromatographic Methods

Diastereomers differ in energy content and therefore have different physical and chemical properties. Enantiomers, by contrast, have identical physical properties and consequently are difficult to separate and quantitate. Chromatographic methods such as thin-layer chromatography, gas–liquid chromatography, and high-pressure liquid chromatography (HPLC) offer distinct advantages over classical techniques in the separation and analysis of stereoisomers, especially enantiomers. These methods show promise for moderate-scale separations of synthetic intermediates as well as for final products. Chiral HPLC is frequently used for large-scale preparation of optical isomers. For large-scale separations and in consideration of the cost of plant-scale resolution processes, the sorption methods offer a substantial increase in efficiency over recrystallization. Large-scale preparative liquid chromatographic systems are already on the market as process units for isolating and purifying chemicals and natural products. A new technology is emerging to address the needs of the industry (discussed later in this chapter).

Various methods can be used for chromatographic separation of enantiomers, but some kind of chiral discriminator or selector is always necessary. Two types of selectors can be distinguished: a chiral additive in the mobile phase or a chiral stationary phase. Another possibility is precolumn derivatization of the sample with a chiral reagent to produce diastereomeric molecules that can be separated by achiral chromatographic methods. Chiral separation methods including validation are (Chapters 6 and 7) extensively discussed in this book, and numerous examples from a variety of fields are included.

Enzyme-Catalyzed Reactions

Enzyme-catalyzed reactions are ideally suited for production of single enantiomers because the interaction of proteins with small chiral molecules, in general, is highly stereospecific and shows clear preference for single enantiomers (Chapter 4). Various classes of enzymes such as oxidoreductases, transferases, hydrolases, lyases, isomerases, and ligases can be used. Hydrolases are by far the most important class of enzymes suitable for preparation of chiral compounds, and their applications have been developed extensively during the past two decades. A number of examples may be found in this book. Preparation of chiral alcohols, chiral acids and esters, amino acids, prochiral diols, and prochiral diesters is also described.

Agricultural Applications

The biological activity of crop protection agents containing one or more stereogenic centers is often strongly related to the absolute configuration of the molecule. Chapter 5 deals with gas chromatographic and liquid chromatographic (HPLC) methods for the analysis of economically significant or soon-to-be-marketed chiral crop protection agents and also with selected experimental compounds currently under development. Compounds are covered under three major headings: weed control, disease control, and insect control.

Classes of compounds discussed include 2-aryloxy propionates, arylalanines, pyrimidines, organophosphates, pyrethroids, benzoyl ureas, and pheromones.

Food and Beverages

Many of the organic components of foods and beverages are chiral molecules, some of which are known to contribute greatly to characteristic tastes and textures. Chiral evaluation and discrimination is also important in odor perception. Therefore enantiomeric separations are useful for evaluating flavor and fragrance components. Enantiomeric purity can be used to evaluate storage, shipping, handling, age, and adulteration of various foods, beverages, and other consumer products (Chapter 9). Discussed at length are amino acids, alcohols, acids, esters, sugars, lactones, aldehydes, ketones, monoterpenes, and vitamins.

Chiral Technology

It can be argued that separation methods can compete with synthetic methods for production of individual enantiomers of interest and are leading to the development of the new chiral technology. This is possible because a number of preparative separation methods have been developed. Undoubtedly these methods will complement synthetic methods and enzyme catalysis methods described earlier. Chapter 10 details the general strategy, available chiral stationary phases, criteria for selecting the chiral stationary phase, choice of chromatographic methods, and improvement of separations. Many applications are described, covering subjects such as productivity in simulated moving bed mode, supercritical fluid chromatographic separations, and resolution of a large number of racemic compounds with a variety of chiral stationary phases.

Usefulness of membrane systems for enantioselective transport of chiral molecules is covered in Chapter 11. Basic treatments of membrane theory, morphology, and transport considerations are included to better acquaint the reader with this mode of separation. Discussed in some detail are different types of enantioselective transport systems, grouped by structural class of the organic carrier molecule. Details of separation conditions have been provided for phenylpropanolamine, ephedrine, metazapine, phenylglycine, albuterol, terbutaline, propranolol, and ibuprofen.

Immobilized enzymes in hollow-fiber membrane reactors may be very useful. In Figure 1, preparation of an L-acid is demonstrated. These hollow-fiber membrane laboratory scale modules have a capacity of resolving 165 kg per year for a compound with a molecular weight of 150. With 20 full-scale modules, the capacity would be 100,000 kg per year. Membrane separations may well provide the basis for chiral technology, which will be hard to match in terms of cost and convenience.

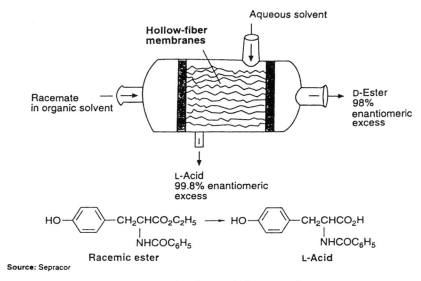

Figure 1. Preparation of an L-acid from a racemic ester.

Regulatory Requirements

Regulatory requirements demand that chiral separations be conducted to study purity and pharmacologic and toxicologic activities of each enantiomer separately. The impact of these regulations is extensively discussed in later chapters. Metabolism of chiral compounds can contribute to their toxicity. In this regard, stereoselective renal elimination of drugs is very important and is discussed at length in this book. Clinical examples are also given, where possible, to illustrate the principles.

References

1. *Chiral Separations by Liquid Chromatography*; Ahuja, S., Ed.; ACS Symposium Series 471; American Chemical Society: Washington, DC, 1991.
2. Ahuja, S. Presented at the 1st International Symposium on Chiral Molecules, Paris, 1988.
3. Blaschke, G.; Kraft, H. P.; Fickentscher, K.; Koehler, F. *Arzneim-Forsch.* **1979**, *29*, 1640.
4. "Top 200 Drugs in 1982", *Pharm. Times*, **1982**, *April*, 25.
5. *USP Dictionary of USAN and International Drug Names*; U.S. Pharmacopeial Convention: Rockville, MD, 1995.
6. *Chem. Eng. News*, **1992**, *Sept.*, **1994**, *Sept.*

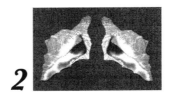

2

Commercial Manufacture of Chiral Pharmaceuticals

Paul van Eikeren

Use of chirality in the structure of pharmaceutical products promises higher drug specificity and improved efficacy. There is an increasing trend toward the use of chirality in drug design and toward the development, marketing, and sale of drugs as single isomers. Given that the cost of bulk active is a significant contributor to the cost of manufacturing a pharmaceutical product, pharmacological and specialty chemical companies have directed efforts toward development of scaleable technologies for the manufacture of single-isomer chiral pharmaceuticals. This chapter uses a case studies approach to examine the technical and economic issues associated with the commercial manufacture of chiral pharmaceuticals.

New Market Forces Affect Profitability of Industry

The worldwide pharmaceutical industry is in the midst of major structural changes. In the 1970s and 1980s the industry introduced several classes of new blockbuster drugs, including nonsteroidal anti-inflammatory drugs, H_2-antagonists, β-blockers, angiotensin-converting enzyme inhibitors, and lipid-lowering drugs. The industry blossomed with the development of large sales forces and the absence of pricing pressures. Sales for the industry increased 10% annually in the 1980s, while operating profit margins increased from 17% in 1980 to 26% in 1990. The early 1980s were the golden era for the industry: Drugs corre-

3407–8/96/0009$16.50/0

sponding to 26% of the current world pharmaceutical market were launched between 1981 and 1986 (*1*).

As the 1990s began, it became evident that new market forces were affecting the industry (Figure 1). Privately managed health care became the dominant

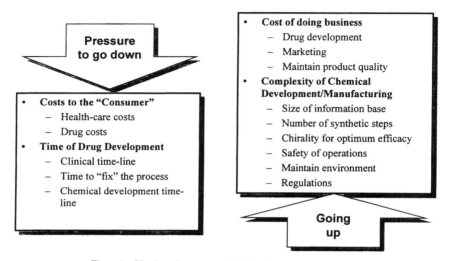

Figure 1. Trends and countertrends in the pharmaceutical industry.

force controlling costs in the United States. In most foreign markets, government-controlled managed health care became increasingly dominant. This trend is driven by insurance groups and corporate and government interest lowering health care costs. These forces have shifted the balance of power from drug manufacturers to drug purchasers and payers with the commensurate pressure on cost control. A principal target for cost control has been the selling price of pharmaceutical products, because they are a clearly identifiable and well-defined component of health care costs.

Because bringing a new pharmaceutical to market requires an investment of about $230 million, pharmaceutical companies are carefully reviewing their strategies for drug development. It requires on average 10 to 12 years to develop a new drug, and patent protection gives the holder a monopoly for only 17 years from the time the patent is issued. Accordingly, pharmaceutical companies must recover their investment in the short span of 5 to 7 years between product launch and patent expiration. To justify taking these large risks, pharmaceutical companies are striving to increase the life span of their product investment by decreasing the time for drug development (aiming for a target of 2500 days). This has important implications for chemical development efforts, including (1) desire for lower risk, proven technologies for manufacture, and (2) earlier development of and commitment to the manufacturing route in anticipation of successful clinical development and to comply with the requirements of the regulatory agencies.

While there are clear demands for lower costs and a shortened development time, the industry is also in the midst of a countertrend. Pharmaceutical companies are finding that the cost of doing business is increasing dramatically for three reasons. First, clinical development is more complex and more extensive because of increasingly rigorous drug approval processes. Second, marketing is consuming an increased fraction of revenues to capture market share in an increasingly competitive marketplace. Third, manufacturing is spending more to satisfy the ever-tightening regulatory constraints under which it must operate.

In addition to this countertrend, the industry has also observed that the complexity and cost of chemical development and manufacturing have been increasing dramatically. This increase results from two principal factors. First, in an ongoing drive to develop more selective pharmaceuticals, medicinal chemists are turning to more complex structures. The increased complexity is reflected in the fraction of modern pharmaceuticals that contain chiral centers (more than 50% of Phase II candidates in clinical development) in comparison with pharmaceuticals of 10 years ago (less than 10%), as well as in the growing number of steps required to synthesize modern pharmaceuticals. Only 5 to 8 steps were usually required 10 years ago, but modern pharmaceuticals require 12 to 15 steps. Second, chemical manufacturing is operating in the midst of increasing regulatory pressure with regard to worker safety, product quality control, environmental quality control, and reporting requirements.

Cost of Bulk Active Is a Significant Contributor to Pricing

To date, approximately 3000 different pharmaceuticals have been approved for human or animal use in the world and are being marketed (*2*). The majority of the market consists of the top 100 drugs that are sold in the principal markets of North America, Western Europe, and Japan. The bulk active chemical present in these formulated pharmaceuticals contributes as little as 5 to 10% (for patent-protected pharmaceuticals) to as much as 30 to 40% (for generic and over-the-counter drugs) to the selling price of a formulated pharmaceutical product. To reduce the cost of pharmaceuticals, the industry is actively pursuing approaches to reduce the cost of bulk active.

The impact of bulk active costs on the industry is illustrated by the distribution of sales and volumes of bulk actives (Figure 2). Although a few products have annual bulk active sales of more than $100 million, most fall in the range of $5 million to $50 million in annual sales. Similarly, although annual production of bulk active ranges from thousands of metric tons (for naproxen, ibuprofen, acetaminophen, and aspirin) to only a few kilograms (for some bulk actives with

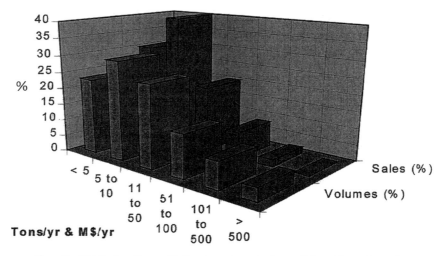

Figure 2. Distribution of annual bulk actives sales and volumes. (Adapted from reference 2.)

exceptionally high specific activities), most bulk active products are produced at annual volumes of 5 to 50 metric tons.

Chirality in Pharmaceutical Development

Chirality Provides Higher Drug Specificity

The majority of compounds currently in the pharmaceutical development pipeline are chiral. Chirality offers a means to increase the specificity of drug action by providing optimal spatial orientation of functionality in the drug. The growing emphasis on chirality is reflected in the number of patents that use the word *chiral*: A search on Derwent Patents identified about 1200 in 1989, compared with only about 250 in 1970.

Chiral drugs when produced by conventional synthetic organic chemistry contain a 50/50 mixture (racemate) of mirror-image isomers known as *enantiomers*. Although enantiomers have identical chemical compositions, their three-dimensional structures are different and often interact differently with cell receptors in living organisms. Recent scientific advances have enabled scientists to understand better the biochemistry underlying the therapeutic action of many drugs. Increased knowledge of the interaction of specific drugs with receptors and enzymes in living organisms has led to the development of an increasing number of chiral drugs, particularly in single-isomer form.

Single-isomer drugs operate by binding to three-dimensional receptors in a highly specific manner. The interaction between the drug and the receptor either stimulates or inhibits a biological function of the receptor and thereby initiates the therapeutic effect. In some cases the desired therapeutic effect is attributed only to one of the enantiomers. The other enantiomer may interact weakly or not at all with the particular receptor or may exert an altogether different effect by binding with another type of receptor, leading to a side effect. But for some drugs, the single isomer has not provided a better alternative to the racemate. For example, the scientific evidence does not appear to support the pharmaceutical development of labetalol (*3*) and etodolac (*4*) as single isomers.

Use of Chirality in Drugs Is Increasing

Historically, chiral drugs were developed and marketed as racemates because suitable single-isomer manufacturing technology was lacking. Figure 3 shows that in 1990, two-thirds of the total number of drugs commercially available were made via chemical synthesis. Of these synthetic drugs, approximately 40% were chiral. Of the more than 500 synthetic chiral drugs commercially available, approximately 90% were sold as racemates, whereas 10% were sold in single-isomer form (*5*).

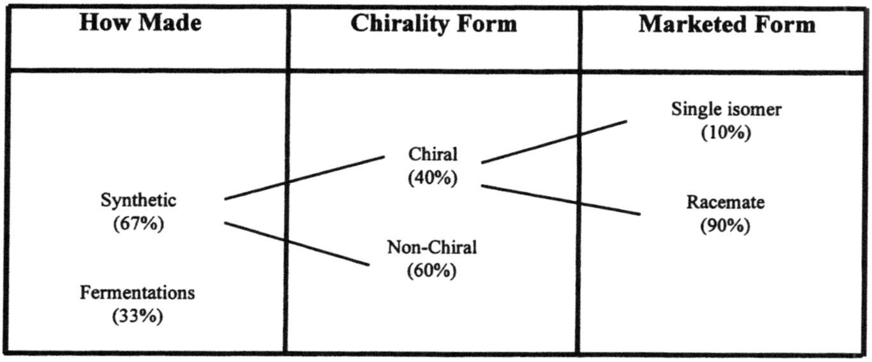

How Made	Chirality Form	Marketed Form
Synthetic (67%)	Chiral (40%)	Single isomer (10%)
Fermentations (33%)	Non-Chiral (60%)	Racemate (90%)

Figure 3. Distribution of drug forms on the market in 1990.

Figure 4 shows that today, of the approximately 400 drugs now in the development pipeline at Phase II human clinical trials, nearly 60% are chiral (compared with 40% of drugs commercially available), of which 40% (compared with 10% of drugs commercially available) are being developed in single-isomer form.

The importance of chirality in modern drug design is illustrated in the development of a group of drugs designed to combat hypertension by inhibiting angiotensin-converting enzyme (ACE), a key regulator of blood pressure. These *ACE inhibitors* bind to the active site of the enzyme and thereby prevent the formation of angiotensin II, a powerful vasoconstrictor. Representative examples of this class of drugs are shown in Figure 5. These drugs have revolutionized the

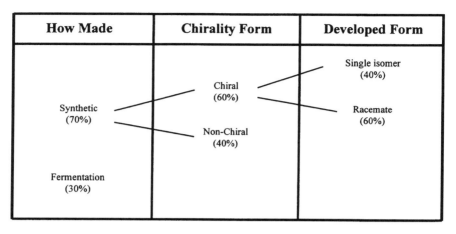

How Made	Chirality Form	Developed Form

Figure 4. Distribution of drug forms currently in Phase II clinical development pipeline.

treatment of hypertension. The first drug, captopril, was launched in 1981, and the second, enalapril, in 1984. Combined, they have enjoyed worldwide sales of formulated product in excess of $1 billion. This class of drugs contains at least two chiral centers, allowing for 4 possible isomers (2^2).[1] The most complex variants, perindopril and ramipril, each contain 5 chiral centers, allowing for 32 possible isomers (2^5). Although multiple isomers are available, without exception these drugs have been developed and marketed as single isomers.

   Producing chiral drugs in single-isomer form is an important task in drug manufacture. This aspect is illustrated in Table I, which shows the top 20 drugs ranked on the basis of 1990 worldwide sales (in bulk dosage form). Ten of the top 20 were marketed as single isomers and represented about 53% of the sales volume ($12 billion of a total of $22.5 billion).

Trend Toward Single-Isomer Drugs to Continue

During the drug discovery process, new chiral chemical entities are generally synthesized and initially tested as mixtures of stereoisomers. If desirable pharmacologic activity is found, a number of issues regarding the potential differences between the enantiomers of the candidate need to be addressed.

---

[1]ACE inhibitors are diastereomers—that is, they contain two or more chiral centers. For example, captopril contains two chiral centers, resulting in four possible isomers representing two enantiomer pairs: (R, S) and (S, R); (R, R) and (S, S). Diastereomers have different physicochemical properties and therefore can frequently be separated by conventional means. Thus, the production of captopril from L-proline produces a mixture of two diastereomers that can be separated by crystallization. Accordingly, formation of separable diastereomers affords a practical way to manufacture single-isomer drugs.

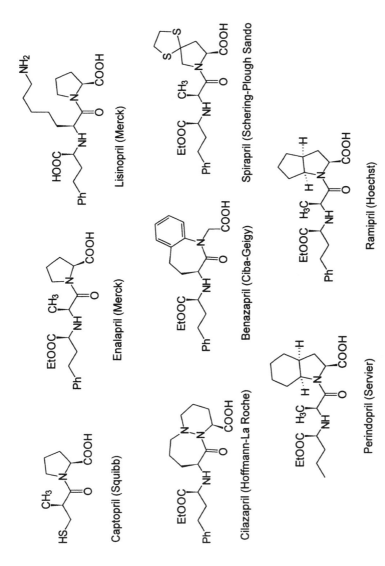

Figure 5. Structures of ACE inhibitors showing use of chirality.

Table I. Ranked Sales of Top 20 Drugs in 1990

Sales ($ million)	Therapeutic Class	Drug Name	Form
2370	Antiulcer	Ranitidine	Achiral
2000	Antibiotic	Amoxicillin	Single isomer
1800	Antibiotic	Ampicillin	Single isomer
1520	ACE inhibitor	Captopril	Single isomer
1500	ACE inhibitor	Enalapril	Single isomer
1400	NSAID	Ibuprofen	Racemate
1290	Calcium antagonist	Nifedipine	Achiral
1040	Antibiotic	Cefaclor	Single isomer
1030	Antiulcer	Cimetidine	Achiral
1020	β-Blocker	Atenolol	Racemate
975	NSAID	Diclofenac	Achiral
960	Calcium antagonist	Diltiazem	Single isomer
950	NSAID	Naproxen	Single isomer
900	Antibiotic	Cephalexin	Single isomer
750	Antihypercholesteremic	Lovastatin	Single isomer
630	Antiulcer	Famotidine	Achiral
620	Contrast medium	Iohexol	Racemate
600	Antibiotic	Ceftriaxone	Single isomer
585	NSAID	Piroxicam	Achiral
555	Bronchodilator	Albuterol	Racemate

1. enantiomer differences in
   - pharmacology
   - toxicology
   - pharmacokinetics
2. development of enantiospecific analytical protocol
3. extent of metabolic inversion
4. extent of species differences
5. therapeutic index of active enantiomer
6. side effects due to the inactive enantiomer
7. viability of large-scale synthesis
8. therapeutic benefit of single isomer
9. marketing benefit of single isomer

The issues center on how the patient is best served. The decision to develop the racemate or single isomer is not automatic. The decision, as with nonchiral drugs, revolves around two principal issues: (1) attainment and demonstration of safety and efficacy of the new drug and (2) development of new pharmaceutical products with optimum use of time and money. Each drug is an individual entity, and the decision must be based on the individual merits of the drug in question (for additional discussions, see reference 6). In light of these issues, development of the racemate is sometimes the appropriate decision. Justifications for development of the racemate are summarized as follows:

1. Pharmacologic activities of combined enantiomers are additive or synergistic.
2. Racemate exhibits high therapeutic index.
3. Inactive enantiomer has exceptionally low toxicity or side effects.
4. Enantiomer undergoes rapid isomerization or metabolic inversion.
5. Physical properties (e.g., solubility) of racemate provide benefit.
6. Racemate offers regulatory or marketing advantages.
7. Drug is indicated for a life-threatening condition.
8. Single isomer cannot be produced cost-effectively at scale.

Nevertheless, it is expected that the trend to single-isomer drugs will continue during the 1990s and beyond because of developing trends that include the following:

- New drugs frequently exhibit a high eudismic ratio (ratio of eutomer activity to distomer activity); that is, the activity resides primarily in only one of the enantiomers.
- A few new drugs have a low therapeutic index, making it difficult to justify developing the racemate.
- For some new drugs, the low-activity enantiomer is toxic.
- For most drugs, metabolic inversion is rare.
- Regulatory authorities in the developed countries are requiring separate testing of the pharmacologic and toxicologic properties of the racemate and the individual enantiomers before any drug based on a racemate can be approved. Concurrent testing of both enantiomers plus the racemate is time-consuming and represents a significant additional cost.
- Technological developments have made manufacturing the single isomer more economically feasible.

The last trend is of special importance. Ten years ago a number of drugs were developed as racemates because manufacturers claimed that the technology for cost-effective production was not available. However, technology has developed significantly over the past 10 years, and today it is difficult for manufacturers to make that claim.

Approaches to Manufacturing

Strategies

Modern chiral drugs are synthesized via a sequence of reactions from smaller, lower molecular weight precursors. Chiral centers are introduced at the appropriate place in the reaction sequence by incorporating chiral precursors available

from the chiral pool or by employing asymmetric reactions or resolution processes (Figure 6).

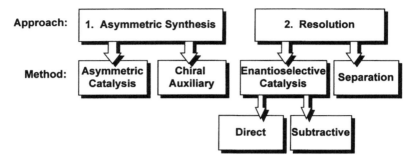

Figure 6. Two basic approaches to manufacture of single isomers.

Asymmetric reactions involve converting a prochiral center to a chiral center
by using a chiral agent, acting in stoichiometric or catalytic fashion, to favor the
formation of the desired enantiomer. Resolution involves separating a racemate
along the reaction pathway and can be achieved by using a chiral agent, acting
again in a stoichiometric or catalytic fashion (enantioselective catalyst), which
can be carried out in two ways: The enantioselective catalyst can directly trans-
form one enantiomer of an intermediate to the desired next intermediate, or the
enantioselective catalyst can destroy the unwanted enantiomer, leaving behind
the desired enantiomer.

Selection of Approach

Practicality and Cost

Asymmetric synthesis is, in principle, the most cost-effective method for producing single-isomer products, because all of the precursor is converted to the desired isomer. The interest in asymmetric approaches is reflected in the number of
new commercial applications that has been reported to be under development
(Table II).

However, the decision to implement an asymmetric synthesis approach
should be based on an assessment of efficiency and cost. Factors to consider include (1) the catalyst efficiency (that is, the number of product molecules produced per molecule of the catalyst); (2) the availability and cost of the catalyst
and the reaction starting materials (especially critical for low-value products); (3)
reaction conditions, such as very low temperatures or high pressures; and (4) reaction kinetics. Other practical considerations include air or moisture sensitivity,
reaction concentration and batch time as they reflect volumetric productivity,
and ease of removal of the catalyst.

Table II. Products of Commercial Interest Manufactured by Asymmetric Synthesis

Ultimate Product	Company	Chiral Intermediate	Production Reaction	Chiral Catalyst Ligand	Catalyst Metal
Cilastatin	Sumitomo	(structure)	Cyclopropanation	Schiff base complex	Cu
Dispalure control (mating disruption)	J.T. Baker	(structure)	Epoxidation	Dialkyl tartrate	Ti
Glycidols	ARCO	(structure)	Epoxidation	Dialkyl tartrate	Ti
Antihypertensive (pilot-scale quantities)	Merck Bristol–Myers Squibb	(structure)	Epoxidation	Bis-salen ligand	Mn
L-735,524 (pilot-scale quantities of this AIDS drug)	Merck Sepracor	(structure)	Epoxidation	Bis-salen ligand	Mn
L-Dopa	Monsanto	(structure)	Hydrogenation	(DIPAMP)COD phosphine ligand	Rh
L-Phenylalanine	Enichem	(structure)	Hydrogenation	PNNP phosphine ligand	Rh
Primaxin	Takasago	(structure)	Hydrogenation	(S)-BINAP binaphthyl phosphine	Ru
Levomenthol	Takasago	(structure)	Rearrangement	(S)-BINAP binaphthyl phosphine	Rh

In today's climate of pressure on pharmaceutical pricing, cost is a major issue. The major factors affecting the cost of manufacturing are shown as follows:

1. cost of substrate (starting material)
2. cost of chiral agent (e.g., resolving agent, chiral auxiliary, or catalyst)
3. cost of reagents, solvents, and processing aids
4. cost of capital equipment and efficiency of using capital equipment (volumetric productivity)
5. cost of specialized in-process controls (analytical equipment and procedures)
6. cost of hazard management and regulatory controls
7. cost of waste treatment
8. total number of steps in reaction, number of handling steps, overall yield
9. position of resolution step in overall synthesis

Enantiomeric Purity

In light of these product quality and cost issues, using catalytic methods has advantages over using separation methods or stoichiometric auxiliaries. These advantages, summarized in Table III, stem from the use of a substoichiometric, rather than stoichiometric, amount of an expensive chiral agent to produce the single isomer.

Table III. Comparison of Separations, Chiral Auxiliaries, and Catalysis for Manufacture of Single-Isomer Chiral Pharmaceuticals

Factor	Separations and Chiral Auxiliaries	Catalysis
Chiral agent	• Chiral resolving agent • General • Stoichiometric	• Chiral catalyst • Specific • Substoichiometric
Concentration	Many dilute streams	One concentrated product stream
Process complexity	Many steps and operations	Few steps and operations
Equipment	Conventional	Sometime specialized
Solvents	Organic	Aqueous and organic
Waste stream	More	Less

Despite its obvious advantages, asymmetric synthesis has an important inherent limitation. This limitation, illustrated in Figure 7, stems from the stringent requirement for enantiomeric purity in the pharmaceutical industry, generally greater than 99% enantiomeric excess (less than 0.5% of the undesired enan-

- **Resolutions give high enantiomeric excess at the expense of yield**
  - subtractive kinetic resolutions
  - successive crystallizations

- **Asymmetric synthesis give high enantiomeric excess only when exceptionally enantioselective**
  - E > 100 to get 98% ee

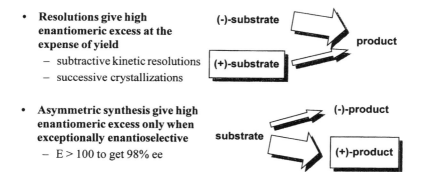

Figure 7. Intrinsic advantage of resolution and intrinsic limitation of asymmetric synthesis.

tiomer). Enantiomeric excess is the excess of predominant enatiomers expressed as percent, that is:

$$\frac{(\text{predominant enantiomer}) - (\text{minor enantiomer})}{(\text{predominant enantiomer}) + (\text{minor enantiomer})} \times 100$$

To achieve these enantiomeric purities requires exceptional enantioselectivity in the catalyst; more than a 200-fold preference for the reaction to the desired enantiomer. Such enantioselectivity is often difficult to achieve. By contrast, resolution methods can, in principle, always achieve the desired enantiomeric purities, albeit at the expense of yield. For example, subtractive resolution (destruction of the undesired enantiomer) can produce product of the needed enantiomeric purity with a catalyst of modest enantioselectivity by simply allowing the reaction to proceed beyond 50% conversion. Similarly, resolutions by diastereomeric crystallization can achieve needed enantiomeric purities by staging the operations— that is, by repeated crystallization.

   In light of these limitations, successful implementation of asymmetric synthesis in a manufacturing scheme requires three elements. First, perform the asymmetric synthesis as early as possible in the synthetic scheme to maximize the throughput. Second, employ an asymmetric synthesis step with good enantioselectivity (e.g., leading to greater than 85% enantiomeric excess), which allows selection of a broad range of reactions and conditions that are efficient, cost-effective, and scalable. Third, employ a subsequent enrichment or purification step (e.g., crystallization of an intermediate whose eutectic composition is below the enantiomeric excess of the product) to reach the desired enantiomeric purity. The availability of this enrichment step reduces the risk of ending up with batches of product that fail to meet specifications, since "failed" batches can be enriched to meet specifications.

Comparing Approaches: Nonsteroidal Anti-inflammatories

Danger of Generalizations

Chirality introduces an extra dimension into the challenge of designing an industrial synthesis, one that is often the overriding factor in choosing a synthesis route. One general conclusion is that it is dangerous to draw general conclusions about the superiority of one technology over the other. Production costs are strongly dependent on the costs of raw materials. The most economically viable route varies from one product to another, even within groups of closely related products. Moreover, the viability of the route is frequently very site-specific— that is, if the company has a unique position in raw materials or has underutilized facilities for carrying out certain types of reactions, then this could have an overriding impact on route selection. Hence, each case must be judged on its own merits.

α-Arylpropionic Acid Anti-inflammatories

The α-arylpropionic acids are a class of nonsteroidal anti-inflammatory drugs (NSAIDs) that also exhibit analgesic and antipyretic activity. They act by inhibiting the enzyme cyclooxygenase, which causes a decrease in prostaglandin synthesis. Their principal side effects are gastrointestinal effects (e.g., gastric ulceration) due to the systemic decrease in prostaglandin levels. The three most important members of this group are the profens: ibuprofen, naproxen, and ketoprofen (Figure 8). Production volumes of ibuprofen and naproxen (Table IV) are very large relative to most drugs, reflecting the breadth of their use for chronic treatments and the large dosages required for effective treatment.

Table IV. Sales and Production Volumes for Ibuprofen, Naproxen, and Ketoprofen

Drug	Worldwide Sales ($ million)		Bulk Active Manufactured (metric tons)	
	1990	1995	1990	1995
Ibuprofen	1400	1680	7700	10,500
Naproxen	990	1000	1290	1800
Ketoprofen	290	340	95	115

In vitro experiments have demonstrated that the (S)-enantiomers of the profens preferentially act as competitive inhibitors to the cyclooxygenase subunit. For example, (S)-naproxen has 28 times the anti-inflammatory activity of (R)-naproxen. Thus the use of the (R)-isomer rather than the racemate allows a sig-

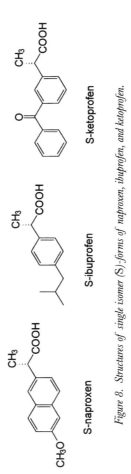

Figure 8. Structures of single isomer (S)-forms of naproxen, ibuprofen, and ketoprofen.

nificantly lower net dosage. Additionally, (*R*)-naproxen has toxic side effects. On this basis, (*S*)-naproxen was developed and is marketed by Syntex in single-isomer form.

(S)-Naproxen Manufacture

Expiration of patent protection on (*S*)-naproxen in Europe in 1988 and the United States in 1993 prompted the development of a generic market for (*S*)-naproxen. Additionally, a joint venture between Syntex and Procter & Gamble obtained approval for use of (*S*)-naproxen in certain indications, albeit at reduced dosages, for the over-the-counter market (under the brand name Aleve). Projected increased production volumes in 1995 relative to 1990 (Table IV) with no significant increase in bulk active sales reflect the typical price erosion that occurs when a patented drug goes generic. In view of the economic importance of this drug in the pharmaceutical marketplace and the expected price pressures on the cost of bulk active with loss of patent protection, industrial and academic laboratories have actively pursued new strategies for the manufacture of (*S*)-naproxen. All the strategies outlined in Figure 4 have been explored. Some of these strategies, including the known commercial manufacturing routes, are summarized below.

Synthesis of Racemic Naproxen. Two principle routes are used on the industrial scale for the synthesis of racemic naproxen. In one, base-catalyzed condensation of 2-acetyl-6-methoxynaphthalene with alkyl chloroacetate (Darzens reaction) provides the α-alkyl naphthylglycidate, which is subsequently converted to racemic naproxen by a series of steps analogous to those in the Boots process for making racemic ibuprofen (described later). In the other, a reaction of 6-methoxy-2-naphthylmagnesium bromide with the salt of α-bromopropionic acid (Grignard reaction) provides racemic naproxen (Figure 9). It is believed that this scheme is the basis for the commercial method used by Syntex to produce racemic naproxen.

Resolution of Racemic Naproxen. Resolution of naproxen is based on the formation of diastereomeric salts with chiral amines including cinchonidine, (−)-α-phenethylamine, and *N*-methyl-D-glucamine. The preferred method involves *N*-methyl-D-glucamine, which results in a resolution process that is practical and cost-effective. This method has been developed and commercialized by the Syntex group. The undesired (*R*)-enantiomer is readily racemized and reused by treatment with base and heating. The chiral amine can be recovered and recycled.

Zambon Chiral Auxiliary Route. Zambon has developed and commercialized a process for the production of (*S*)-naproxen employing a stoichiometric quantity of a chiral auxiliary. The process, illustrated in Figure 10, is based on the enantioselective 1,2-aryl-shift rearrangement of alkylarylketals catalyzed by Lewis

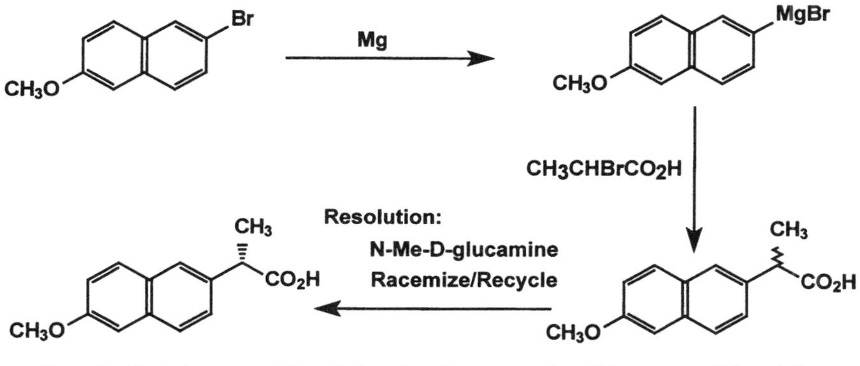

Figure 9. Basis for commercial method used by Syntex to produce (S)-naproxen. (Adapted from references 7 and 8.)

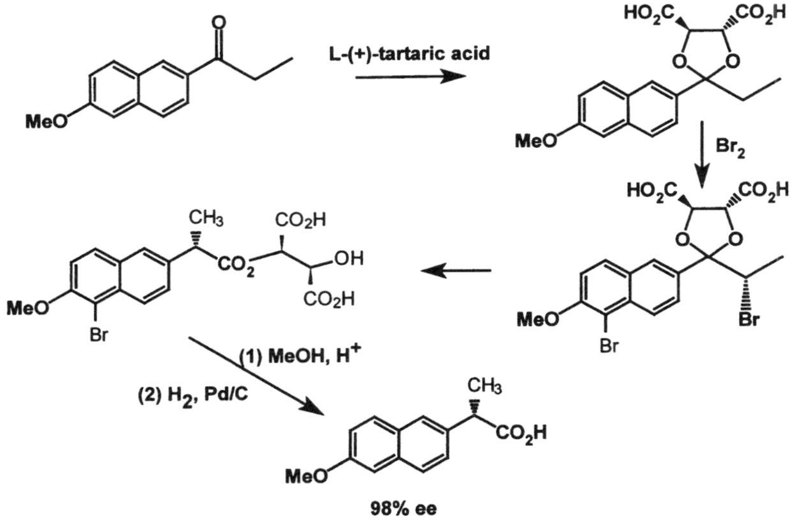

Figure 10. Zambon process for the manufacture of (S)-naproxen (10).

acids. The chiral auxiliary, a dialkyl ester of $(2R,3R)$-tartaric acid, performs two functions: (1) It promotes a highly diastereoselective bromination, and (2) it increases the enantiomeric excess of the isolated propionic acid ester beyond the diastereomeric excess of the bromoketal (a kinetic resolution). At the end of the reaction $(2R,3R)$-tartaric acid is recovered and reused. According to Villa and Pozzoli (9), the process is carried out in standard steel- and glass-lined equipment, exhibits high volumetric productivity, and provides an overall yield of (S)-naproxen in excess of 75% based on starting ketone.

Catalytic Asymmetric Synthesis Routes. The Monsanto and DuPont groups have reported asymmetric synthesis routes employing asymmetric catalysts. The Monsanto route, illustrated in Figure 11, is based on asymmetric hydrogenation in the presence of a catalytic amount of enantiomerically pure ruthenium (2,2′-bis(di-arylphosphino)-1,1′-binaphthyl)dihalide, a method first reported by Noyori and coworkers. A novel aspect of the process is the electrolytic production of the α-hydroxy acid, which upon dehydration produces the requisite 2-arylacrylic acid precursor. The DuPont route, illustrated in Figure 12, is based on the asymmetric hydrocyanation of 2-arylstyrene using an enantiomerically pure nickel catalyst. Although these processes appear economically promising relative to Syntex's racemic resolution route, neither process has yet been commercialized. This fact reflects higher costs of starting materials and the difficulty of bringing a new process to market in an environment where the competing process has enjoyed many years of development and optimization and uses manufacturing facilities that have been fully depreciated.

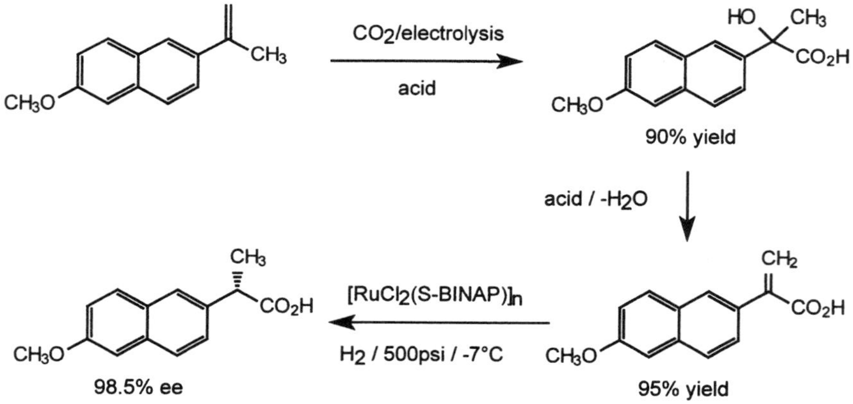

Figure 11. *Monsanto asymmetric synthesis route to (S)-naproxen (11).*

(S)-Ibuprofen Manufacture

It has been shown that the (R)-enantiomer of ibuprofen undergoes metabolic inversion in vivo, presumably through the mechanism illustrated in Figure 13. Because the process goes through the coenzyme-A thiolester of ibuprofen, an activated substance, all of the (R)-enantiomer is converted to the physiologically active (S)-enantiomer. These results argued for the development and marketing of ibuprofen as a racemate.

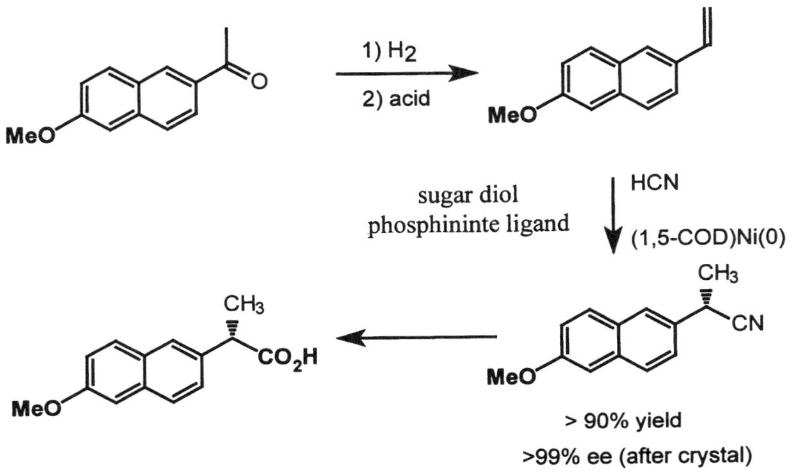

Figure 12. *DuPont asymmetric synthesis route to (S)-naproxen (12).*

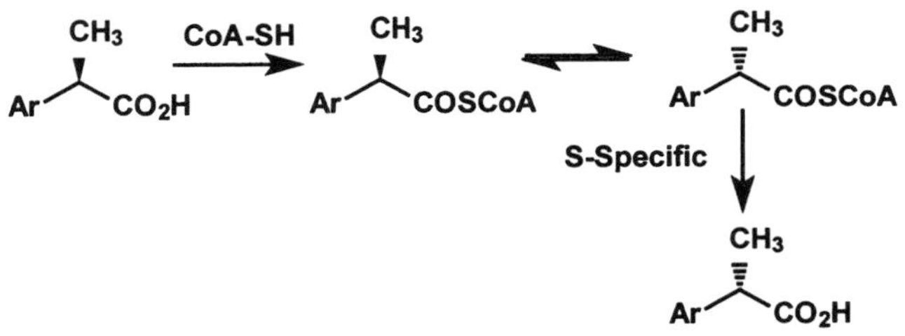

Figure 13. *Proposed route of metabolic inversion of profens.*

Recently, however, a convincing case has been made for the development of ibuprofen as a single isomer. The kinetics of metabolic inversion are slow, and thus administration of the single isomer provides more rapid onset of pharmacologic action. An (S)-ibuprofen product is now being marketed in Austria. In the United States, a joint venture between Johnson & Johnson and Merck has filed a new drug application with the Food and Drug Administration seeking approval for (S)-ibuprofen in the over-the-counter market. Approval is pending.

Synthesis of Racemic Ibuprofen. Commercial production of racemic ibuprofen starts with isobutylbenzene. In the Boots process, illustrated in Figure 14, isobutylbenzene is carried through a Friedel–Crafts acylation to make isobutylacetophenone. The ketone is transformed to the glycidate ester via a Darzens condensation with ethyl chloroacetate. Acid-catalyzed rearrangement of the gly-

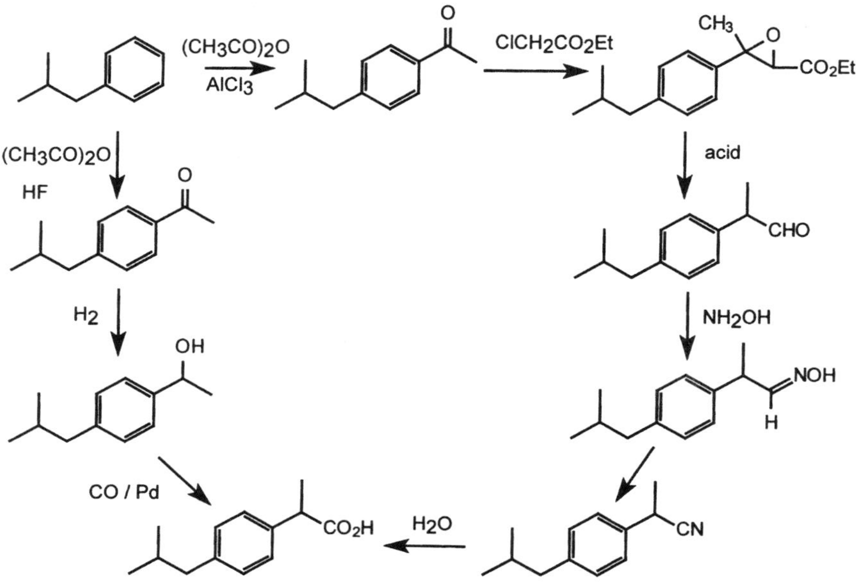

Figure 14. Hoechst–Celanese and Boots processes for manufacture of racemic ibuprofen.

cidate ester yields the aldehyde, and conversion to the nitrile via the oxime, followed by hydrolysis of the nitrile, produces racemic ibuprofen. This six-step scheme has been supplanted by a three-step scheme developed by the Hoechst–Celanese group. The scheme employs a technically challenging but much improved hydrogen fluoride–catalyzed acylation of isobutylbenzene. The resulting isobutylacetophenone is reduced to the alcohol by catalytic hydrogenation and allowed to undergo direct hydroformylation to racemic ibuprofen in the presence of a novel catalyst. The new route represents a milestone in pharmaceutical manufacturing because of its use of high-technology reactions that demand specialized conditions and equipment. The large capital investment needed to put this route in place (in excess of $50 million) is justified on the basis of the dramatic improvement in volumetric productivity coupled to large-volume product demands.

Resolution of Racemic Ibuprofen. Racemic ibuprofen is a relatively inexpensive pharmaceutical commanding in volume prices of about $10 to $15 per kilogram of bulk active. It is produced on a large scale in the United States by Ethyl Corporation and Hoechst–Celanese and in India by several companies. Given that (S)-ibuprofen is expected to exhibit at most twice the potency of the racemate, (S)-ibuprofen bulk active can command a higher price than the racemate but not an exceptional premium. A rule of thumb in these circumstances is that the single-isomer bulk active commands a price 2 to 2.5 times the racemate bulk active price. Because the price difference is small, manufacturing methods for (S)-ibuprofen have to be simple (not capital-intensive) and very efficient to be commercially viable. Accordingly, this argues strongly for a resolution route with recycling of the (R)-enantiomer, especially if carried out by the companies that have highly efficient and fully depreciated processes for the racemate. The patent literature suggests that the principal manufacturing route of (S)-ibuprofen as practiced by Ethyl Corporation is via resolution with (–)-α-phenethylamine.

Asymmetric Synthesis. The Merck group has developed an innovative asymmetric synthesis of (R)-ibuprofen using a readily available chiral auxiliary (Figure 15). The process uses the diastereoselective reaction between a chiral alcohol and the prochiral precursor, ibuprofen ketene. Use of inexpensive (R)-pantolactone produced the ester with high diastereoselectivity. However, the process has never been commercialized, reflecting the tight constraints on process economics and the requirement for the more expensive (S)-pantolactone to produce (S)-ibuprofen.

Case Studies of Pharmaceutical Manufacturing

Diastereomeric Resolution: β-Lactam Antibiotics

Diastereomeric crystallization is based on the interaction of a racemic product with the single isomer of a chiral material to give two diastereomeric derivatives, which are usually salts. The salts formed are diastereomers with different physical properties, which can be separated by physical means. Generally, the most efficient method of separating the diastereomers is crystallization, which employs differences in solubilities. The approach has the advantage of simplicity and can be accomplished with standard equipment. The method is well suited to batch production, which is the norm in the pharmaceutical industry.

Despite the view that it is a low-technology approach, the separation of diastereomeric salts by crystallization (classical resolution) is a widely used manufacturing method that provides a large fraction of single-isomer drugs produced by synthetic means. A study of a representative group of market drugs shows that more than 65% are manufactured by methods involving diastereomer salt crystallization (*14*).

Amoxicillin, ampicillin, cefaclor, and cephalexin are antibiotic products that are rated, on the basis of total sales, in the top 10 products sold in single-isomer form (Table III). These products contain either D-phenylglycine OR D-hydroxyphenylglycine as chiral elements, which command sales prices of about $15 and

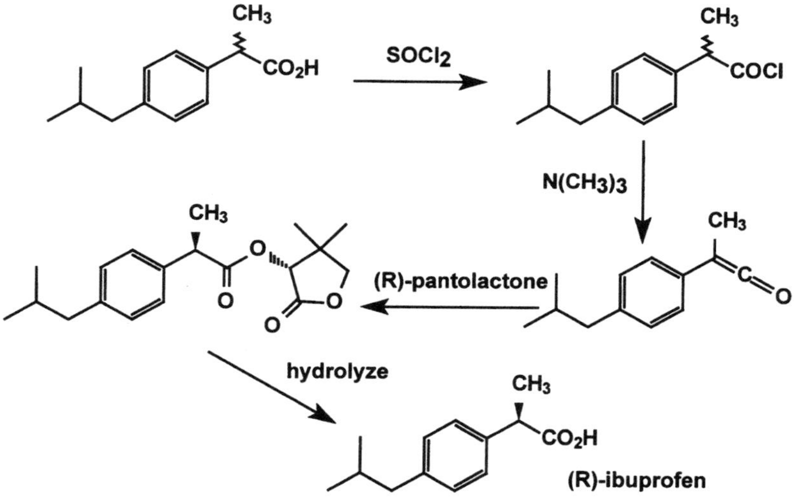

Figure 15. Merck chiral auxiliary route to single-isomer ibuprofen (13).

$45 per kilogram, respectively. These two amino acids are produced by classical resolution processes at a scale of more than 1000 metric tons per year using (+)-10-camphorsulfonic acid and the corresponding bromocamphorsulfonic acid as resolving agents (15). (+)-10-Camphorsulfonic acid is a readily available chiral acid made from natural camphor that in large quantities costs about $15 per kilogram. Racemic phenylglycine is produced by Strecker reaction of benzaldehyde with hydrogen cyanide followed by hydrolysis of the cyanohydrin. The undesired L-phenylglycine is racemized by reaction with benzaldehyde to form the Schiff base, followed by treatment with base.

Enzymatic Resolution: Diltiazem

Diltiazem is a highly successful cardiovascular product developed by Tanabe Seiyaku Co. (Osaka, Japan) for the treatment of hypertension. It is marketed in the United States through Marion Merrell Dow (Kansas City, MO) under the trade names Cardizem and Cardizem CD. Diltiazem, whose structure is shown in Figure 16, belongs to the group of cardiovascular agents known as calcium channel blockers.

The original synthesis developed by Tanabe, illustrated on the left side of Figure 16, uses a tin-catalyzed epoxide ring opening of racemic methyl *p*-methoxyphenylglycidate with *o*-nitrothiophenol. Elaboration of the adduct yields an α-hydroxy acid derivative, which is resolved by using L-lysine. Since there is no practical way to recycle the unwanted enantiomer, it is discarded. Reduction of the nitro group followed by further elaboration yields diltiazem. The sequence

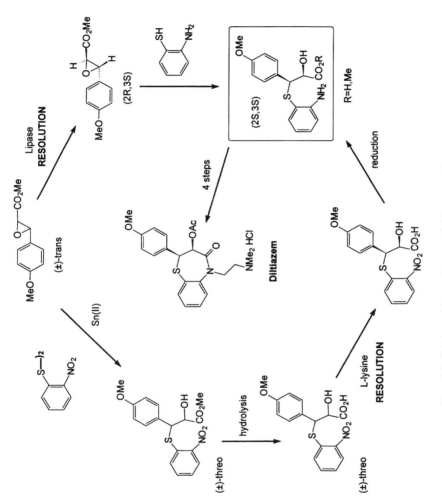

Figure 16. Existing and improved routes to Diltiazem manufacture.

is characterized as a racemic synthesis with the resolution occurring relatively late. At manufacturing scale, late resolutions are inefficient because equal amounts of undesired enantiomers are being carried through the early steps of the process. Since manufacturing of early intermediates is carried out at larger volumes than later intermediates (to compensate for yield losses in a sequential synthesis), late resolution dramatically reduces the throughput of product through a reactor train. This increases the cost of the process. Accordingly, it is generally desirable to carry out the resolution as early as possible in the sequence.

The old route is being supplanted by a new route developed by Tanabe in conjunction with Synthelabo (Gargenville, France), Sepracor (Marlborough, MA), and Andeno (Venlo, Netherlands). Development of the new route was motivated partly by the need to reduce bulk active costs in anticipation of price erosion accompanying patent expiration. The new route, illustrated on the right side of Figure 14, is two steps shorter and carries out the resolution process in the first step. This strategy has increased throughput and allowed more effective use of existing equipment.

A key element in the new route is the efficient resolution of racemic methyl *p*-methoxyphenylglycidate, which is readily produced by Darzens condensation of inexpensive *p*-methoxybenzaldehyde with methyl chloroacetate. The process developed by Sepracor and illustrated in Figure 17 is based on the discovery of a series of lipases that preferentially catalyze the hydrolysis of the unwanted $(2S,3R)$-glycidate ester. An interesting feature of the reaction is that the hydrolysis product undergoes spontaneous decarboxylation and rearrangement to the aldehyde, which is a powerful inhibitor of the enzyme. To overcome this drawback, sodium bisulfite is added to the reaction mixture to complex the aldehyde and prevent the inhibition. In the practical implementation of this process at scale, two additional issues needed to be addressed. First, the enzyme had to be recov-

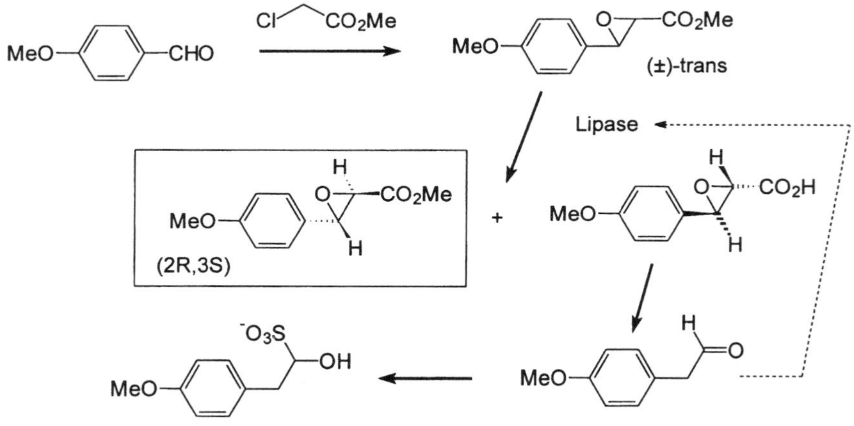

*Figure 17. Sepracor process for methyl-*p*-methoxyphenylglycidate resolution.*

ered and reused, since it was relatively expensive. Second, the reaction mixture consists of two liquid phases—an organic ester-containing phase and an aqueous bisulfite adduct–containing phase—which must be separated at the end of the reaction.

To address these needs, Sepracor developed a novel enzyme-membrane reactor for the commercial implementation of the enzymatic resolution process. As illustrated in Figure 18, the reactor is based on porous hollow fibers having large open pores on the outside and small pores on the inside. Enzyme is introduced into the spongy matrix of the fiber by simple filtration through the membrane and trapped in the membrane by the ultrafiltration film. Enzyme is retained in the membrane by flowing the ester-containing toluene phase, in which the enzyme is not soluble, on the outside. Thus the membrane provides a simple way to immobilize the enzyme and offers a stable high surface area interface for rapid mass transfer between phases. The membranes are assembled into bundles and fabricated into stainless steel production modules having approximately 60 m² of surface area. Since the beginning of 1993, Tanabe has used this process to manufacture methyl (2R,3S)-p-methoxyphenylglycidate at a scale in excess of 50 metric tons per year.

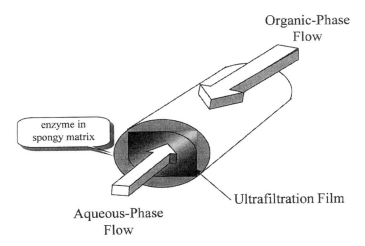

Figure 18. Supported enzyme in Sepracor two-phase membrane bioreactor.

Concluding Remarks

The central theme of this chapter is that control of production cost is playing an increasingly important role. The industry's response to cost control has been to implement new, more efficient technologies. Although the benefits of implementation are readily appreciated, the costs of implementation are not. Successful development of a new industrial-scale synthesis of a modern chiral pharmaceutical may take a team of chemists 6 to 9 months of laboratory synthesis, and a large team of chemists and chemical engineers may require 3 years of process development work to achieve a commercial process. The cost of development work alone is in the tens of millions of dollars, but the cost of construction of dedicated facilities can be hundreds of millions of dollars. Consequently, commitments to new technology are very costly and must be justified by adequate potential return on capital.

Pharmaceutical companies have historically done all manufacturing in house to maintain control over all aspects of the manufacturing process. In light of the restructuring in the industry, a new trend is developing for the pharmaceutical company to purchase advanced intermediates and to carry out only the last few steps in the synthesis. This trend has led to new business opportunities for fine chemical manufacturers. However, the cost and risk of developing new technologies for improved manufacturing have been shifted from the pharmaceutical companies to the fine chemical manufacturers. What are the business implications? As shown in Figure 2, the sales of intermediates are generally in the volume range of 1 to 100 metric tons per year, with the majority at the lower end. Sale prices of chiral intermediates are generally in the range of $50 to $1000 per kilogram, where the higher prices apply to only a limited number of low-volume complex molecules. And development of a novel chemistry for the commercial production of a chiral intermediate from conception to commercialization could take 2 to 3 years at a cost of $2 million to $3 million. What business conditions justify this level of investment? If we assume a 10-year payback period for a development program costing $2.5 million and aim for a gross margin of 40% with 50% profit, then $5.4 million in sales per year is required to pay back the investment. This means that for a product costing $50 per kilogram, 108 metric tons per year in sales would be required to support the investment. For a product costing $1000 per kilogram, annual sales of approximately 5.4 metric tons would be required, but as pointed out previously, there are very few opportunities at the high-price end. In addition, the analysis has not considered the capital investment required to implement the newly developed technology, which can be staggering.

This analysis suggests that successful implementation of new technology requires that the technology have several favorable characteristics:

- The process can be carried out in conventional multipurpose equipment to utilize existing capacity and avoid constructing new facilities.
- The process uses a catalyst that is inexpensive and readily available (e.g., biological catalysts such as enzymes that are commercially available in bulk or synthetic catalysts such as Jacobsen's catalyst that can be prepared from inexpensive raw materials).
- The process is generic and can be used to make a range of products. This spreads the cost and risk of development over several products.

Acknowledgments

The author acknowledges his former colleagues at Sepracor Inc. who performed the diltiazem work described earlier. Special thanks go to Roger Pettman and Robert L. Bratzler who provided many helpful comments during the preparation of the manuscript.

References

1. Lehman Brothers; "PharmaPipelines: Implications of Structural Changes for Returns on Pharmaceutical R&D"; Lehman Brothers: New York, April 12, 1994.
2. Bernhagen, W. *Chim Oggi.* **1993,** *March/April,* 9–14.
3. Jamali, F. In *Drug Stereochemistry: Analytical Methods and Pharmacology,* 2nd ed.; Wainer, I. W.; Drayer, D., Eds.; Marcel Dekker: New York, 1993; pp 375–384.
4. Wright, M. R.; Davies, N.; Jamali, F. *J. Pharm. Sci.* **1994,** *83,* 911–912.
5. Millership, J. S.; Fitzpatrick, A. *Chirality* **1993,** *5,* 573–576.
6. Cayen, M. N. *Chirality* **1991,** *3,* 94–98.
7. Felder, E.; Pitre, D.; Zutter, H. U.S. Patent 4 246 164, 1981.
8. Holton, P. G. U.S. Patent 4 515 811, 1985.
9. Villa, M.; Pozzoli, C. In *Proceedings of Chiral '95 USA Symposium;* Spring Innovations: Stockport, England, 1995.
10. Giordano, C.; Castaldi, G.; Cavicchioli, S.; Villa, M. *Tetrahedron* 1989, *45,* 4243–4252.
11. Wagenknecht, J. H. U.S. Patent 4 601 797, 1986.
12. Nugent, W. A.; RajanBabu, T. V.; Burke, M. J. *Science (Washington, D.C.)* **1993,** *259,* 479–483.
13. Corley, E. G.; Larsen, R. D.; Grabowski, E. J.; Reider, P. U.S. Patent 4 940 813, 1990.
14. Roth, H. J.; Kleeman, A.; Beissen, T. *Pharmaceutical Chemistry;* Hornwood: Chichester, England, 1988; Vol. 1.
15. Kamphuis, J.; Boesten, W. H. J.; Kapstein, B.; Hermes, H. F. M.; Sonke, T.; Broxterman, Q. B.; van den Tweel, W. J. J.; Shoemaker, H. E. In *Chirality in Industry;* Collins, A. N.; Sheldrake, G. N.; Crosby, J., Eds.; Wiley: Chichester, England, 1992.

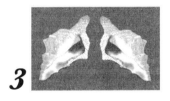

3

Studies on Enantioselective Synthesis

E. J. Corey

One of the most interesting and crucial areas of chemical research at present is the development of new catalysts and reagents for enantioselective synthesis. Many laboratories have contributed to the recent flood of important new results in this area. Through their research the science of chemical synthesis has been profoundly altered and enriched. This paper reviews certain aspects of our work in this area subsequent to that which was described at the 31st National Organic Chemistry Symposium in 1989 (*1*). Specifically, the first part describes research on catalytic enantioselective Diels–Alder reactions and the second part deals with enantioselective addition of carbon to carbonyl groups.

Catalytic Enantioselective Diels–Alder Reactions

Earlier research in our group on the use of the 8-phenylmenthyl group as a chiral controller for Diels–Alder reaction of acrylate esters (*2*) nurtured the idea that neighboring aromatic π-groups could influence transition-state energies in such a way as to enforce high stereoselectivity (*3*). Thus, the AlCl₃-catalyzed Diels–Alder reaction shown in Figure 1 afforded a major adduct (*ca.* 97:3 diastereoselectivity) that could be transformed into the optically pure lactonic prostaglandin precursor **1b** with recovery of 8-phenylmenthol. We supposed

3407–8/96/0037$15.50/0
© 1996 American Chemical Society

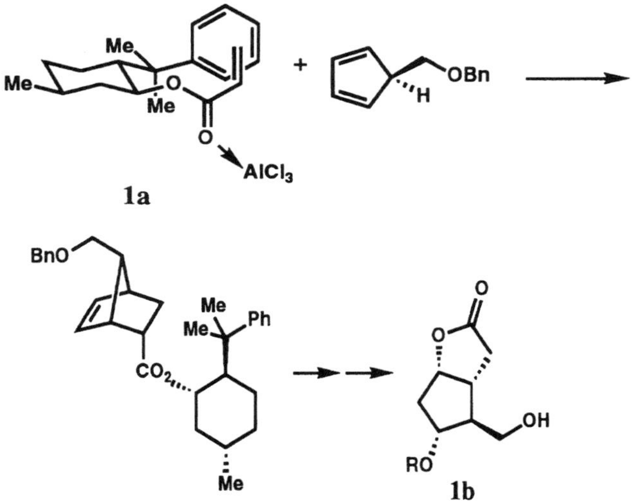

Figure 1. Enantioselective route to prostaglandins using the 8-phenylmenthol controller.

that the catalytically active species is the *s-trans* ester–AlCl$_3$ complex **1a** and that the aromatic ring stabilizes both the complex and the transition state by proximity with the electron deficient carbonyl carbon; acrylate face selectivity is then a consequence of the steric screening by the aromatic ring. It is important to recognize that electron deficiency in the Lewis-acid complexed carbonyl is preserved even in the Diels–Alder product, so it must exist in the transition state for the reaction, as well as in the initial complex of AlCl$_3$ with the dienophile. The idea that π-aromatic units could be used both to stabilize specific transition-state geometries and to provide stereoselectivity through facial screening seems to be a promising heuristic for the design of enantioselective Diels–Alder catalysts. Some support for the value of this speculation has recently been obtained in several different systems.

The bistriflamide of *S,S*-1,2-diphenyl-1,2-diaminoethane reacts with trimethylaluminum to form a crystalline diazaaluminolidine that is a useful catalyst for enantioselective Diels–Alder reactions (*1, 4–8*), as exemplified by the reaction shown in Figure 2, which can be applied to the synthesis of prostanoids. In the solid state the diazaaluminolidine exists as a dimer whose structure was determined by X-ray crystallography to be shown in Figure 3 (*8*). The ^{1}H, ^{13}C, and ^{19}F NMR spectra of this diazaaluminolidine show that it also exists as a dimer in CH$_2$Cl$_2$ solution (*8*). However, in the presence of one or more equivalents of 3-acryloyl-1,3-oxazolidine-2-one, the Diels–Alder catalyst forms a 1:1 complex whose preferred geometry was revealed by low-temperature ^{1}H and ^{13}C NMR analysis (including NOE) to be that shown in Figure 4 (**4a**). It is of interest that this geometry of the 1:1 complex places the metal-complexed carbonyl group and an aromatic ring in proximity and leads to the possible transition-state as-

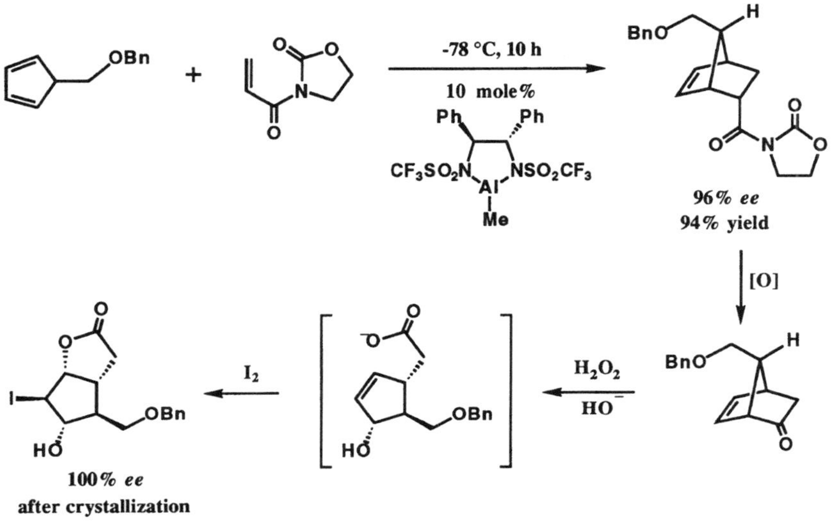

Figure 2. Catalytic enantioselective Diels–Alder reaction for the general synthesis of prostaglandins (ee is enantiomeric excess).

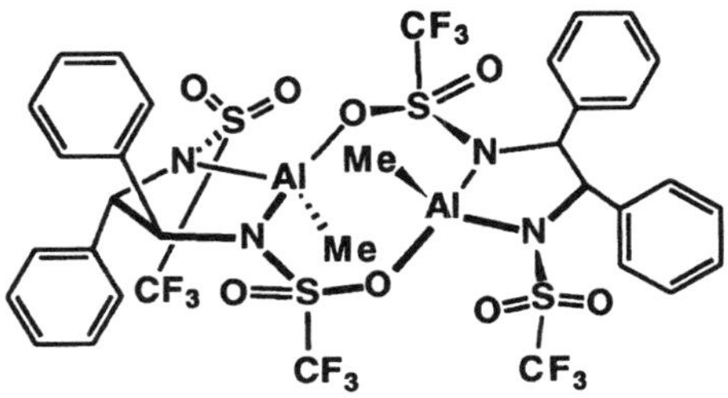

Figure 3. Dimeric structure of the chiral diazaaluminolidine catalyst in solution (CH₂Cl₂) and in the solid state.

sembly shown in Figure 4 (**4b**) that is consistent with the observed enantioselectivity. An alternative transition-state assembly that cannot be excluded is shown in Figure 4 (**4c**). Despite the NMR evidence indicating geometry **4a** to be favored, it is entirely possible that the main reaction course involves a minor component of the system such as **4c**. This sort of issue invariably recurs in the analysis of mechanistic pathways for enantioselective reactions and, indeed, is a central problem in the design of new enantioselective catalysts and reagents. It is one of the most interesting problems in the field of mechanistic chemistry. In the case of the reaction shown in Figure 2, we believe that the matter of the geometry of the transition-state assembly can be resolved by further experimentation.

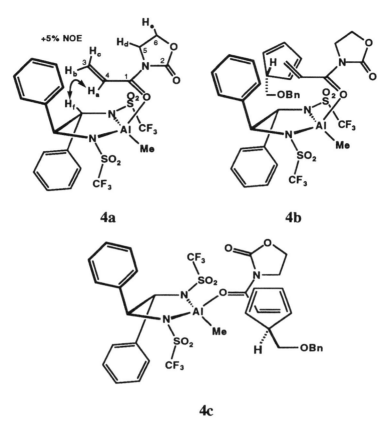

Figure 4. Structures of the dienophile–catalyst complex (**a**) and Diels–Alder transition-state assembly (**b**). Alternative transition-state assembly (**c**).

At present, the strongest evidence in favor of transition-state assembly **4a** is the fact that the replacement of the two phenyl groups of the catalyst by *tert*-butyl results in a catalytic reaction that is not enantioselective. This observation seems inconsistent with reaction via **4c**, but is easily understood in terms of **4b** and the crucial role of aromatic π-neighboring group stabilization and screening in the transition state.

The high enantioselection observed in catalytic Diels–Alder reactions such as that in Figure 2 does not extend to reactions of carbonyl addition, such as the reaction of aldehydes with trimethylsilylcyanide. This fact is also consistent with the crucial role of the phenyl groups discussed above. In the case of carbonyl addition such neighboring π-aromatic interactions would be expected to lower the energy of the reactive aldehyde–catalyst complex more than that of the transition state.

The use of chiral oxazaborolidines as catalysts for the enantioselective reduction of ketones by boranes has proven to be a valuable and general new synthetic method (*1, 7, 9*). The oxazaborolidine system is clearly of great interest as a basis for enantioselective catalysis of other important synthetic reactions. Recently, the application of an *S*-tryptophan-derived oxazaborolidine to catalysis of the Diels–Alder reaction with acrolein derivatives has been described (*10*). The preparation of the catalyst (**5a**) and an exciting application are summarized in Figure 5. Just 5 mol% of catalyst **5a** suffices for the acceleration and control of the addition of cyclopentadiene and 2-bromoacrolein to form enantioselectively the (2*R*)-adduct **5b** in very high yield. Equally good results are obtained with 2-chloroacrolein. The reaction of 2-methylacrolein with cyclopentadiene proceeds with somewhat lower enantioselectivity (*R/S* 98/2). In each of these reactions the adduct with the *exo* arrangement of the formyl group is obtained with high selectivity (*ca.* 96/4). In contrast, the corresponding reaction of cyclopentadiene with acrolein proceeds to form the *endo* formyl adduct with opposite and low enantiofacial selectivity (30:70) with respect to the enal.

The catalytic Diels–Alder process outlined in Figure 5 has been harnessed for practical purposes. The small amount of *endo* aldehyde contaminating **5b** is easily removed by stirring at room temperature with the equivalent amount of aqueous silver nitrate since it is much more reactive than **5b** and affords water soluble products. The bromo aldehyde **5b** can be converted to a wide variety of useful products (Figure 6). Two very efficient methods have been developed for the transformation of **5b** into bicyclo[2.2.1]heptene-2-one of high (>99%) enantiomeric purity (Figure 7). Catalyst **5a** is also useful for the promotion of highly enantioselective Diels–Alder reactions of 2-chloro- or 2-bromoacrolein with furan, 2-methylbutadiene, or 2-triisopropylsilyloxybutadiene.

The application of catalyst *ent*-**5a** to the catalytic enantioselective synthesis of prostaglandins is summarized in Figure 8. Although this process has not been optimized (the results shown are from the initial experiment), it is apparent that this approach is of great interest for practical production. The catalyst precursor, *R*-*N*-tosyltryptophan, is cheap and easily recovered for reuse.

The concept that led to the discovery of the enantioselective Diels–Alder catalyst **5a** (**9a** in Figure 9) can be summarized as follows. If the dienophilic α,β-enal coordinates to the face of boron in **9a** which is *cis* to the 3-indolylmethyl substituent, the π-basic indole unit can stabilize the π-acidic coordinated dienophile through internal π-complexation as shown in Figure 9. In the complex the indole and α,β-enal moieties can assume a parallel orientation at the ideal separation (*ca.* 3 Å) for π-interaction. Steric screening by the indole ring would serve to block one face of the α,β-enal from attack by the diene. The transition-state assembly **9b** for the reaction of the *s-cis*-α,β-enal complex with cyclopentadiene successfully predicts the correct absolute configuration of the product **9c**. Some support for this model is provided by the following experimental results (*11*). (1) Addition of 2-bromoacrolein to catalyst **9a** in CH$_2$Cl$_2$ at −78 °C results in an immediate orange-red coloration (similar to that of the indole-

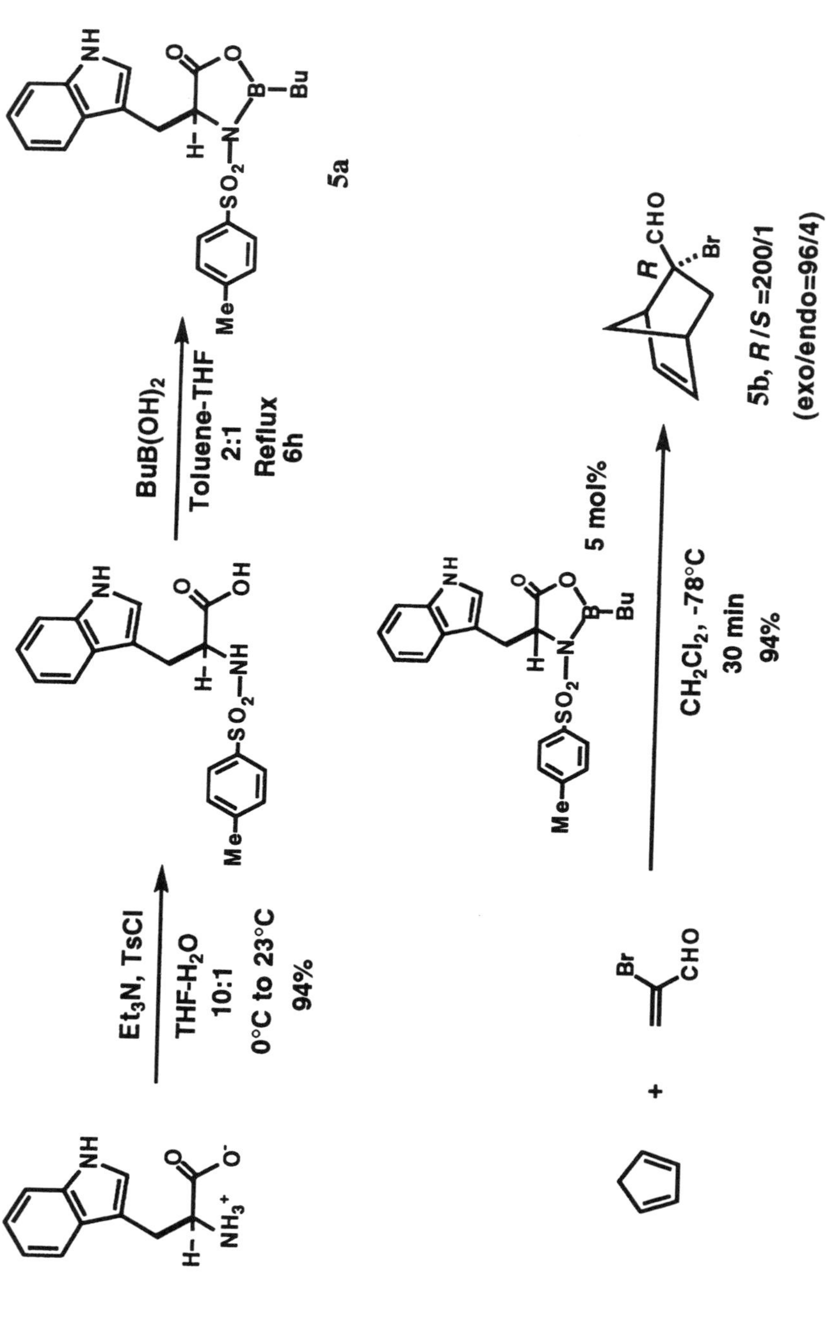

Figure 5. Preparation from S-tryptophan of oxazaborolidine **5a** and its application as an enantioselective Diels–Alder catalyst.

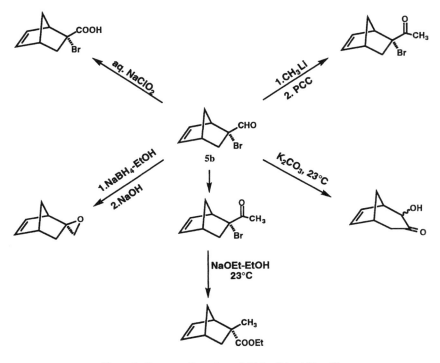

*Figure 6. Some transformations of Diels–Alder Adduct **5b**.*

1,3,5-trinitrobenzene mixture in CH$_2$Cl$_2$), indicating a π-acceptor–donor inter-action. (2) Catalysts in which the 3-indolylmethyl group of **9a** is replaced by phenyl or isopropyl are effective in catalyzing the Diels–Alder reaction, but are only weakly enantioselective and actually favor the *enantiomer* of **9c**. (3) The cata-lyst analogous to **9a** that is obtained from *N*-tosyl-(α,*S*, β*R*)-β-methyltryptophan is as effective as **9a** in accelerating and controlling the formation of adduct **9c**, as expected for transition-state assembly **9b.** It is even somewhat more enantiose-lective than **9a** as a catalyst for the reaction of cyclopentadiene and 2-methyl-acrolein. The geometry of the complex of the β-methyltryptophan-derived cata-lyst (**10a**, Figure 10) with 2-methylacrolein has been probed by NMR studies, especially low-temperature ¹H 2-D NOESY measurements (*11*). The NMR data demonstrate that the catalyst–aldehyde complex is conformationally fairly rigid and that the preferred molecular geometry approximates that depicted in **10b** of Figure 10 (*ca.* 55° HC$_\alpha$C$_\beta$H dihedral angle). The aldehyde complexation is rapidly reversible on the NMR time scale at 210 K as indicated by the chemical shifts for the aldehyde moiety (e.g., as compared to the static methylacrolein–BF$_3$ complex) and its sharp spectrum. That 2-methylacrolein is complexed to the face of boron which is proximate to the indole ring as indicated by the bright orange-

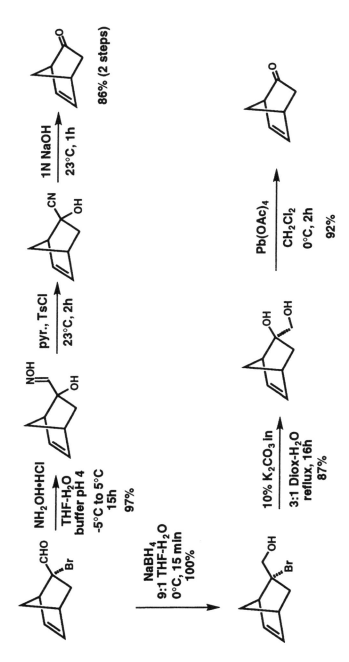

Figure 7. Efficient syntheses of chiral bicyclo[2.2.1]hepten-2-one.

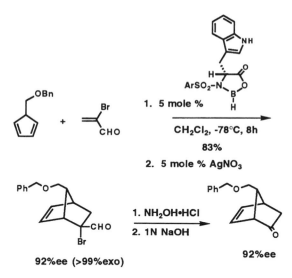

Figure 8. Enantioselective catalytic synthesis of a key prostaglandin intermediate.

9a

9b

9c

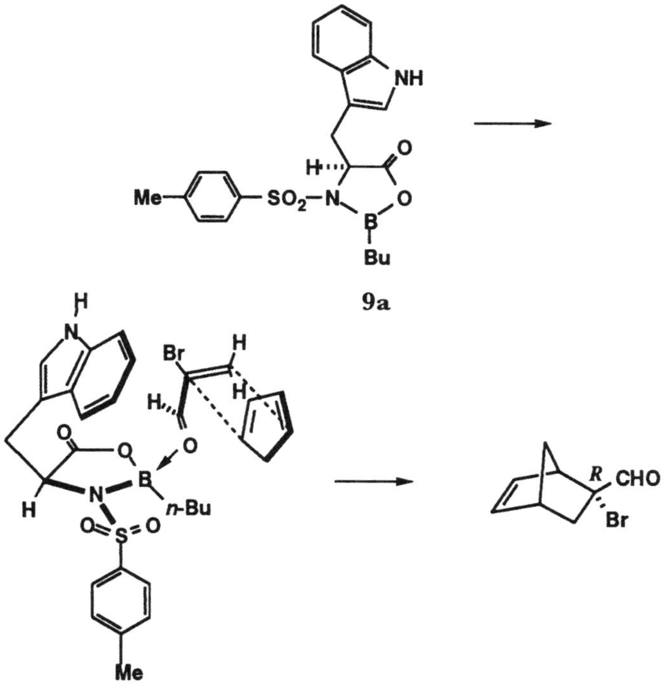

*Figure 9. Proposed transition-state assembly for the enantioselective addition of 2-bromoacrolein and cyclopentadiene catalyzed by **9a**.*

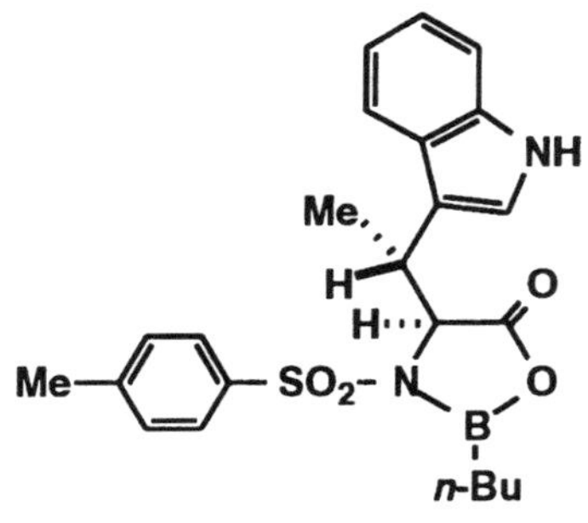

10a

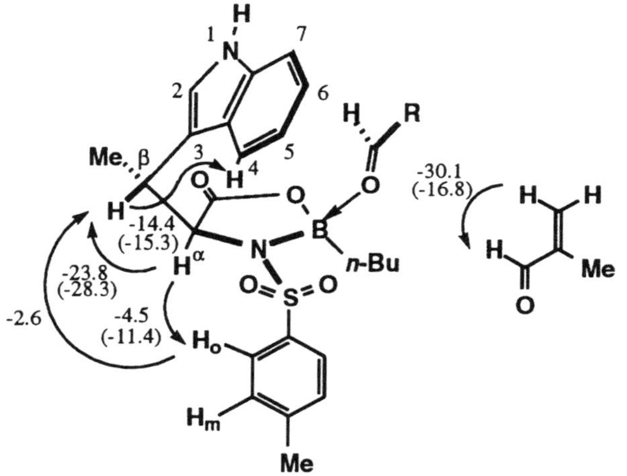

10b

*Figure 10. NOE data on the complex of catalyst **10a** with 2-methylacrolein.*

red color of the complex at 210 K, which fades upon warming to 250 K and reappears on cooling. This color, which corresponds to a broad absorption band in the 400 to 600 nm region, indicates π-complexation between the π-donor indole ring and the coordinated aldehyde, consistent with the sort of arrangement shown in **10b** and with a *ca.* 3 Å spacing between donor and acceptor elements. Further evidence for the proximity of the coordinated aldehyde and indole subunits in the complex derives from the fact that the ^{1}H NMR peak due to the CH$_3$ of 2-methylacrolein in the 1:1 mixture with **10a** moves *upfield* with decreasing temperature (from 1.79 δ at 262 K to 1.46 δ at 188 K in CD$_2$Cl$_2$) in contrast to the *downfield* methyl shift of a 1:1 mixture of 2-methylacrolein and BF$_3$ with decreasing temperature.

Since the NMR data indicate that the association of α,β-enal and catalyst is rapidly reversible on the NMR time scale, it is clear that complexes of both *s-cis* and *s-trans* forms of the α,β-enal are in rapid equilibrium and that enantioselection arises from a faster reaction from the *s-cis* complex **11a** than from the *s-trans* complex **11b** in Figure 11. One apparent reason for a preferred pathway via **11a** is that the transition state for addition of cyclopentadiene involves less steric repulsion than that from **11b.** Specifically, the sp^2 to sp^3 transformation of the α and β carbons in the transition state from **11b** entails serious repulsion between the α-bromine substituent and the indole ring, which maintains proximity because of the attraction with the electron deficient formyl carbon in the transition state. The transition state from **11a** is free from this repulsion involving bromine. The nearest indole atom neighbor of the formyl carbon in complex **11a** or **11b** is the nitrogen, a geometry which corresponds electronically (in the HOMO–LUMO sense) to that observed in the crystalline indole-1,3,5-trinitrobenzene complex (*12*), and which is therefore ideal for maintaining proximity between the formyl and indole subunits. In summary, the *s-cis* and *s-trans* complexes **11a** and **11b** expose opposite faces of the dienophile to attack by the diene, and the enantioselectivity of the reaction is a consequence of the considerable difference in transition-state energies, with that corresponding to reaction from the *s-cis* complex **11a** being the lower. It is of considerable interest in connection with these ideas that, as we have shown for the solid state (by X-ray diffraction) and for solution in CH$_2$Cl$_2$ (by low temperature ^{13}C and ^{1}H NMR studies), the only observed form of the BF$_3$-2-methylacrolein complex is that with *s-trans* geometry (*13*). The fact that the **9a** (or **5a**) -catalyzed reaction of cyclopentadiene and acrolein is not very enantioselective is consistent with this analysis for **11a** and **11b** pathways.

Using the ideas just described it has been possible to design a totally different oxazaborolidine system (derived from *S*-proline and *S*-naproxen) that catalyzes preferentially Diels–Alder addition to the *s-trans* form of 2-bromoacrolein or 2-methylacrolein (*14*). Detailed studies of this new system are still underway, but the results obtained thus far are in good agreement with the transition-state analysis described for the **11a, 11b** cases.

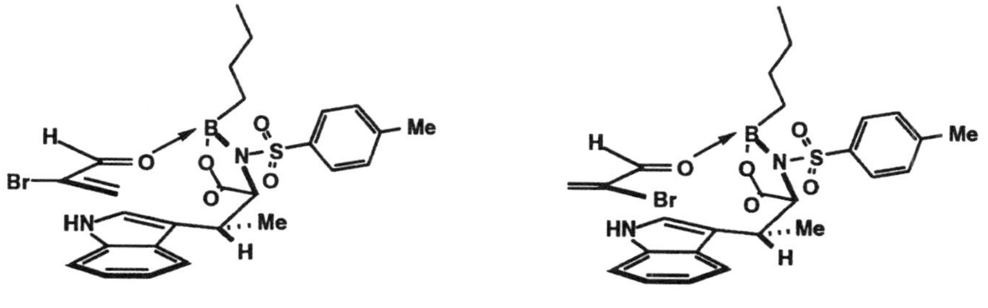

11a   **11b**

Figure 11. S-cis *and* s-trans *complexes of 2-bromoacrolein with Diels–Alder catalyst* **10a**.

Another effective catalyst for enantioselective Diels–Alder reactions has been developed which is based on the readily available chiral bisoxazolines **12a** (*15*) and **12b** (*16*) (Figure 12). Metal complexes of these rigid ligands provide catalysts of well defined molecular geometry which are very promising for a number of important reactions (*17*). The complex of **12b** with magnesium iodide or magnesium tetraphenylborate catalyzes the reaction of cyclopentadiene with 3-acryloyl-1,3-oxazolidin-2-one (**13a**) in CH$_2$Cl$_2$ at –50 °C to form the adduct **13c** (Figure 13) with 95.5/4.5 enantioselectivity and 98/2 *endo/exo* selectivity (84% yield) (*16*). The absolute stereochemistry of this reaction agrees with the expectation of a favored transition-state assembly **13b** in which tetrahedral magnesium binds and activates the bidentate dienophile and phenyl screening provides the basis for face-selective addition to the *s-cis* form of the dienophile (*16*). The use of the complex of **12b** with Cu(II) to catalyze the same reaction leads to fairly selective formation of the *enantiomer* of **13c**, as expected for a process via square planar complex of **12b·**Cu(II) with the dienophile **13a**. The complex of **12a** with ferric iodide is also an effective catalyst for the conversion of **13a** to **13c** probably via an octahedral cationic complex (**12a·**FeI$_2$**·13a**)$^+$ (*15*). An interesting feature of these three catalytic systems is that a cationic form of the metal complex is required to accelerate the reaction. Neutral species are clearly not sufficiently Lewis acidic because of the electron donating nature of the bidentate bisoxazoline ligand. These findings emphasize the important influence of ligand and charge on catalytic effectiveness.

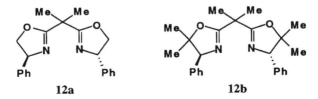

Figure 12. Chiral bisoxazoline ligands for enantioselective metal catalysis.

Narasaka and co-workers have discovered and applied the fascinating catalytic Diels–Alder summarized in Figure 14 (*18*). The structure of the catalyst (**14b**), which is supported by ^{1}H NMR data (*18a, 18b*), and the bidentate nature of the dienophile **14a** strongly suggest that the catalytic process involves the addition of the diene to an octahedral complex of **14a** and **14b**. Since there are *ten* possible diastereomeric octahedral complexes, and twice as many if both *s-cis* and *s-trans* conformers of the dienophile are considered, it is quite remarkable that substantial enantioselectivity was observed. Because of the importance of understanding the mechanistic–structural basis for such enantioselectivity, and our interest in the theory of catalytic enantioselective Diels–Alder processes, we extended our investigations to the Narasaka system with interesting results (*19*).

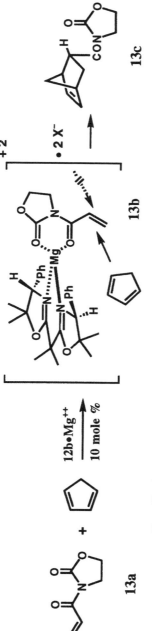

Figure 13. An enantioselective Diels–Alder reaction catalyzed by the magnesium bisoxazoline complex 12b.

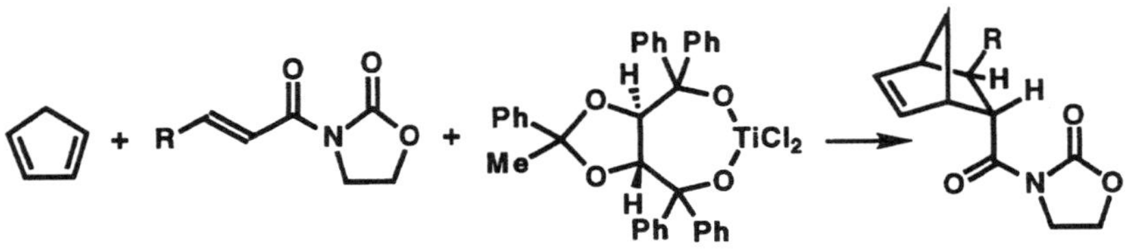

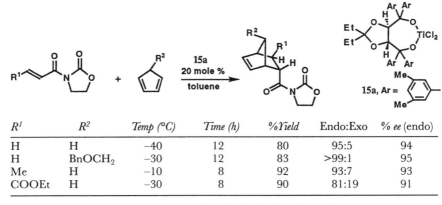

14a   **14b, 0.1 equiv**   **14c**

14c, R = H: 93% yield, 96 : 4 *endo : exo*, 64% ee (*endo*)
14c, R = Me: 87% yield, 92 : 8 *endo : exo*, 91% ee (*endo*)
14c, R = Ph: 72% yield, 88 : 12 *endo : exo*, 64% ee (*endo*)

Figure 14. The catalytic Diels–Alder system introduced by K. Narasaka (Tokyo University).

As is indicated in Figure 15, the replacement of phenyl in the Narasaka catalyst by 3,5-dimethylphenyl, which is more π-basic, produced a very superior catalyst (**15a**) that afforded greatly improved enantioselectivities. The importance of π-basicity in the aryl substituent is further indicated by the results summarized in Figure 16 with a series of catalysts **16a** in which the aryl group was varied (*20*). The superiority of catalyst **16a**, Ar = 3,5-dimethylphenyl, over **16a**, Ar = 3,5-bistrifluoromethylphenyl, or **16a**, Ar = 3,5-dichlorophenyl, provides convincing experimental evidence that the aryl group enhances enantioselectivity as a neighboring π-electron-rich substituent that simultaneously serves a stabilizing and steric screening function in the transition state. One attractive formulation of the transition-state assembly for the enantioselective pathway in the Narasaka system is depicted in Figure 17 (*19*).

R^1	R^2	Temp (°C)	Time (h)	%Yield	Endo:Exo	% ee (endo)
H	H	−40	12	80	95:5	94
H	$BnOCH_2$	−30	12	83	>99:1	95
Me	H	−10	8	92	93:7	93
COOEt	H	−30	8	90	81:19	91

Figure 15. An improved catalyst for the Narasaka reaction.

Catalyst 16a:

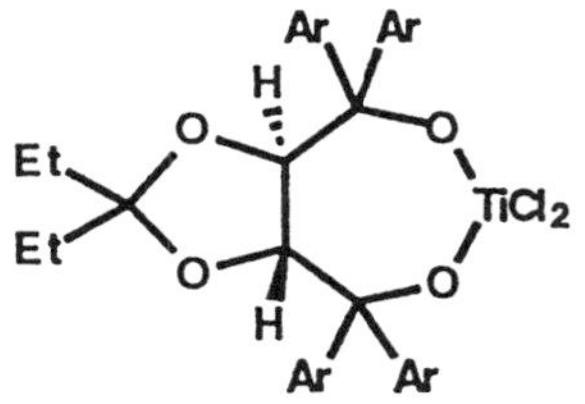

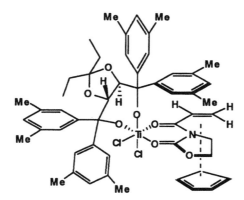

Catalyst, Ar	Temp (°C)	Time (h)	%Yield	Endo:Exo	% ee (endo)
Ph[a]	−50	12	70	93:7	73
β-Naphth[a]	−50	36	86	93:7	68
6-MeO-β-Naphth[a]	−50	40	76	90:10	82
3,5-diMe-Ph[a]	−40	12	80	95:5	94
3,5-diCF₃-Ph[b]	−50	20	81	92:8	68
3,5-diCl-Ph[b]	−50	18	75	93:7	44

[a]Catalyst prepared by Method A: (1) diol + Ti(O i-Pr)₄; (2) SiCl₄.
[b]Catalyst prepared from diol and TiCl₄ in ether, removal of solvent in vacuo, addition of toluene, and

Figure 16. The effect of variation of the aromatic substituent in the Narasaka system on catalytic enantioselectivity.

Figure 17. Possible transition-state assembly for the $(R^*O)_2TiCl_2$-catalyzed Diels–Alder reaction in the Narasaka system.

Enantioselective Addition of Carbon to Carbonyl Groups

The reduction of achiral ketones, R_LCOR_S, to chiral secondary alcohols, R_LCHOHR_S, by means of catalytic amount of oxazaborolidine **18a** (Figure 18) (or enantiomer, 5-10 mol%) with borane or catecholborane as the stoichiometric reductant provides a practical route to many useful chiral starting materials and reagents for synthesis (7, 9). Since the chiral precursor of catalyst **18a** can be re-

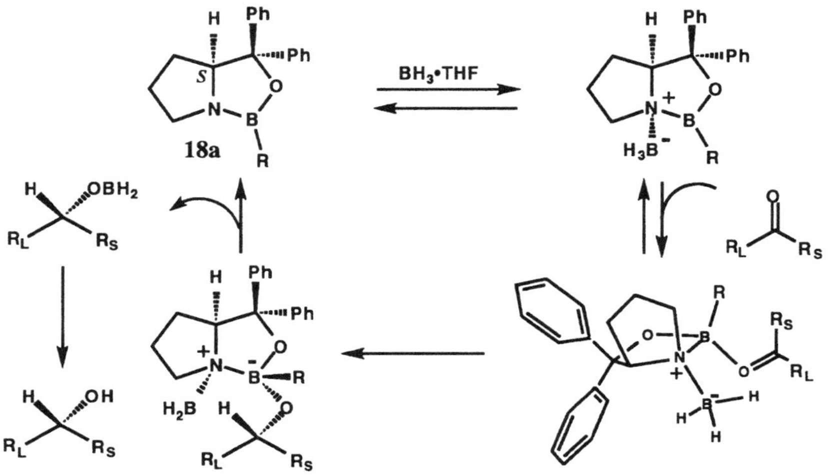

Figure 18. Enantioselective oxazaborolidine-catalyzed reduction of carbonyl compounds.

covered efficiently for reuse and since the stereochemistry of the reduction product is predictable from the mechanistic pathway shown in Figure 18, the application of this methodology has expanded rapidly. Catalyst **18a** or the enantiomer
has been used in the synthesis of natural products such as ginkgolides and
forskolin, and for making chiral α-amino and α-hydroxy acids, for example. Recent research in our laboratory has further broadened the utility of oxazaborolidine **18a** to include the catalysis of enantioselective nucleophilic ethynylation of
aldehydes.

The new enantioselective alkynylation process is outlined in Figure 19 (*20*).
The alkynylation reagent is either an alkynyldimethylborane or an alkynyldifluo-
roborane, and the addition reaction is accelerated and controlled sterically by the
proline-derived oxazaborolidine **19a**, which is employed in stoichiometric
amount at −78 °C in toluene solution. Thus the adduct of cyclohexane carboxaldehyde and phenylacetylene, (*R*)-phenylethynylcyclohexylcarbinol, is obtained
in 65% yield with 95/5 *R/S* enantioselectivity. Even better results have been obtained with the (4*S*,5*R*)-4,5-diphenyl-2-*n*-butyl-1,3,2-oxazaborolidine (**20a**) as the
chiral reagent for the process summarized in Figure 20. The oxazaborolidine
20a affords better yields and higher enantioselectivities in the ethynylation reaction than does the oxazaborolidine **19a**. A further advantage of **20a** is that it can
be used in sub-stoichiometric amount, which is quite practical since the catalyst
precursor, 1,2-diphenyl-2-aminoethanol, can be recovered efficiently from these
reactions for reuse. Although this new method is still under development, it
shows promise of exceptional utility. Some of the data obtained thus far are summarized in Figure 21. In a related development, the enantioselective vinylation of

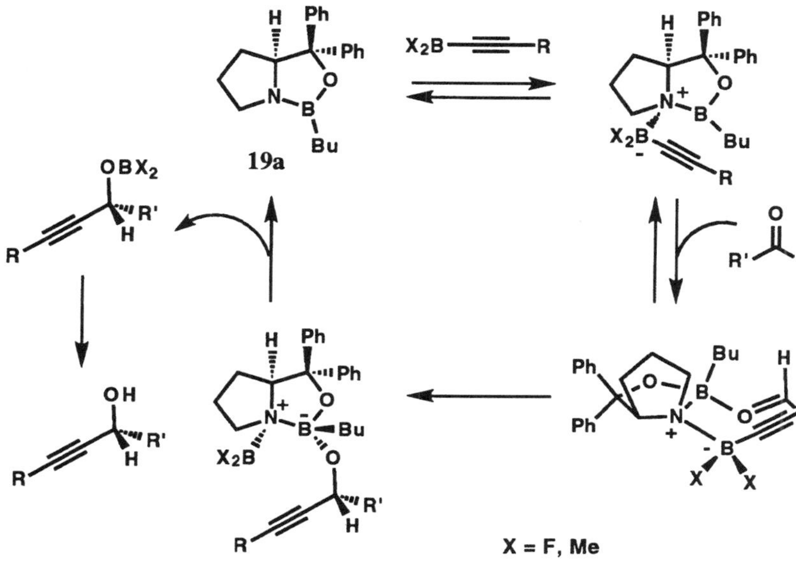

Figure 19. Stoichiometric ethynylation of aldehydes promoted by a proline-derived oxazaborolidine.

aldehydes using E-vinylzinc compounds and (−)-3-*exo*-(dimethylamino)isoborneol as catalyst has recently been described by Oppolzer (*21*).

The chiral, C_2-symmetric bisoxazoline system described earlier in connection with catalysis of enantioselective Diels–Alder reactions (*15, 16*), has considerable potential for application to other enantioselective processes (*17*). In this section a novel approach to the catalytic synthesis of chiral cyanohydrins is presented that features a pair of chiral catalysts. One catalyst is effectively a chiral Lewis acid that binds and activates an aldehyde, while the chiral co-catalyst binds and activates HCN, effectively to form a kind of chiral cyanide ion. This process is of practical as well as theoretical interest because of the versatility of cyanohydrins as synthetic intermediates (Figure 22). The importance of chiral cyanohydrins in synthesis is further attested to by the growing list of laboratories concerned with the problem of efficient enantioselective synthesis of this class (*22, 23*).

The workings of the new system are summarized in Figure 23 (*24*). The (S,S)-bisoxazoline magnesium complex **23a** is used as the catalyst for electrophilic activation and binding of the aldehyde component. The bisoxazoline **23b** is employed as co-catalyst for binding and activating HCN, which is present in only catalytic (trace) amount with trimethylsilylcyanide being used as the stoichiometric reagent. Using this system with 20 mol% of Lewis acid **23a** and 12 mol% of bidentate base **23b** in propionitrile-methylene chloride at −78 °C,

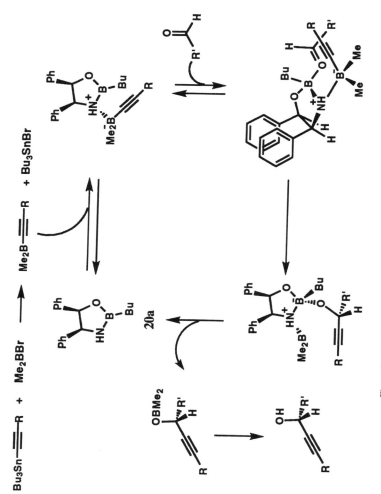

Figure 20. Catalytic enantioselective ethynylation of aldehydes.

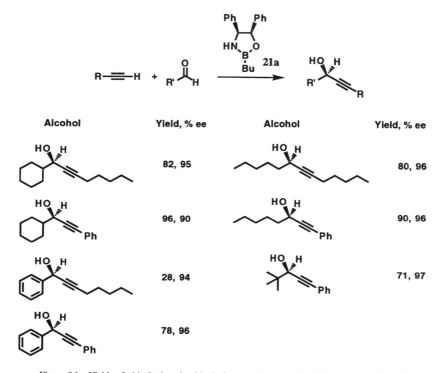

Figure 21. *Yields of chiral ethynylcarbinols from reactions catalyzed by oxazaborolidine* **21a**.

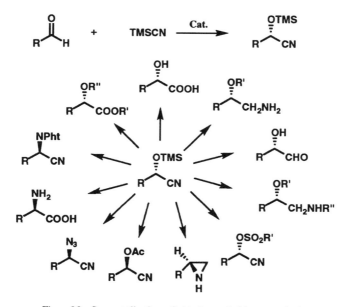

Figure 22. *Some applications of chiral cyanohydrins to synthesis.*

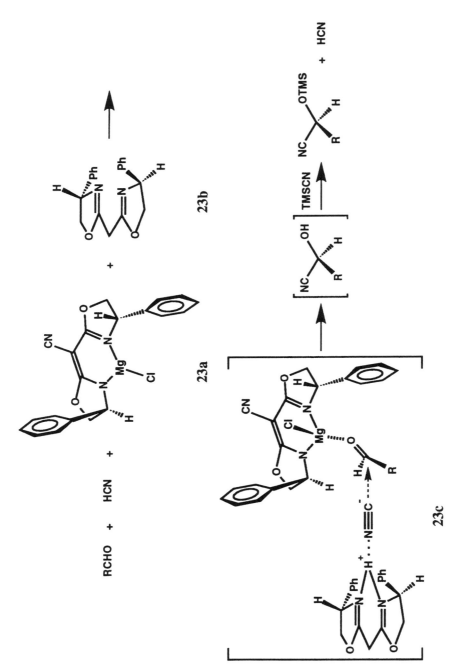

Figure 23. Enantioselective, catalytic synthesis of cyanohydrin trimethylsilyl ethers from aldehydes and trimethylsilylcyanide using a dual chiral catalyst system, **23a** + **23b**.

enantioselectivities for (*S*)-cyanohydrin formation in the range 97/3 to 95/5 are obtained for a range of aldehyde substrates including *n*-heptanal, cyclohexane carboxaldehyde, and trimethylacetaldehyde. The predominance of (*S*)-cyanohydrin in each case is consistent with the pathway shown in Figure 23. The steric screening of the aldehyde by the phenyl groups of the magnesium complex **23a** and the dissymmetry of the bisoxazoline-complexed HCN combine to favor transition-state assembly **23c** quite strongly. With the less reactive conjugated and aromatic aldehydes, lower enantioselectivities are observed.

Support for the pathway for enantioselective cyanohydrin formation shown in Figure 23 is provided by experiments using cyclohexane carboxaldehyde in which the catalytic components have been varied. Using the electrophilic catalyst **23a** alone with trimethylsilylcyanide and no bisoxazoline co-catalyst **23b** resulted in a much slower reaction and (*S*)-cyanohydrin product of only 65% ee. Further, the reaction of cyclohexane carboxaldehyde and trimethylsilylcyanide in the presence of 20 mol % of **23a** and 12 mol % of the *enantiomer* of bisoxazoline **23b** produced silylated (*S*)-cyanohydrin (90% yield, −78 °C, standard condition) of only 38% ee. These data indicate a synergistic action of catalyst **23a** and co-catalyst **23b**, which clearly argues for the intervention of the four-component ensemble shown **23c**, or equivalent. Although only a trace of free HCN is necessary for the process to proceed, a faster rate of reaction is observed when additional HCN is generated by the addition of, for example, 0.05 mol % of methanol to the reaction mixture. These data clearly favor ensemble **23c**, but the case for this pathway requires additional study. Nonetheless, **23c** is intriguing as a harbinger of more complex and sophisticated catalytic ensembles that remain to be discovered.

Acknowledgment

I am deeply grateful to the extremely talented and devoted group of co-workers who have carried out the studies described herein. It has been a great pleasure and privilege to be associated with this group, which includes Yi Bin Xiang, Stanislaw Pikul, Nobuyuki Imai, Sepehr Sarshar, Teck-Peng Loh, Thomas D. Roper, Mark C. Noe, Mihai Azimioara, Hung-Yue Zhang, Kazuaki Ishihara, Yasushi Matsumura, Karlene A. Cimprich, Graham B. Jones, and Zhe Wang. This research was assisted financially by grants from the National Institutes of Health, the National Science Foundation, and Pfizer Inc.

References

1. Corey, E. J. *Abstracts of Papers*, 31st National Org. Symposium, 1989; 1.
2. Corey, E. J.; Ensley, H. E. *J. Am. Chem. Soc.* **1975**, *97*, 6908.
3. *See also* Corey, E. J.; Becker, K. B.; Varma, R. K. *J. Am. Chem. Soc.* **1972**, *94*, 8616.

4. Corey, E. J.; Imwinkelried, R.; Pikul, S.; Xiang, Y. B. *J. Am. Chem. Soc.* **1989**, *111*, 5493.
5. Corey, E. J.; Imai, N.; Pikul, S. *Tetrahedron Lett.* **1991**, *32*, 7517.
6. Pikul, S.; Corey, E. J. *Org. Synth.* **1992**, *71*, 22, 30.
7. Corey, E. J. *Pure Appl. Chem.* **1990**, *62*, 1209.
8. Corey, E. J.; Sarshar, S.; Bordner, J. *J. Am. Chem. Soc.* **1992**, *114*, 7938.
9. *See* Corey, E. J.; Link, J. O. *J. Am. Chem. Soc.* **1992**, *114*, 1906; Corey, E. J.; Link, J. O. *Tetrahedron Lett.* **1992**, *33*, 3431.
10. Corey, E. J.; Loh, T.-P. *J. Am. Chem. Soc.* **1991**, *113*, 8966.
11. Corey, E. J.; Loh, T.-P.; Roper, T. D.; Azimioara, M. D.; Noe, M. C. *J. Am. Chem. Soc.* **1992**, *114*, 8290.
12. Hanson, A. W. *Acta Crystallogr.* **1964**, *17*, 559.
13. Corey, E. J.; Loh, T.-P.; Sarshar, S.; Azimioara, M. *Tetrahedron Lett.* **1992**, *33*, 6945.
14. Corey, E. J.; Rohde, J. J., unpublished results.
15. Corey, E. J.; Imai, N.; Zhang, H.-Y. *J. Am. Chem. Soc.* **1991**, *113*, 728.
16. Corey, E. J.; Ishihara, K. *Tetrahedron Lett.* **1992**, *33*, 6807.
17. For other studies of chiral bis(oxazolines) for enantioselective catalysis *see* (a) Lowenthal, R. E.; Abiko, A.; Masamune, S. *Tetrahedron Lett.* **1990**, 31, 6005; (b) Evans, D. A.; Woerpel, K. A.; Hinman, M.; Faul, M. M. *J. Am. Chem. Soc.* **1991**, *113*, 726; (c) Nishiyama, H.; Kondo, M.; Nakamura, T.; Itoh, K. *Organometallics* **1991**, *10*, 500; (d) Lowenthal, R. E.; Masamune, S. *Tetrahedron Lett.* **1991**, *32*, 7373.
18. (a) Narasaka, K.; Iwasawa, N.; Inoue, M.; Yamada, T.; Nakashima, M.; Sugimori, J. *J. Am. Chem. Soc.* **1989**, *111*, 5340; (b) Iwasawa, N.; Hayashi, Y.; Sakurai, H.; Narasaka, K. *Chem. Lett.* **1989**, 1581; (c) Iwasawa, N.; Sugimori, J.; Kawase, Y.; Narasaka, K. *Chem. Lett.* **1989**, 1947; (d) Narasaka, K.; Inoue, M.; Yamada, T.; Sugimori, J.; Iwasawa, N. *Chem. Lett.* **1987**, 2409. (e) Narasaka, K.; Inoue, M.; Yamada, T. *Chem. Lett.* **1986**, 1967. (f) Narasaka, K.; Inoue, M.; Okada, N. *Chem. Lett.* 1986, 1109.
19. Corey, E. J.; Matsumura, Y. *Tetrahedron Lett.* **1991**, *32*, 6289.
20. Corey, E. J.; Cimprich, K. A.; Jones, G. B., unpublished results.
21. Oppolzer, W.; Radinov, R. N. *Helv. Chim. Acta* **1992**, *75*, 170.
22. (a) Reetz, M. T.; Kunish, F.; Heitmann, P. *Tetrahedron Lett.* **1986**, *27*, 4721; (b) Narasaka, K.; Yamada, T.; Minamikawa, H. *Chem. Lett.* **1987**, 2073; (c) Minamikawa, H.; Hayakawa, S.; Yamada, T.; Narasaka, K. *Bull. Chem. Soc. Jpn.* **1988**, *61*, 4379; (d) Kobayashi, S.; Tsuchiya, Y.; Mukaiyama, T. *Chem. Lett.* **1991**, 541. (e) Hayashi, M.; Matsuda, T.; Oguni, N. *J. Chem. Soc., Chem. Commun.* **1990**, 1364; (f) Nitta, H.; Yu, D.; Kudo, M.; Inoue, S. *J. Am. Chem. Soc.* **1992**, *114*, 7969.
23. For enzymatic methods *see* (a) Ognyanov, V. I.; Datcheva, V. K.; Kyler, K. S. *J. Am. Chem. Soc.* **1991**, *113*, 6992; (b) Inagaki, M.; Hiratake, J.; Nishioka, T.; Oda, J. *J. Am. Chem. Soc.* **1991**, *113*, 9360.
24. Corey, E. J.; Wang, Z. *Tetrahedron Lett.* **1993**, *34*, 4001.

4

Enzyme-Catalyzed Reactions

David L. Coffen

The application of enzyme catalyzed reactions for the preparation of chiral compounds in enantiomerically pure (or enriched) form is reviewed, with emphasis on practical aspects and experimental techniques. Relevant aspects of enzymology are discussed, including preferred types of enzymes, scope and mechanisms of enzyme catalysis, and the complementarity with microbial transformation of organic compounds. Discussion of experimental details covers addressable functional groups, planning a biocatalytic step in a multi-step synthesis, commercial enzyme sources, reaction medium, pH and temperature control, in-process analysis and control, product isolation, stereochemical product analysis, and scale-up issues such as enzyme reuse and recycle of "wrong" enantiomer. The chapter concludes with a table of 66 typical examples covering several compound classes, presented with structural formulas and an extensive bibliography.

The several thousand bioactive substances used in medicine and agriculture are obtained by extraction and purification from natural sources, by semisynthesis from naturally derived materials, or by total synthesis from (primarily) petroleum- and coal-based raw materials. When present, stereochemical features in materials obtained by the first and second methods are treated as essential components of molecular structure. However, stereochemical features in materials produced by the third method (total synthesis) are treated with pragmatic ambivalence. To be considered pure, a compound with double bonds must be *cis* or *trans*, not a mixture of both isomers. When two or more chiral centers are present, the prevailing standards in chemical technology assert that a pure compound must be a single diastereomer, but chiral compounds in general can be either racemic or single enantiomers. About 85% of commercially produced synthetic bioactive chiral compounds are racemates.

Molecular biology and biochemistry have established that the bioactivity of bioactive substances is universally expressed in chiral microenvironments created by the chirality of amino acids—or of ribose in a few instances. When a bioactive substance is itself chiral, it is highly unlikely (and rarely observed) that the two enantiomers will be pharmacologically and toxicologically equivalent. The interactions of proteins with small chiral molecules are, in general, highly stereospecific, giving rise to both a clear preference for single enantiomers when producing chiral drugs and a sound basis for the use of enzyme-catalyzed reactions for their production in single-enantiomer form.

It is worth noting that a new impetus for the production of chiral bioactive substances in single-enantiomer form has arisen, one that is unrelated to their interactions with proteins. The growing sophistication of drug discovery methods results in increasingly complex drug molecules, frequently with multiple chiral centers. Particularly for compounds that have modular structures assembled from two or more chiral components, production as single enantiomers is both technically and economically preferable to production as racemates.

In principle, the scope of enzyme catalysis in organic synthesis is unbounded—a perception borne out by the seemingly endless diversity of structures for natural products. In practice, however, the range of laboratory and plant applications is highly restricted by practical realities, despite rapid growth in this field over the past two decades. To understand both the utility and the current limitations of enzyme-catalyzed reactions (*1, 2, 3*), a brief look at some relevant aspects of enzymology is helpful.

Relevant Aspects of Enzymology

Enzymologists recognize six classes of enzymes:

- oxidoreductases
- transferases
- hydrolases
- lyases
- isomerases
- ligases

In some respects, oxidoreductases should be the premier class of enzymes for use in organic synthesis. Transformations such as ketone reduction and remote hydroxylation frequently entail the enantiospecific generation of new chiral centers. However, the reductases and oxidases involved in such reactions are strongly coupled with cofactor redox cycles. Moreover, pure or semipure preparations of such enzymes are scarce. Live-organism processes using microbial transformation offer more practical alternatives. Microbial transformation as a technique in organic synthesis is generally complementary to enzyme catalysis and is outside the scope of this chapter. Comprehensive reviews are available in various monographs (*3, 4*).

Transferases, particularly aminotransferases, show considerable promise in organic synthesis (*5*), as described later in this chapter.

Hydrolases are by far the most important class of enzymes for chiral separations and technology, and their applications in this field have been developed extensively during the past two decades. Lyases, isomerases, and ligases, while relevant to certain aspects of organic synthesis, have not been extensively applied to chiral separations.

The hydrolases useful as catalysts in organic synthesis and chiral separations are generally serine proteases, esterases, and lipases. Members of these three groups have a wide diversity of substrate types but share an important feature in their catalytic domains: the catalytic triad of discontiguous residues made up of aspartate, histidine, and serine (DHS) (*6*). Glutamate replaces aspartate in some instances (e.g., in acetylcholine esterase) (*7*), and cysteine replaces serine in papain.

The position of the DHS triad in the catalytic site allows facile proton exchange among the three side chains, and the deprotonated serine side chain is acylated by the substrate in the key step of hydrolysis (Scheme I). Overlaid on this constant feature of the catalytic site is a remarkable variation in binding-pocket topography that allows a wide range of substrate structures to be processed by hydrolases of different types and from different sources. Functional

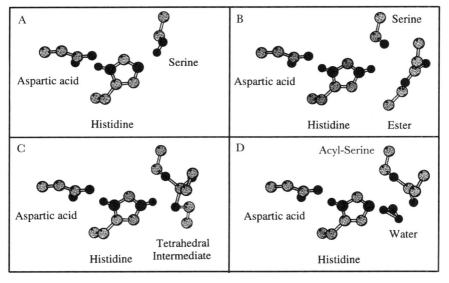

Scheme I.

recognition of stereochemistry in both the alcohol and the acid moieties of an ester can be maintained with little regard for other structural features of the substrate.

One important feature of this process that is evident from the mechanism depicted is that catalytic activity cannot be maintained under acidic conditions. At pH values much below 7, the aspartate side chain is protonated, so the catalytic hydrolysis is self-regulating. At high pH values, uncatalyzed hydrolysis may intervene, causing loss of the desired selectivity. Therefore, to maintain an adequate rate of enzyme-catalyzed hydrolysis, pH must be carefully controlled, preferably in the 7.0–8.0 region. With small-scale reactions, pH can be controlled with a suitably buffered reaction medium. For larger scale reactions, continuous addition of base using a pump controlled by the output signal of a pH meter is a preferred method, as discussed later.

A second important feature of the process stems from the reversibility of all the component steps. The process is generally more responsive to the law of mass action than to the laws of thermodynamics and, through judicious choice of experimental conditions, can be driven in either the synthetic direction or the hydrolytic direction—much as the balance between Fisher esterification and acid-catalyzed ester hydrolysis is controlled by experimental conditions.

A third feature of the catalytic process, one that is not implied in Scheme I, is that the topography of the catalytic binding pocket can recognize and exhibit kinetic bias toward stereochemical features in *both* the acid and the alcohol moieties of the ester substrate—in some cases simultaneously. The example shown in Scheme II illustrates this point and also exemplifies the reversible nature of the reaction. It also illustrates yet another dimension of hydrolase catalysis: transesterification.

A thorough understanding of stereochemical outcomes such as that in Scheme II requires precise knowledge of the protein–substrate–product binding interactions and kinetics. Armed with such information, one can predict the results for a particular enzyme–substrate combination—for example, (R) or (S) product. However, very little information of this kind exists. X-ray crystallographic protein structures of a number of serine proteases have been determined, largely in the context of their relevance to key metabolic or disease processes. But the esterases and lipases of greatest utility in organic chemistry have only recently begun to attract the attention of molecular biologists and protein crystallographers. Human pancreatic lipase and a few fungal lipases are among those whose structures have been determined (9, 10).

Nevertheless, the abundance of examples of enzyme-catalyzed stereoselective hydrolyses (and esterifications) accumulating in the literature has made it possible to draw some predictively useful inferences about the protein–substrate binding interactions involved (11), particularly when a specific enzyme is applied to a closely related series of substrates (12). For example, with 18 of 28 3-substituted dimethyl glutarates tabulated by Ohno and Otsuka (11), hydrolysis catalyzed by pig liver esterase favored the enantiomer depicted in Scheme III.

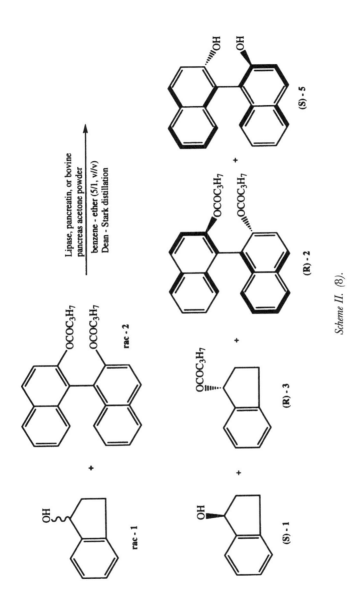

Lipase, pancreatin, or bovine
pancreas acetone powder
benzene - ether (5/1, v/v)
Dean - Stark distillation

Scheme II. (8).

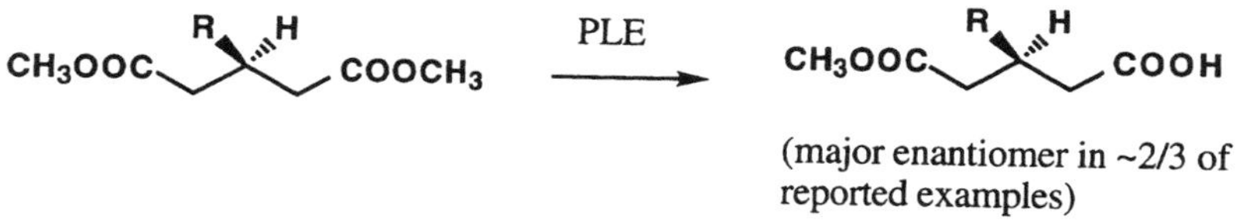

Scheme III. (11).

Predictive tools (*13*) must be used with caution when changes in enzyme type or source are involved or when a chiral moiety is transposed from the alcohol to the acid part of an ester. The series of *d*-biotin precursors shown in Scheme IV illustrates this point. The product in each instance has been converted to the lactone shown in Scheme V, which is an established synthetic precursor of *d*-biotin.

Planning a Biocatalytic Step

When biocatalysis is contemplated as a component of a synthesis scheme for obtaining an enantiomerically pure product, several questions must be answered.

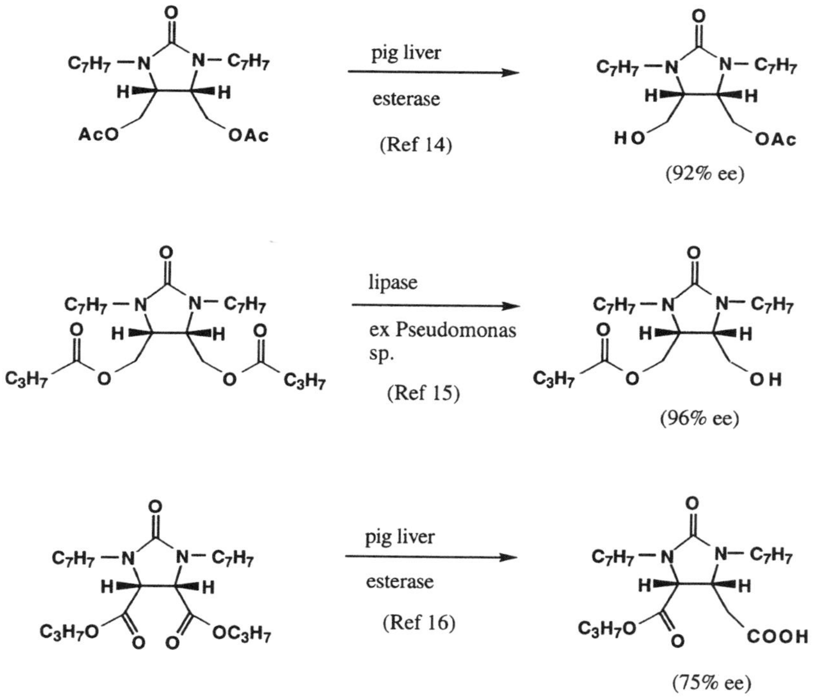

Scheme IV.

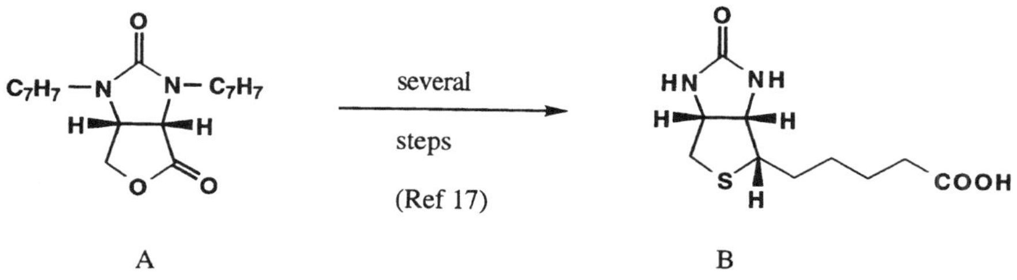

Scheme V. (17).

One of the first is whether to use biotransformation by intact organisms or kinetic resolution catalyzed by an isolated enzyme. If the latter is selected, a choice must then be made between the hydrolytic and synthetic modes of carrying out such reactions. A third question is at what point in a multistep synthesis biocatalysis can be most expediently employed.

Biocatalysis is a highly empirical activity, and the best answers to such questions are determined by the specific application. Nevertheless, some useful generalizations can be drawn. If biocatalysis is to be applied in the step that creates the chiral center of interest, biotransformation often serves better than isolated-enzyme catalysis. This includes steps in which trigonal carbon atoms become tetrahedral, such as additions to olefins and ketone reductions, and steps in which a methylene group is hydroxylated, but not steps in which one of two identical functional groups in a prochiral or *meso* setting is manipulated to create the chiral center. Often, however, microbial fermentation is used simply to provide convenient access to a highly enantioselective hydrolase that is not available in isolated form from commercial sources.

Problems such as enzyme availability and cofactor cycling can frequently be handled more adeptly by microorganisms than by chemists. However, the chemist is then faced with another set of problems:

- Which organism to use?
- Where to obtain an inoculum of the organism selected?
- How to culture it?
- How to sterilize media and equipment and exclude other microbes?
- How to bring the substrate into the medium?
- How much of the substrate will be consumed as a nutrient?
- How to isolate the product from a complex mixture of nutrients, growth factors, and other metabolites?

Conversely, if biocatalysis is applied at a stage when the chiral center already exists, the isolated enzyme is often preferable. In this case, the planning process should take into account the range of functional groups addressable with enzymes and the types and sources of enzymes available. The proximity of the chiral center to the addressable group and the physicochemical properties (e.g., solubility) of the intermediates must also be considered. Amines, amides, alcohols, carboxylic acids, esters, and nitriles are the best functional groups to work with.

After a suitable combination of enzyme and functional group is selected, the merits of synthetic versus hydrolytic modes must be evaluated. Beginning with the work of Klibanov (*18*), the use of hydrolases in organic solvents to synthesize esters from alcohols and carboxylic acids by direct esterification has become widespread. Scheme VI shows an example of this method in a case for which a number of hydrolytic schemes failed. The unreacted alcohol had all the functional and stereochemical features required for incorporation into the side chain of the vitamin D_2 metabolite shown.

We now come to the third question: At what point in a multistep synthesis is biocatalysis best applied? Different considerations give different answers.

At or near the beginning means that the maximum amount of material has to be processed in the biocatalytic step. This is a disadvantage if the biocatalytic step is complicated to carry out or requires an expensive enzyme, or if the product isolation procedure is difficult. The distinct advantage is that only half as much material has to be taken through the downstream steps. If the downstream chemistry puts the integrity of a key chiral center at risk (e.g., if the center is α to a carbonyl group), a late-stage resolution is preferable. When highly crystalline intermediates appear in downstream steps, the opportunity to raise enantiomeric purity by recrystallization favors early-stage resolution.

At or near the end means that the minimum amount of material has to be processed in the biocatalytic step. For this to be an advantage, it must override the fact that (at least) two moles of material must be taken through the entire synthesis for each mole of final product desired and that (at least) half of the high-value substance synthesized must be discarded.

When the contemplated scale is large enough for this question to matter (e.g., in commercial drug production), the best answer can be found by determining where in a multistep synthesis a realistic opportunity to recycle (reracemize or resymmetrize) the unwanted enantiomer occurs.

To summarize, the best placement of a biocatalytic step in a multistep synthesis should represent the best compromise of four issues:

- optimized throughput in the catalytic step
- optimized throughput in the overall process
- optimal choice of a functional group to serve as a "resolution handle"
- opportunity to recycle unwanted enantiomers

To assist with planning, a list of published examples covering each of the addressable functional groups mentioned earlier is provided (Scheme VII).

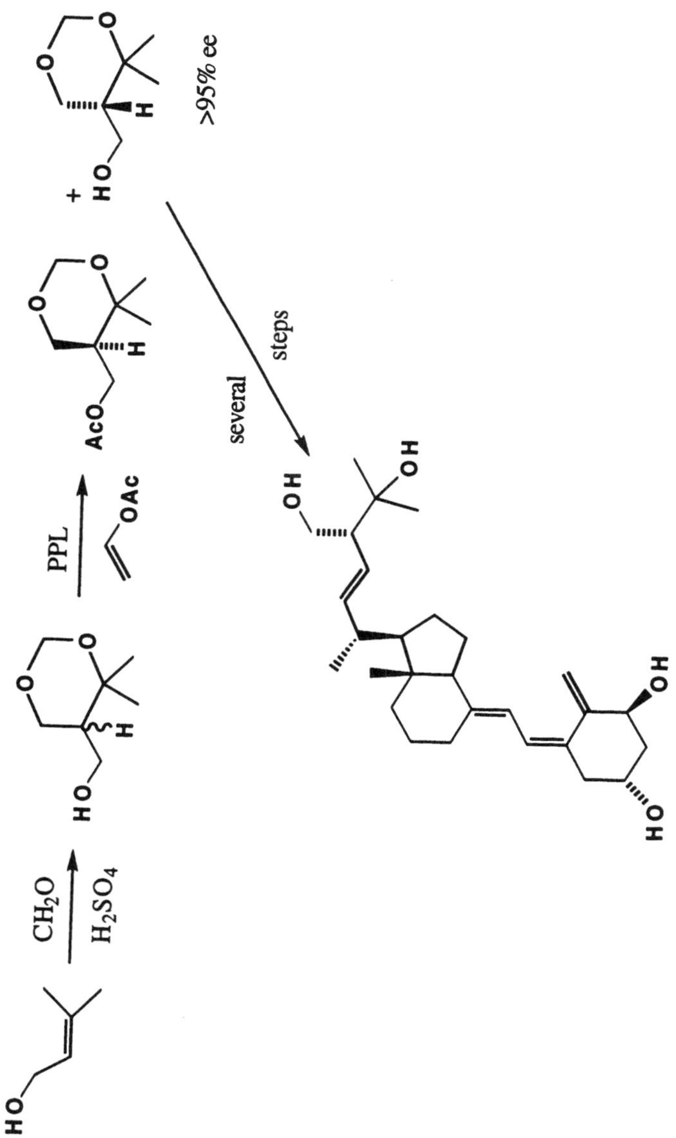

Scheme VI. (19).

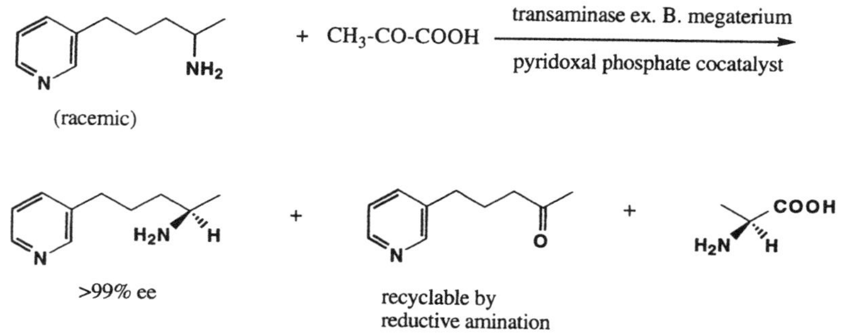

Scheme VIIa. Amines (20).

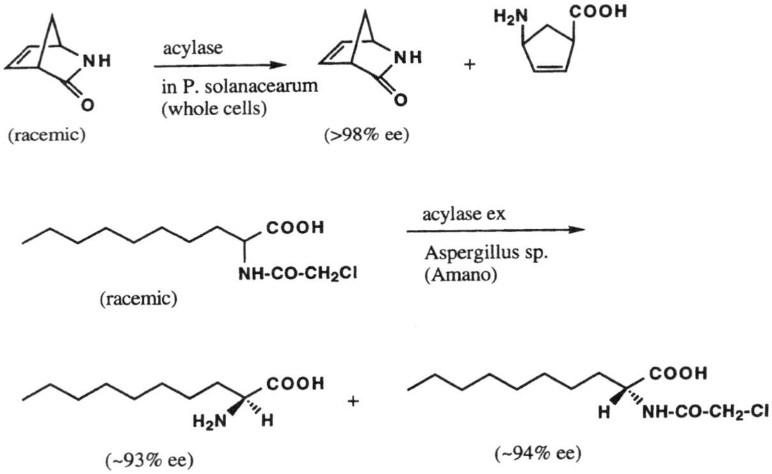

Scheme VIIb. Amides (21, 22).

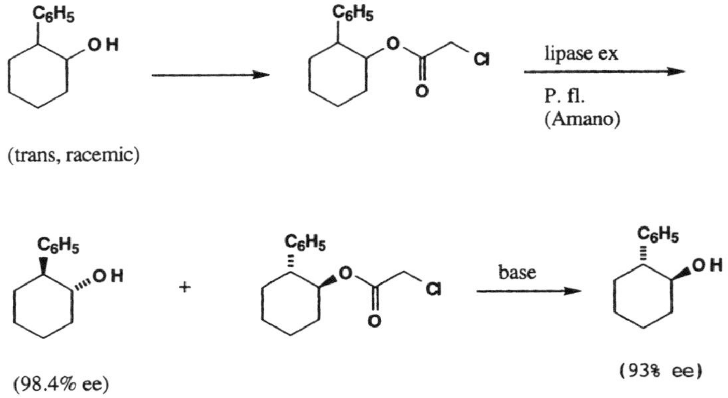

Scheme VIIc. Alcohols (23).

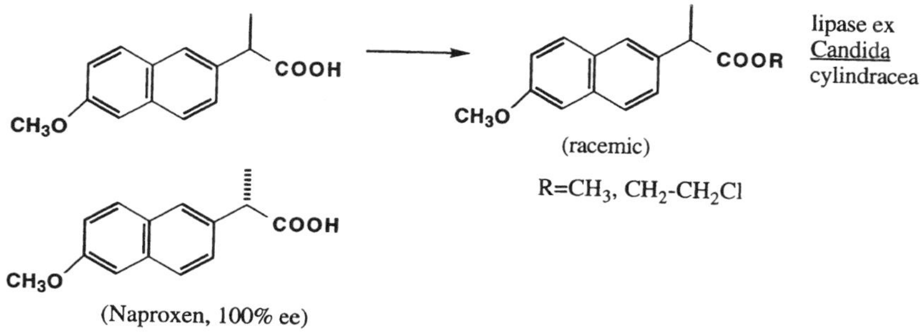
COOH
CH3O
COOR
lipase ex
Candida
cylindracea
(racemic)
R=CH3, CH2-CH2Cl
(Naproxen, 100% ee)

Scheme VIId. Carboxylic acids (24, 25).

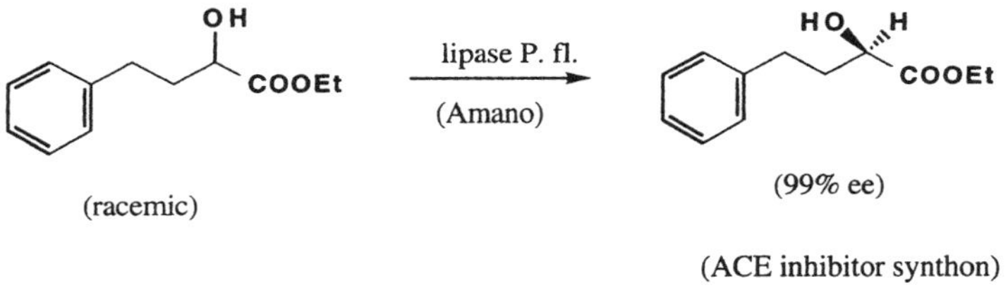
OH
COOEt
lipase P. fl.
(Amano)
HO
H
(racemic)
(99% ee)
(ACE inhibitor synthon)

Scheme VIIe. Esters (26).

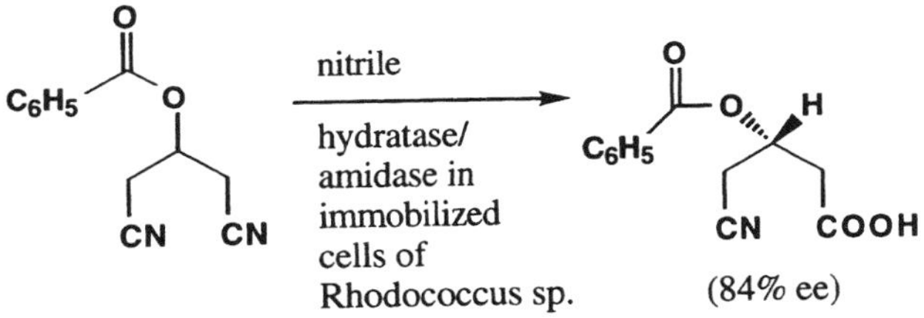
O
C6H5
CN
COOH
nitrile
hydratase/
amidase in
immobilized
cells of
Rhodococcus sp.
H
(84% ee)

Scheme VIIf. Nitriles (27).

Running an Enzyme-Catalyzed Reaction

In any organic synthesis project, once an overall synthesis strategy has been formulated, the planning of tactical details is best done on the basis of closest known analogies for each step. Many enzyme-catalyzed reaction procedures have been published in precise and reliable form, and they provide useful analogies for deriving experimental conditions for a new reaction. Recent volumes of *Organic Syntheses* contain several such examples (see additional examples at end of this chapter) (*23, 28–30*), and the practical compendium edited by Roberts, Wiggins, and Casy (*31*) offers many more.

Ten important experimental concerns are discussed in the subsections that follow.

Types and Sources of Enzymes

To carry out their intended functions, certain enzymes must be secreted from the cells in which they are expressed. Harvesting such enzymes from a production broth generally requires nothing more than precipitation from the medium after removal of the cells. Expression levels are also frequently higher for secreted enzymes, so they are generally cheaper and more readily available than intracellular enzymes.

Some of the enzymes frequently used in chiral separations are listed in Table I, together with commercial sources. Many other companies also provide enzymes of use in this context, and those listed offer many additional products of this type.

Organic Reaction Medium

The general conditions developed by Klibanov and subsequent workers for applying hydrolases in a synthetic sense require aprotic organic solvents with low, but nonzero, water content. Complete dehydration can cause protein denatura-

Table I. Enzymes Used in Chiral Separations

Enzyme	Source	Substrates
Subtilisin (ex *Bacillus* sp.)	A, B, N, S	amides, esters
α-Chymotrypsin	B, S	amides, esters
Papain	A, B, S	amides, esters
Acetylcholine esterase (AChE)	S	esters
Pig liver esterase (PLE)	A, S	esters
Pig liver acetone powder (crude PLE)	S	esters
Cholesterol esterase (various sources)	A, B, S	esters
Porcine pancreatic lipase (PPL)	A, S	esters
Steapsin (crude PPL)	S	esters
Microbial lipase ex *Candida cylindracea*	A, B, S	esters
Microbial lipase ex *Pseudomonas cepacia* (or *fluorescens*)	A, S	esters
Microbial lipase ex *Mucor* sp.	A, N	esters
Microbial lipase ex *Candida antarctica*	N	esters
Transaminases	B, S	amines, ketones
Alcohol dehydrogenases	B, S	alcohols, including prochiral diols
Aldolases	B, S	specific ketone or aldehyde aldol condensations

Notes: A is Amano International Enzyme Co., P.O. Box 1000, Troy, VA 22974 (tel. 800-446-7652); B is Boehringer Mannheim Corp., 9115 Hague Road, P.O. Box 50414, Indianapolis, IN 46250-0414 (tel. 800-428-5433); N is Novo Nordisk A/S, Novo Allé, 2880 Bagsvaerd, Denmark (tel. t45 4444 8888); S is Sigma Chemical Company, P.O. Box 14508, St. Louis, MO 63178-9916 (tel. 800-325-5832).

tion and loss of catalytic activity. Toluene, benzene–ether, and hexane are among the solvents used. The enzyme may be contained in a reverse micellar dispersion through which reactants and products freely migrate. Since products and reactants (when transesterification is used) are neutral, pH control is not important. Excess water and volatile by-products that may impede the progress of the reaction can be removed by azeotropic codistillation with solvent under reduced pressure (*8*).

Aqueous Reaction Medium

Water is, in many respects, the ideal medium for an enzyme-catalyzed reaction; the more obvious reasons include the facts that water is environmentally preferred for all solution-phase chemistry and that enzymes evolved to function in essentially aqueous systems. Measurement and control of pH are also more straightforward in water. Unfortunately, the vast majority of organic compounds has limited solubility in water. The preference for water as a reaction medium is usually reconciled with the poor water solubility of many organic compounds in one of two general ways. Water-miscible organic cosolvents can be used at levels that greatly enhance substrate solubility without denaturing the enzyme. Experience suggests that tetrahydrofuran (THF) and dimethylformamide (DMF) should be lead candidates in the *empirical* selection of a suitable cosolvent, but other water-miscible solvents can be used. The presence of a cosolvent can affect the enantioselectivity (*32–34*). The second method is to focus on lipases in selecting an enzyme catalyst for the reaction. As their name implies, lipases were evolved to process lipids in living organisms. Therefore, a substrate for a lipase need only be presented to the enzyme as a liquid; it need not be dissolved in the medium. Lipases preferentially act on the surface micelles in a lipid suspension while the enzyme remains in the aqueous phase. This characteristic can be used to an advantage for water-insoluble solid substrates. The solid can be dissolved in THF at high concentration to create a water-immiscible liquid that can, in many cases, then be added to the aqueous preparation of enzyme and buffer with high agitation to create an emulsion or stable suspension without inducing crystallization of the substrate. Such a preparation behaves like a lipid suspension.

Reactions using lipases and liquid suspensions of substrates will frequently perform better if a surfactant is added to make the system a more stable emulsion. Surfactants of choice include salts of taurocholic acid and various nonionic detergents such as Triton X-100. The addition of calcium salts promotes adsorption of enzymes to micelle surfaces.

The ionic strength of an aqueous reaction medium is another factor that can influence both the catalytic activity and the substrate specificity of hydrolases. This too requires empirical assessment in the optimization stage of an enzymatic resolution, but the effect of changing ionic strength can be significant, as illustrated in Scheme VIII.

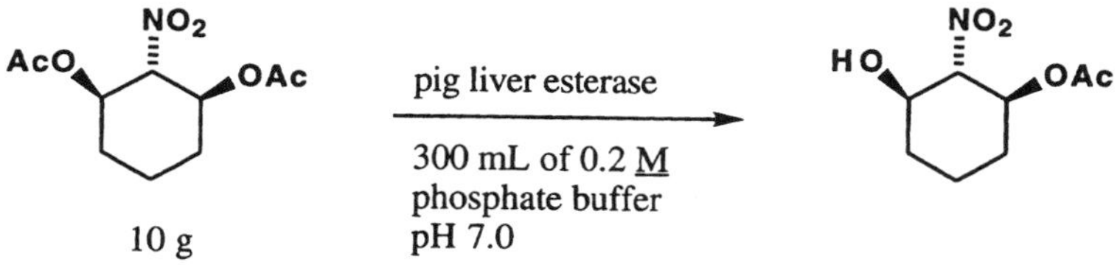

Scheme VIII. (32).

pH Control

As mentioned, a hydrolytic enzyme using the DHS triad cannot function at pH levels below that at which the carboxylate of the aspartate side chain is protonated. Effective pH control must therefore be maintained during an enzymatic hydrolysis. For small-scale reactions, up to a few grams of substrate, pH can be controlled by using a medium buffered with a mixture of di- and trisodium phosphate, as in Scheme IX.

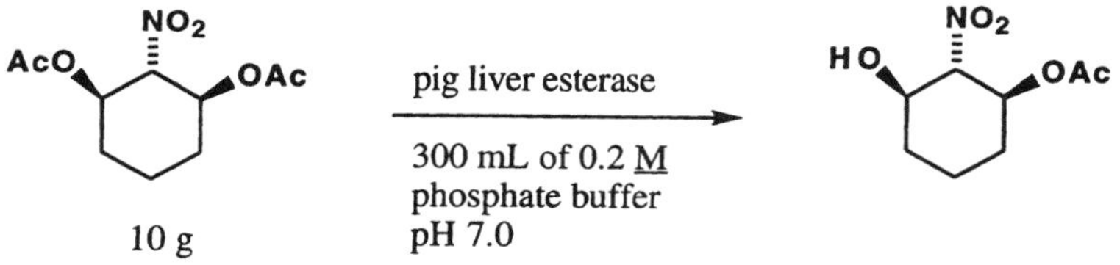

Scheme IX. (28).

For large-scale reactions, aqueous base is pumped into the reaction mixture using an electric pump that is switched on by the output signal of a pH meter whenever the pH drops below a set point. A suitable buffer is also used in these reactions to prevent pH spiking. Additional control of the reaction progress is achieved by using a standardized solution of base (e.g., 2 N NaOH) and recording the volume added, which allows a direct calculation of the molar quantity of substrate hydrolyzed.

A typical laboratory setup incorporating these features is depicted in Figure 1. Multikilogram batches using production-size fixed equipment can be produced in the same manner.

Temperature Control

Temperature control is needed for two reasons. When the temperature is too low, the reaction is too slow, and substrate solubility may also be decreased. When the temperature is too high, the enzyme denatures, losing catalytic activity, but loss of substrate specificity may occur before the denaturation temperature is reached.

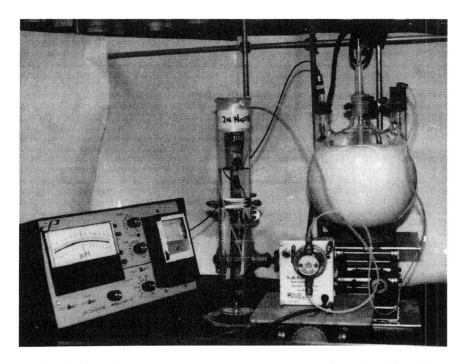

Figure 1. Typical laboratory setup for an enzyme-catalyzed hydrolysis, showing (from left to right) pH meter, reservoir for solution of base, peristaltic pump for adding base, and reaction flask fitted with pH electrode and stirrer.

Enzymatic resolutions are rarely run below 20 °C, and few published examples are run above 60 °C. However, within these limits, there is usually an optimum temperature for a reaction. In cases where enantiomeric excess (ee) of product close to 100% is being realized in reactions that take several days to reach 50% conversion of racemic substrate, operating temperatures of 40 to 55 °C should be routinely tried.

In-Process Control

In-process control is the application of suitable analytical methods to determine how far the enzymatic reaction has proceeded and to measure the ee of products and reactants at suitable intervals. Only in ideal cases will an enzyme exhibit total enantioselectivity. In such cases, conversion of racemic starting material stops at 50%, and both product and remaining reactant have 100% ee. In more typical cases, the enzyme catalyzes reaction of both enantiomers but at usefully different rates. In these more general cases, the reaction must be stopped short of 50%

conversion if high-ee product is desired, or taken past 50% conversion if high-ee reactant is preferred.

Suitable analytical methods are usually gas chromatography (GC) or high-pressure liquid chromatography (HPLC), with conventional columns to check the extent of reaction and chiral columns for ee assessments. Chiral column ee analysis usually entails an investment of time and effort in analytical method development, but this is essential if an enzymatic resolution is to be properly developed. Screening of enzymes and reaction conditions is greatly facilitated by appropriate analytical methods. If the substrate contains a second chiral center of fixed absolute configuration, the reactant and product stereoisomers are epimers, and diastereomeric excess (de) analyses can be carried out on conventional columns. Such was the case with the d-α-tocopherol side chain synthon in Scheme X.

Product Isolation

The underlying premise of any kinetic resolution is that the product from the reactive stereoisomer will be separable from the unreactive isomer by conventional isolation and purification techniques. This condition must be met if a resolution is to succeed.

When the chiral entity of interest is a carboxylic acid and the enzyme-catalyzed reaction involves making or hydrolyzing an ester or amide bond, one of the reaction components will be a free carboxylic acid. In these cases, the water solubility of the acid at basic pH provides an obvious and effective separation technique. Very different solubility characteristics will often be encountered, even in the absence of a carboxylic acid group. For example, the alcohol product and ester starting material of the enzymatic hydrolysis in Scheme XI were completely separated by partitioning between water and hexane (*37*).

The ketone and amine in the reaction mixture of the aminotransferase-catalyzed amine resolution in Scheme VII were separated by taking advantage of the pH dependence of their water solubility (Scheme XII).

When selective extraction from an aqueous medium is not feasible, fractional distillation or column chromatography must be used. In some cases, selective crystallization works (Scheme XIII).

Fractional distillation through a 30-cm Goodloe column served to separate the C_{13} ester and alcohol in the reaction mixture of Scheme X.

Since the physical properties that are pertinent to conventional separation methods are the same for both racemic and resolved materials, the feasibility of the necessary separation can be assessed at the outset, even before the resolution is achieved.

Product Analysis

The chemical purity of a resolved product must first be established, using the methods and criteria applicable to any newly synthesized material. Analysis for

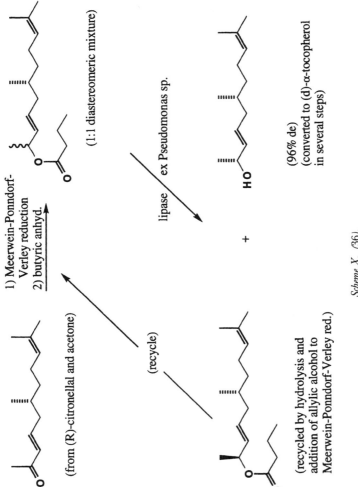

Scheme X. (36).

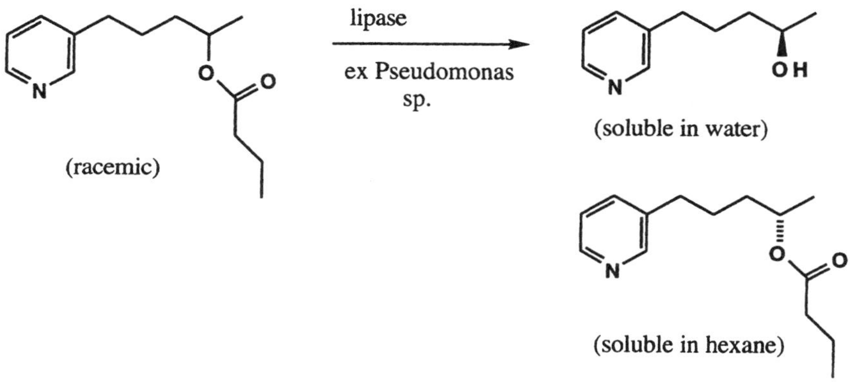

lipase

ex Pseudomonas
sp.

(racemic)

(soluble in water)

(soluble in hexane)

Scheme XI.

Extractable from
water at pH 7.5

Extractable from
water at pH >13

Scheme XII.

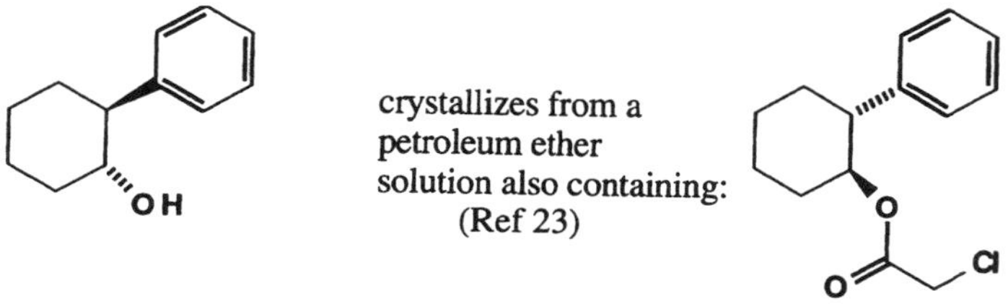

crystallizes from a
petroleum ether
solution also containing:
(Ref 23)

Scheme XIII.

stereochemical purity requires specialized methods, of which optical rotation is probably the least desirable. Except when specific rotations are very high, optical rotation measurements are neither sensitive nor reproducible enough to detect variations in the ±1% range.

The analytical methods used must be sensitive enough to ensure the high stereochemical purity required for many end uses; for example, chiral drug substances should be more than 99% enantiomerically pure.

Two types of analytical methods meet these requirements. Diastereomeric derivatization renders the minor enantiomer separable from the major enantiomer and detectable by sensitive HPLC and GC methods. For alcohols and amines, the well-known Mosher reagent (*38*) (*R*)- or (*S*)-α-methoxy-α-(trifluoromethyl)phenylacetic acid is usually employed. Carboxylic acids can be derivatized with the α-methyl-*p*-nitrobenzylamines (*39*). Enantiomerically pure reagents are essential for diastereomeric derivatization. To verify their purity, either the vendor or the user should analyze the reagents by a method of the type described later. It is also important to verify that derivatization of both enantiomers goes to completion under the conditions used. This can be conveniently done with the racemate. The second category of reliable methods has as its basis a point made in the introduction of this chapter. Just as the microenvironment in which small chiral molecules interact with proteins is responsive to stereochemical features, chromatographic column packings can be constructed from chiral materials and used to separate enantiomers, the relative adsorption and desorption rates being diastereomerically biased by such materials.

The pioneering work of Pirkle first reduced this principle to practice, and Chapter 11 gives a full account of the method. The same type of technology, covalently bonding suitable enantiomerically pure polar compounds to silica and other types of HPLC column-packing material, has more recently advanced into the complementary domain of GC. Excellent separations of enantiomers are achievable by using equipment such as permethylated β-cyclodextrin capillary GC columns, many of which are now commercially available. A short list of typical columns and vendors is provided in Table II.

The sensitivity and reproducibility of such GC and HPLC methods are extremely high, making enantiomeric purity assays of 99.9% and above fairly routine.

Absolute Configuration

Some methods for catalytic kinetic resolution (e.g., the Sharpless epoxidation of racemic allylic alcohols) come with reliable predictors for establishing the configuration of the resolved product (*40*). However, at the current stage of development, any such rules associated with enzyme-catalyzed resolutions should be applied with some skepticism. A fact as important as the absolute configuration of the product must be established by a method that is independent of the product genesis. This is a standard problem in synthetic organic chemistry, and standard solutions entail X-ray crystallography (preferably with sulfur or some heavier element present) or correlation with a substance of known absolute configuration.

Scale-Up

The principal issues to be addressed during scale-up from laboratory glassware to fixed equipment in a pilot or production plant are the use of explosion-proof electrical equipment for pH control and economic concerns involving enzyme utilization and recycling of the undesired stereoisomer.

Table II. Chiral Columns

Type	Product Name	Vendor
Gas chromatography		
Chirasil-Val	Chirasil-Val III	Alltech Associates, Inc. Deerfield, IL
XE-60-(S)-valine-(S)-α-phenylethylamide	—	Chrompack, Inc. Raritan, NY
Cyclodextrin	—	Kupper & Co. Bonaduz, Switzerland
Cyclodextrin	Lipodex	Macherey–Nagel Duren, Germany
Liquid chromatography		
Pirkle	Pirkle-concept chiral selectors	Regis Technologies, Inc. Morton Grove, IL
Cyclodextrin	Cyclobond	Advanced Separation Technologies Whippany, NJ
Cellulose	Chiralcel	J. T. Baker Phillipsburgh, NJ
Protein	Ultron ES-OVM	MAC-MOD Chadds Ford, PA

Note: — means data are not available.

In my experience, direct scale-up from a 5-L flask to a 100-gal reactor gave rise to no unusual problems; moreover, the potential for runaway exotherms, prevalent with many processes during the reaction or quench phases, simply does not exist with enzymatic processes.

Concerns about escalating enzyme cost with scale-up can be addressed through enzyme immobilization on a solid support (*41*). This renders the enzyme recoverable for use in the next batch, but the process must be redeveloped in its entirety. Choosing the cheapest commercial enzyme available that is consistent with the cost and quality required for the final product is the best plan. The enzyme charge used in each batch can then be discarded.

Recycling the wrong stereoisomer becomes more of a concern in large-scale operations, and as mentioned earlier, this issue can be resolved only by carefully planning where in the overall synthetic scheme enzyme catalysis is best applied.

Additional Examples

Selected examples are provided in Appendix I for use in the design of new enzymatic catalysis applications by analogy with existing ones.

Appendix I. **Enzyme-Catalyzed Chiral Syntheses** *(continued)*

Substrate	Enzyme	Products	Yield[a] (%)	ee (%)	Ref.
Chiral Alcohols					
(racemic)	lipase ex *Aspergillis niger* (Amano A)		47%	>99%	40
		+	53%	85%	
(racemic)	lipase ex *Pseudomonas* sp.* (Amano P)		30%	100%	41
		+	43%	100%	
(racemic) + lauric acid in hexane	lipase ex *Candid cylindracea* (Lipase My)		44%	98%	42
(cis, racemic)	cholesterol esterase from bovine pancrease (Sigma)		~40%	93%	43
(racemic)	lipase ex *Pseudomonas* sp.		~40%	97%	44
(racemic)	porcine pancreatic lipase (PPL)		~36%	>94%	45

Appendix I. Enzyme-Catalyzed Chiral Syntheses *(continued)*

Substrate	Enzyme	Products	Yield[a] (%)	ee (%)	Ref.
(racemic) + in heptane (at 60 °C)	PPL	+	~45% ~45%	99.5% (recryst.) 99.5%	46
(racemic) in isooctane sat'd with H_2O	lipase ex *Candida cylindracea* (lipase OF-360) on Celite	+	44& 40%	93% 94%	47
+ $(CH_3CO)_2O$ in benzene	lipase ex *Pseudomonas* sp. (on Celite)	+	43% 39%	>95% >95%	48
	lipase ex *Pseudomonas* sp.	+	48% 48%	>99% 99%	49
+ in (i-Pr)$_2$O	lipase ex *Pseudomonas* sp.	+	48% 50%	93% 87%	50

Appendix I. Enzyme-Catalyzed Chiral Syntheses *(continued)*

Substrate	Enzyme	Products	Yield[a] (%)	ee (%)	Ref.
	lipase (PPL, (Type II)		~40%	92%	51
in CHCl$_3$	lipase ex *Pseudomonas* sp.		55%	62.4%	52
			40%	94%	
	lipase ex *Pseudomonas* sp. (Amano P)		21%	>95%	53
			43%	>95%	
	lipase ex *Pseudomonas* sp. (Amano AK)		41%	>95%	54
			43%	>95%	
with anion-exchange resin and in (i-Pr)$_2$O	lipase ex *Pseudomonas* sp. (Amano M-12-33)		81%	91%	55
(diastereomeric)	PPL		35.5%	>95%	56

Appendix I. Enzyme-Catalyzed Chiral Syntheses *(continued)*

Substrate	Enzyme	Products	Yield[a] (%)	ee (%)	Ref.
			36.5%	>95%	
 (racemic) + lauric acid in cyclohexane	lipase ex *Candida rugosa* (Amano AY30) (2 cycles)		45.8%	98.4%	57
 (racemic)	pancreatin		36%	94%	58
			36%	90%	
 + trichloroethyl butyrate *or* trifluoroethyl laurate, in ether	PPL		43%	>97%	59 60
			37.5%	90%	

Chiral Acids and Esters

Substrate	Enzyme	Products	Yield[a] (%)	ee (%)	Ref.
 (racemic)	papain		~50%	>97%	61
	lipase ex *Candida* *cylindrica*		27% (R=Methyl) 45% (R=octyl)	94% ~96%	62 63
	incubation with whole cells, *Corynebacterium* *equi* IFO 3730			>99%	64

Appendix I. Enzyme-Catalyzed Chiral Syntheses *(continued)*

Substrate	Enzyme	Products	Yield[a] (%)	ee (%)	Ref.
	protease ex *Streptomyces griseus* (Sigma Type XXI)		~50%	~96%	65
	PPL (on Celite)		40%	>99%	66
(racemic)	pig liver (PLE)	+	40% 48%	100% 80%	67
	lipase ex *Pseudomonas* sp. (Amano P)	+	34% 53%	99% 68%	26 108
	lipase ex *Pseudomonas* sp. (Amano P)	+	45% 50%	99% 75%	26
(racemic)	α-chymotrypsin (Sigma)	+	27% 48%	>95% —	68
	PPL	+	35–40%	75%	69

Appendix I. Enzyme-Catalyzed Chiral Syntheses (continued)

Substrate	Enzyme	Products	Yield[a] (%)	ee (%)	Ref.
			—	—	
(racemic)	lipase P, ex *Pseudomonas* sp.	+	50% / 44%	95% / 98%	70
(98% ee)	lipase ex *Candida cylindracea*		92%	98%	
(ibuprofen ester) + HN$_3$ in t-butanol	lipase ex *Candida antarctica* SP 435	+	~56%	— / 96%	71
	lipase ex *Candida cylindracea*				72

Aminoacids

Substrate	Enzyme	Products	Yield[a] (%)	ee (%)	Ref.
	acylase I (porcine kidney, Sigma)	+	35% / 40%	>99% / >99%	73
	lipase ex *Aspergillis niger*		37%	94%	74

Appendix I. Enzyme-Catalyzed Chiral Syntheses *(continued)*

Substrate	Enzyme	Products	Yield[a] (%)	ee (%)	Ref.
(threo, racemic)	acylase ex *Aspergillis* sp. (Amano)		27.5%	98%	75
	esterase ex *Pseudomonas* sp. (DMF 5/8)		43%	90%	76
+ NH_3	β-methylaspartase (ammonia lyase ex *Clostridium tetanomorphum*)		60%	—	77
Prochiral Diols					
(meso, n=1,2,3,4)	PPl		74–94%	72–96%	78 79 80
(meso)	PPL		96%	>99%	79
(meso)	PPL		78%	94%	81
	PPL		75%	99%	82 83
+	lipase *Candida antarctica* (Novo Nordisk SP 435)		48%	>99%	84
(meso)	lipase ex *Pseudomonas fluorescens*		39%	>99%	85

Appendix I. Enzyme-Catalyzed Chiral Syntheses *(continued)*

Substrate	Enzyme	Products	Yield[a] (%)	ee (%)	Ref.
R = H, CH₃	lipase ex *Mucor javanicus* (Fluka)		75% (R=H) 60% (R=CH₃)	>99% >99%	86
(meso)	PLE		80–90%	>95%	87
	acetylcholinesterase ex electric eel (Sigma)		79%	>95%	88
	lipase ex *Pseudomonas fluorescens* (Amano)		92%	96%	89
(ex triol)	lipase ex *Pseudomonas fluorescens* (Amano)		~50% ~50%	>99% >99%	90
+	lipase ex *Pseudomonas* sp. (Sigma Type XIII)		53%	96%	91
	PPL		45%	88%	92
(meso)	lipase ex *Pseudomonas fluorescens* (Fluka)		98%	>98%	93

Appendix I. Enzyme-Catalyzed Chiral Syntheses *(continued)*

Substrate	Enzyme	Products	Yield[a] (%)	ee (%)	Ref.
	PPL	(config. tentative)	66%	95%	94
C_6H_5 ∼ C_6H_5 AcO OAc (racemic)	incubation with whole cells, *Trichoderma viride* (IFO 9065)	C_6H_5 ∼ C_6H_5 HO OH	31%	100%	95
(racemic)	a) PPL, b) bovine pancreas acetone powder or incubation c) with *Bacillus* sp. L-75 followed by hydrolysis of (S)-diester		30–50%	95–100%	96 97 98 109

Prochiral Diesters

CH_3OOC ⟋⟍ $COOCH_3$ (meso)	PLE	$HOOC$ ⟋⟍ $COOCH_3$	82%	34%	99
	PLE		61% (recryst.)	98%	100
	PLE		99.6%	80%	101
CH_3OOC $COOCH_3$ / C_6H_5	PLE	CH_3OOC $COOH$ H / C_6H_5	100%	93%	102
OAc / $EtOOC$ $COOEt$	α-chymotrypsin (Fluka)	H OAc / $HOOC$ $COOEt$	84%	95%	103
F / $EtOOC$ $COOEt$	lipase-MY ex *Candida cylindricea*	F / $EtOOC$ $COOH$	87%	91%	104

Appendix I. Enzyme-Catalyzed Chiral Syntheses *(continued)*

Substrate	Enzyme	Products	Yield[a] (%)	ee (%)	Ref.
	lipase ex *Pseudomonas* sp. (Amano P-30) Sigma type XIII)		89–93%	98.5%	105
	PPL		~100%	>98%	106
+ CH_3OH in $(i\text{-}Pr)_2O$	lipase ex *Pseudomonas* sp. (Amano P-30)		92%	87%	107

[a]Yields are based on moles of product per mole of substrate and so cannot exceed 50% in most cases.
[b]Identified as *P. fluorescens* in earlier literature and more recently as *P. cepacia*.

References

1. *Biotransformations in Organic Chemistry*; Faber, K., Ed.; Springer-Verlag: New York, 1992.
2. *Enzymes in Synthetic Organic Chemistry*; Wong, C.-H.; Whitesides, G. M., Eds. Tetrahedron Organic Chemistry Series; Pergamon-Elsevier: Oxford, England, 1994; Vol. 12.
3. *Biotransformations in Preparative Organic Chemistry*; Davis, H. G.; Green, R. H.; Kelly, D. R.; Roberts, S. M., Eds.; Academic: London, 1989.
4. Halgas, J. *Biocatalysts in Organic Synthesis*; Studies in Organic Chemistry; Elsevier: Amsterdam, Netherlands, 1992; Vol. 26.
5. Sterling, D. I. *Chirality in Industry*; Collins, A. N.; Sheldrake, G. N.; Crosby, J., Eds.; Wiley: New York, 1992; pp 209–222.
6. Blow, D. M.; Birktoft, J. J.; Hartley, B. S. *Nature (London)* **1969**, *221*, 337.
7. Sussman, J. L.; Harel, M.; Frolow, F.; Oefner, C.; Goldman, A.; Toker, L.; Silman, I. *Science (Washington, DC)* **1991**, *253*, 872.
8. Lin, G.; Liu, S-H.; Chen, S.-J.; Wu, F.-C.; Sun, H.-L. *Tetrahedron Lett.* **1993**, *34*, 6057.
9. Derewenda, Z. S.; Sharp, A. M. *Trends Biochem. Sci.* **1993**, *18*, 20.
10. Derewenda, U.; Swenson, L.; Green, R.; Wei, Y.; Yamaguchi, S.; Joerger, R.; Haas, M.J.; Derewenda, Z. S. *Protein Eng.* **1994**, *7*, 551.
11. Ohno, M.; Otsuka, M. *Org. React. N.Y.* **1989**, *37*, 16.
12. Toone, E. J.; Werth, M. J.; Jones, J. B. *J. Am. Chem. Soc.* **1990**, *112*, 4946.

13. Jones, J. B. *Aldrichimica Acta* **1993**, *26* (4), 105.
14. Wang, V.-F.; Sih, C. J. *Tetrahedron Lett.* **1984**, *25*, 4999.
15. Unpublished results with Kalaritis, P.; Regenye, R. W.
16. Iriuchijima, S.; Hasegawa, K.; Tsuchihashi, G.-i. *Agric. Biol. Chem.* **1982**, *46*, 1907.
17. Gerecke, M., Zimmermann, J. P.; Aschwanden, W. *Helv. Chim. Acta* **1970**, *53*, 991.
18. Klibanov, A. M. *Acc. Chem. Res.* **1990**, *23*, 114.
19. Choudhry, S. C.; Belica, P. S.; Coffen, D. L.; Focella, A.; Maehr, H.; Manchand, P. S.; Serico, L.; Yang, R. T. *J. Org. Chem.* **1993**, *58*, 1496.
20. Coffen, D. L.; Okabe, M.; Sun, R. C.; Lee, S.; Matcham, G. W. *J. Bioorg. Med. Chem.* **1994**, *2*, 411.
21. Taylor, S. J. C.; Sutherland, A. G.; Lee, C.; Wisdom, R.; Thomas, S.; Roberts, S. M.; Evans, C. *J. Chem. Soc., Chem. Commun.* **1990**, 1120.
22. Sugai, T.; Mori, K. *Agric. Biol. Chem.* **1984**, *48*, 2497.
23. Schwartz, A.; Madan, P.; Whitesell, J. K.; Lawrence, R. M. *Org. Synth.* **1990**, *69*, 1.
24. Gu, D. M.; Chen, C. S.; Sih, C. J. *Tetrahedron Lett.* **1986**, *27*, 1763.
25. Wu, S.-H.; Guo, Z.-W.; Sih, C. J. *J. Am. Chem. Soc.* **1990**, *112*, 1990.
26. Kalaritis, P.; Regenye, R. W.; Partridge, J. J.; Coffen, D. L. *J. Org. Chem.* **1990**, *55*, 812.
27. Crosby, J. A.; Parratt, J. S.; Turner, N. J. *Tetrahedron Asym.* **1992**, *3*, 1547.
28. Eberle, M.; Missbach, M.; Seebach, D. *Org. Synth.* **1990**, *69*, 19.
29. Kalaritis, P.; Regenye, R. W. *Org. Synth.* **1990**, *69*, 10.
30. Kazlauskas, R. J. *Org. Synth.* **1992**, *70*, 60.
31. *Preparative Biotransformations, Whole Cell and Isolated Enzymes in Organic Synthesis*; Roberts, S. M.; Wiggins, K.; Casy, G., Eds.; Wiley: New York, 1993.
32. Guanti, G.; Banfi, L.; Narisano, E.; Riva, R.; Thea, S. *Tetrahedron Lett.* **1986**, *27*, 4639.
33. Björkling, F.; Boutelje, J.; Hjalmarsson, M.; Hult, K.; Norin, T. J. *Chem. Soc., Chem. Commun.* **1987**, 1041.
34. Parida, S.; Dordick, J. S. *J. Org. Chem.* **1993**, *58*, 3238.
35. Okabe, M.; Sun, R.-C.; Zenchoff, G. B. *J. Org. Chem.* **1991**, *56*, 4392.
36. Coffen, D. L.; Cohen, N.; Pico, A. M.; Schmid, R.; Sebastian, M. J.; Wong, F. *Heterocycles* **1994**, *39*, 527.
37. Manchand, P. S.; Schwartz, A.; Wolff, S.; Belica, P. S.; Madan, P.; Patel, P.; Saposnik, S. J. *Heterocycles* **1993**, *35*, 1351.
38. Dale, J. A.; Dull, D. L.; Mosher, H. S. *J. Org. Chem.* **1969**, *34*, 2543.
39. Valentine, D.; Chan, K. K.; Scott, C. G.; Johnson, K. K.; Toth, K.; Saucy, G. *J. Org. Chem.* **1976**, *41*, 62.
40. Rossiter, B. E. In *Asymmetric Synthesis*; Morrison J. D., Ed.; Academic: Orlando, FL, 1985; Vol. 5, p 194.
41. *Immobilized Cells and Enzymes: A Practical Approach*; Woodward, J., Ed.; IRL: Washington, DC, 1985.
42. Akita, H.; Matsukura. H.; Oishi, T. *Tetrahedron Lett.* **1986**, *27*, 5241.
43. Kutsuki, H.; Sawa, I.; Hasegawa, J.; Watanabe, K. *Agric. Biol. Chem.* **1986**, 50,2369.
44. Langrand, G.; Secchi, M.; Buono, G.; Baratti, J.; Triantaphylides, C. *Tetrahedron Lett.* **1985**, *26*, 1857.
45. Pawlak, J. L.; Berchtold, G. A. *J. Org. Chem.* **1987**, *52*, 1765.
46. Oberhauser, T.; Bodenteich, M.; Faber, K.; Penn, G.; Griengl, H. *Tetrahedron* **1987**, *43*, 3931.
47. Gutman, A. L.; Zuobi, K.; Boltansky, A. *Tetrahedron Lett.* **1987**, *28*, 3861.
48. Theisen, P. D.; Heathcock, C. H. *J. Org. Chem.* **1988**, *53*, 2374.
49. Hoshino, O.; Itoh, K.; Umezawa, B.; Akita, H.; Oishi, T. *Tetrahedron Lett.* **1988**, *29*, 567.

50. Bianchi, D.; Cesti, P.; Battistel, E. *J. Org. Chem.* **1988**, *53*, 5531.
51. Laumen, K.; Schneider, M. P. *J. Chem. Soc., Chem. Commun.* **1988**, 598.
52. Hiratake, J.; Inagaki, M.; Nishioka, T.; Oda, J.-i. *J. Org. Chem.* **1988**, *53*, 6130.
53. Ladner, W. E.; Whitesides, G. M. *J. Am. Chem. Soc.* **1984**, *106*, 7250.
54. Pallavicini, M.; Valoti, E.; Villa, L.; Piccolo, O. *J. Org. Chem.* **1994**, *59*, 1751.
55. Washausen, P.; Grebe, H.; Kieslich, K.; Winterfeldt, E. *Tetrahedron Lett.* **1989**, *30*, 3777.
56. Burgess, K.; Jennings, L. D. *J. Org. Chem.* **1990**, *55*, 1138.
57. Inagaki, M.; Hiratake, J.; Nishioka, T.; Oda, J.-i. *J. Am. Chem. Soc.* **1991**, *113*, 9360.
58. Mulzer, J.; Greifenberg, S.; Beckstett, A.; Gottwald, M. *Liebigs Ann. Chem.* **1992**, 1131.
59. Comins, D. L.; Salvador, J. M. *J. Org. Chem.* **1993**, *58*, 4656.
60. Francalanci, F.; Cesti, P.; Cabri, W.; Bianchi, D.; Martinengo, T.; Foa, M. *J. Org. Chem.* **1987**, *52*, 5079.
61. Belan, A.; Bolte, J.; Fauve, A.; Gourcey, J. G.; Veschambre, H. *J. Org. Chem.* **1987**, *52*, 256.
62. Stokes, T. M.; Oehlschlager, A. C. *Tetrahedron Lett.* **1987**, *28*, 2091.
63. Drueckhammer, D. G.; Barbas, C. F.; Nozaki, K.; Wong, C.-H.; Wood, C. Y.; Ciufolini, M. A. *J. Org. Chem.* **1988**, *53*, 1607.
64. Dahod, S. K. U.S. Patent 4 668 624, 1987.
65. Cambou, B.; Klibanov, A. M. *Appl. Biochem. Biotechnol.* **1984**, *9*, 255.
66. Kato, Y.; Ohta, H.; Tsuchihashi, G.-i. *Tetrahedron Lett.* **1987**, *28*, 1303.
67. Fülling, G.; Sih, C. J. *J. Am. Chem. Soc.* **1987**, *109*, 2845.
68. Pottie, M.; Van der Eycken, J.; Vandewalle, M. *Tetrahedron Lett.* **1989**, *30*, 5319.
69. Klunder, A. J. H.; van Gastel, F. J. C.; Zwanenburg, B. *Tetrahedron Lett.* **1988**, *29*, 2697.
70. Lalonde, J. J.; Bergbreiter, D. E.; Wong, C.-H. *J. Org. Chem.* **1988**, *53*, 2323.
71. Blanco, L.; Guibé-Jampel, E.; Rousseau, G. *Tetrahedron Lett.* **1988**, *29*, 1915.
72. Delinck, D. L.; Margolin, A. L. *Tetrahedron Lett.* **1990**, *31*, 6797.
73. deZoete, M. C.; Kock-van Dalen, A. C.; van Rantwijk, F.; Sheldon, R. A. *J. Chem. Soc., Chem. Commun.* **1993**, 1831.
74. Gebler, J. C.; Adamczyk, M.; Chen, Y.-Y.; Fishpaugh, J. R. *Abstracts of Papers*, 206th National Meeting of the American Chemical Society, Chicago, IL; American Chemical Society: Washington, DC, 1993; ORGN 200.
75. Chenault, H. K.; Mahn-Joo, M.; Akiyama, A.; Miyazawa, T.; Simon, E. S.; Whitesides, G. M. *J. Org. Chem.* *1987*, *52*, 2608.
76. Miyazawa, T.; Takitani, T.; Ueji, S.; Yamada, T.; Kuwata, S. *J. Chem. Soc., Chem. Commun.* **1988**, 1214.
77. Mori, K.; Iwasawa, H. *Tetrahedron* **1980**, *36*, 2209.
78. Ramos Tombo, G. M.; Schär, H.-P.; Ghisalba, O. *Agric. Biol. Chem.* **1987**, *51*, 1833.
79. Akhtar, M.; Cohen, M. A.; Gani, D. *J. Chem. Soc., Chem. Commun.* **1986**, 1290.
80. Kasel, W.; Hultin, P. G.; Jones, J. B. *J. Chem. Soc., Chem. Commun.* **1985**, 1563.
81. Laumen, K.; Schneider, M. *Tetrahedron Lett.* **1985**, *26*, 2073.
82. Ader, U.; Breitgoff, D.; Klein, P.; Laumen, K. E.; Schneider, M. P. *Tetrahedron Lett.* **1989**, *30*, 1793.
83. Hemmerle, H.; Gais, H.-J. *Tetrahedron Lett.* **1987**, *28*, 3471.
84. Laumen, K.; Schneider, M. P. *J. Chem. Soc., Chem. Commun.* **1986**, 1298.
85. Sugai, T.; Mori, K. *Synthesis* **1988**, 19.
86. Johnson, C. R.; Bis, S. J. *Tetrahedron Lett.* **1992**, *33*, 7287.
87. Xie, Z.-F.; Suemune, H.; Sakai, K. *J. Chem. Soc., Chem. Commun.* **1987**, 838.
88. Estermann, H.; Prasad, K.; Shapiro, M. J.; Repic, O.; Hardtmann, G. E. *Tetrahedron Lett.* **1990**, *31*, 445.

89. Eberle, M.; Egli, M.; Seebach, D. *Helv. Chim. Acta* **1988**, *71*, 1.
90. Johnson, C. R.; Senanayake, C. H. *J. Org. Chem.* **1989**, *54*, 735.
91. Wirz, B.; Schmid, R.; Walther, W. *Biocatalysis* **1990**, *3*, 159.
92. Leuenberger, H. G.; Wirz, B. *Chimia* **1993**, *47*, 82.
93. Wang, Y.-F.; Lalonde, J. J.; Momongan, M.; Bergbreiter, D. E.; Wong, C.-H. *J. Am. Chem. Soc.* **1988**, *110*, 7200.
94. Kerscher, V.; Kreiser, W. *Tetrahedron Lett.* **1987**, *28*, 531.
95. Bonini, C.; Racioppi, R.; Righi, G.; Viggiani, L. *J. Org. Chem.* **1993**, *58*, 802.
96. Guanti, G.; Banfi, L.; Brusco, S.; Riva, R. *Tetrahedron Lett.* **1993**, *34*, 8549.
97. Yamamoto, K.; Ando, H.; Chikamatsu, H. *J. Chem. Soc., Chem. Commun.* **1987**, 334.
98. Miyano, S.; Kawahara, K.; Inoue, Y.; Hashimoto, H. *Chem. Lett.* **1987**, 355.
99. Kazlauskas, R. J. *J. Am. Chem. Soc.* **1989**, *111*, 4953.
100. Fujimoto, Y.; Iwadate, H.; Ikekawa, N. *J. Chem. Soc., Chem. Commun.* **1985**, 1333.
101. Jones, J. B.; Hinks, R. S.; Hultin, P. G. *Can. J. Chem.* **1985**, *63*, 452.
102. Bloch, R.; Guibe-Jampel, E.; Girard, C. *Tetrahedron Lett.* **1985**, *26*, 4087.
103. Arita, M.; Adachi, K.; Ito, Y.; Sawai, H.; Ohno, M. *J. Am. Chem. Soc.* **1983**, *105*, 4049.
104. Nakada, M.; Kobayashi, S.; Ohno, M.; Iwasaki, S.; Okuda, S. *Tetrahedron Lett.* **1988**, *29*, 3951.
105. Santaniello, E.; Chiari, M.; Ferraboschi, P.; Trave, S. *J. Org. Chem.* **1988**, *53*, 1567.
106. Kitazume, T.; Sato, T.; Kobayashi, T.; Lin, J. T. *J. Org. Chem.* **1986**, *51*, 1003.
107. Hughes, D. L.; Bergan, J. J.; Amoto, J. S.; Bhupathy, M.; Leazer, J. L.; McNamara, J. M.; Sidler, D. R.; Reider, P. J.; Grabowski, E. J. J. *J. Org. Chem.* **1990**, *55*, 6252.
108. Gutman, A. L.; Bravdo, T. *J. Org. Chem.* **1989**, *54*, 4263.
109. Yamamoto, K.; Nishioka, T.; Oda, J.-i.; Yamamoto, Y. *Tetrahedron Lett.* **1988**, *29*, 1717.

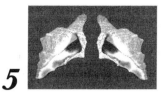

5

Stereoselective Analysis in Crop Protection

Hans-Peter Buser and Eric Francotte

In the past 15 years, more and more chemical research has become focused on stereoselective synthesis. Such stereochemical investigations call for efficient chiral analytical methods to determine optical purities of the synthesized stereoisomers of biologically active compounds, their synthetic precursors, and their metabolites. The introduction of separation techniques based upon HPLC and GC on chiral stationary phases (CSPs) has made the immediate determination of enantiomeric purities without substrate derivatization and with the required high degree of accuracy possible. The present chapter reviews published chiral HPLC- and GC-analytics of economically significant or soon to be marketed chiral crop-protection agents and documents our own work in this fascinating field of research with reference to selected experimental compounds developed in our laboratories.

It is now well established that the biological activity of most drugs and crop protection agents containing one or more stereogenic centers is strongly related to the absolute configuration of the molecule (*1–4*). Over the past 15 years, an ever-growing number of research groups have focused their efforts on stereoselective synthesis, which has led to dramatic improvements in this field of chemistry (*5*). These advances have provided the scientific tools for a systematic investigation of the implications of stereoisomerism as one of the many determinants of biological activity. With these tools at hand, the question whether to develop a biologically active compound as the single stereoisomer or as an isomer mixture

3407–8/96/0093$21.50/0

has become important for the pharmaceutical and agrochemical industries (6, 7). This importance is reflected in an increasing number of scientific publications discussing technical aspects of stereoselective synthesis in industry (8–10).

Such stereochemical investigations call for efficient chiral analytical methods to determine the optical purities of the synthesized stereoisomers of biologically active compounds, their synthetic precursors, and their metabolites. Accurate detection at very low levels is required. The classical method of polarimetry—until about 15 years ago the only method available—is clearly no longer satisfactory. It has largely been replaced by modern, highly sensitive chromatographic and spectroscopic methods, such as high-performance liquid chromatography (HPLC), gas chromatography (GC), nuclear magnetic resonance (NMR) techniques, and, more recently, supercritical fluid chromatography (SFC) and capillary zone electrophoresis (CZE).

The introduction of separation techniques based on HPLC and GC with chiral stationary phases (CSPs) has made possible the immediate determination of enantiomeric purities without substrate derivatization and with the required high accuracy. Over the past 10 years, these new tools have become more and more versatile, and they are now the methods of choice for the chiral analysis of enantiomeric compositions of nonracemic mixtures.

A comprehensive review of the principles of stereoselective analysis of pesticides has already been published (11). This chapter focuses on reviewing published chiral HPLC and GC analytics of economically significant or soon-to-be marketed chiral crop protection agents and then documents our own work in this fascinating field with reference to selected experimental compounds developed in our laboratories. All the chiral stationary phases mentioned in the text are summarized in an appendix.

Weed Control

The structural variability of compounds possessing herbicidal activity is very large (12). Among the important groups of herbicides are such structurally different classes as triazines, ureas and sulfonylureas, 2-aryloxypropionates, arylalanines, and chloracetanilides. While the representatives of some of these groups, such as the triazines, the ureas, and the sulfonylureas, are mostly achiral, other groups, such as the 2-aryloxypropionates and the arylalanines, consist largely of chiral compounds. For herbicides containing stereogenic centers, the relationship between stereoisomerism and biological activity is generally not well understood. During the past 10 to 15 years, however, much attention has been focused on the investigation of differences in the biological activities of stereoisomers. In the course of this work, many chiral separation methods for herbicides have been worked out. Nevertheless, for quite a number of economically important chiral herbicides, very little has been published about the chromatographic resolution of their enantiomers.

2-Aryloxypropionic Acid Derivatives

Substituted 2-aryloxypropionic acids and their ester derivatives are widely used as selective herbicides and are produced in hundreds of thousands of tons annually. Because of the presence of an asymmetric carbon atom, they consist of two enantiomers. In recent years, the differences in biological activities of the enantiomers have received considerable attention, and it has been shown that, in a number of cases, the herbicidal activity arises almost exclusively from the (*R*)-enantiomer (*13–16*). Not surprisingly, the 2-aryloxypropionic acid derivatives are the best documented class of herbicides in terms of chiral analysis (Figure 1).

(*RS*)-2-(4-Chloro-2-methylphenoxy)propionic acid (**1**) (mecoprop; CMPP) and (*RS*)-2-(2,4-dichlorophenoxy)propionic acid (**2**) (dichlorprop; 2,4-DC) are two important members of this class of compounds that were originally introduced in racemic form. Environmental considerations have recently led to the production and marketing of their pure (*R*)-enantiomers.

The need for accurate analytical methods for monitoring their optical isomers explains the great number of methods for the chiral chromatographic analysis of these two compounds in the literature. The first reported direct separation of the enantiomers of acid **1** was carried out on the first generation of an α_1-acid glycoprotein CSP (EnantioPac) (*17*). With this method, samples with optical purities of up to 100% enantiomeric excess (ee) were analyzed. The same article described the chiral separation of the diphenylamide derivative of CMPP on an ionic *N*-(3,5-dinitrobenzoyl)phenylglycine (DNBPG) CSP. The same two methods were then tested for the chiral analysis of phenoxy herbicide mixtures containing CMPP and 2,4-DC (*18*). The EnantioPac column gave poor separation of the enantiomers of 2,4-DC, but the ionic DNBPG column proved to be well suited for the analysis of commercial blends of CMPP and 2,4-DC after conversion of the free acids to their diphenylamide derivatives.

On the second-generation α_1-acid glycoprotein CSP (Chiral-AGP), baseline

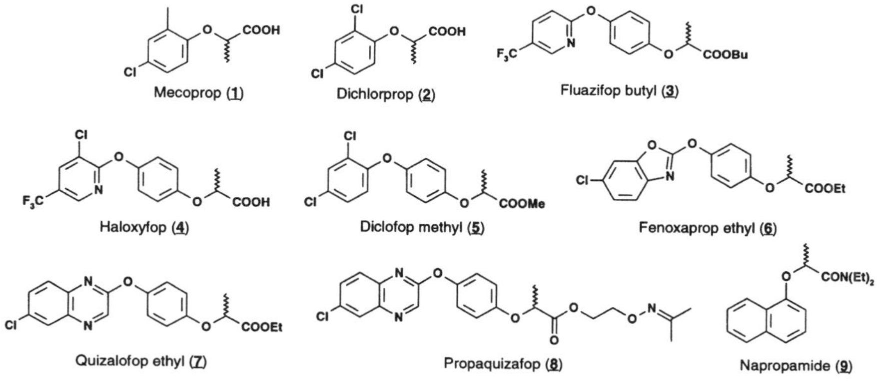

Figure 1. Structure of some commercial chiral 2-aryloxypropionic acid derivatives.

separation of 2,4-DC was achieved (19). This method was applied to demonstrate that marine microorganisms degrade only the (R)-enantiomer of 2,4-DC and leave the (S)-enantiomer unaffected (20).

Two completely different types of CSPs—one based on tartaric acid–phenyl-ethylamine–modified silica (Chiral-2) (21) and the other derived from (R,R)-(N,N'-bis(3,5-dinitrobenzoyl))-trans-1,2-diaminocyclohexane (DACH-DNB) (22)—gave good chiral resolution of the methyl esters of CMPP and 2,4-DC. These methods allow the determination of the optical purities in commercial formulations. DACH-DNB was recently used for the direct chiral analysis of 2,4-DC (23).

Not only HPLC but also capillary GC was applied for the chiral analysis of the methyl esters of both CMPP and 2,4-DC. Baseline separations were achieved with modified cyclodextrin CSPs (Figure 2) (24–26).

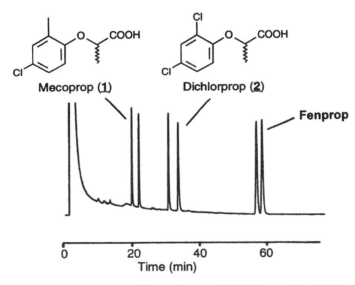

Figure 2. Separation of the enantiomers of mecoprop (1), dichlorprop (2), and fenoprop by GC on pyrex glass capillary column coated with a 1:1 mixture of per-O-pentylated and per-O-methylated β-cyclodextrin (carrier gas: hydrogen, 0.5 bar). (Reproduced with permission from reference 24. Copyright 1991.)

Another 2-aryloxypropionate, butyl (RS)-2-[4,(5-trifluoromethyl-2-pyridyl-oxy)phenoxy]propionate (3) (fluazifop butyl), is a highly selective postemergence herbicide for use against annual or perennial grasses in cotton, soybeans, and other broad-leaved crops. The racemate, which was introduced first, was later re-placed by the (R)-enantiomer (fluazifop-P butyl), which has about twice the her-bicidal activity of the racemate. The enantiomers of both 3 and its methyl ester derivative were first separated on an ionic DNBPG column (27). This has been the standard CSP for the determination of the optical purities of 3 at the production site. During method development, the critical effect of temperature was noticed and used to advantage (the optimum is 5 °C) (*28*). Another Pirkle-type CSP, a covalent DNBPG (*29*), and a Chiral-2 column (*21*) were successfully employed for the stereoselective analysis of **3**. The enantiomers of the free acid of **3,** however, were separated on an Altex Ultrasphere IP reverse-phase column, eluting with a chiral mobile phase containing the Ni(II) complex of L-proline-octylamide at pH 7 to 8 (*27, 30*). With suitable adjustments to the solvent ratio, that method was also applicable to separating the enantiomers of CMPP and 2,4-DC. The enantiomers of the free acid of **3** were separated with a Chiral AGP column, but no experimental details were given (*31*).

(*RS*)-2-[4-(3-Chloro-5-trifluoromethyl-2-pyridyloxy)phenoxy]propionic acid (**4**) (haloxyfop), developed and marketed as its methyl and 2-ethoxyethyl esters (haloxyfop methyl and haloxyfop etotyl), is another selective herbicide used pre- and postemergence for control of annual and perennial grasses in sugar beets, potatoes, leaf vegetables, soybeans, and other crops. For the first reported separation of the enantiomers of **4** ($\alpha = 1.12$), acetylquinine covalently bonded to silica was used as a chiral selector in reverse-phase chromatography (*32*) (acetylquinine-II-silica). Recently, the same authors achieved a much better resolution ($\alpha = 2.35$) using chiral ion-pair chromatography on porous carbon (Hypercarb) with quinine as chiral counterion to discriminate the acid enantiomers (*33*). The 2-ethoxyethyl ester of **4,** haloxyfop etotyl, was resolved on a Chiral-2 column (*21*).

Methyl   (*RS*)-2-(4-(2,4-dichlorophenoxy)phenoxy)propionate   (**5**)   (diclofop methyl) and ethyl (*RS*)-2-[4-(6-chlorobenzoxazol-2-yloxy)phenoxy]propionate (**6**) (fenoxaprop ethyl) are both selective herbicides developed and marketed as racemates for control of annual grasses in wheat. The enantiomers of both compounds can be baseline-separated on a commercial Chiralcel OK CSP (cellulose tricinnamate absorbed into macroporous silica gel) (Figure 3) (*34*). This method was applied to investigate the enantioselective transformation of **5** and **6** in soil.

Two other separation methods for **5** and its derivatives are reported in the literature. The Chiral-2 CSP gave good chiral resolution of **5** itself (*21*), and the Chiral AGP column worked well for its free acid derivative (no details given) (*31*).

The chiral analysis of one more selective 2-aryloxypropionate herbicide, ethyl (*RS*)-2-[4-(6-chloroquinoxalin-2-yloxy)phenoxy]propionate (**7**) (quizalofop ethyl), was also mentioned in one of the papers already cited (*21*). Quizalofop ethyl was launched as a racemate; later, the ethyl and the (±)-tetrahydrofurfuryl esters of its (*R*)-enantiomer (quizalofop-P ethyl and quizalofop-P tefuryl) were introduced as postemergency herbicides to control grasses in cotton, flax, peanuts, sunflowers, soybeans, potatoes, and other vegetables. The enantiomers can be separated on a Chiral-2 column (*21*). A very good chiral resolution of the enantiomers of **7** was achieved on a Chiralcel OB CSP (cellulose tribenzoate coated on silica gel) (Figure 4) (*35*). With this CSP, optical purities of synthetic samples with ee values of 95% were determined.

The structurally closely related 2-(isopropylideneaminooxy)ethyl 2-[4-(6-chloroquinoxalin-2-yloxy)phenoxy]propionate (**8**) (propaquizafop), launched in

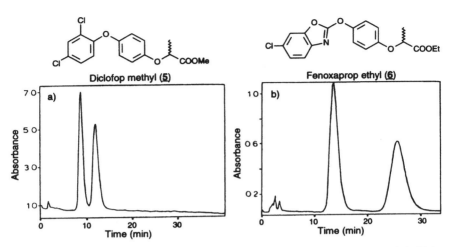

Figure 3. Separation of the enantiomers of diclofop methyl (5) and fenoxaprop ethyl (6) by HPLC on Chiralcel OK (eluent: hexane–2-propanol 7:3). (Reproduced with permission from reference 34. Copyright 1988.)

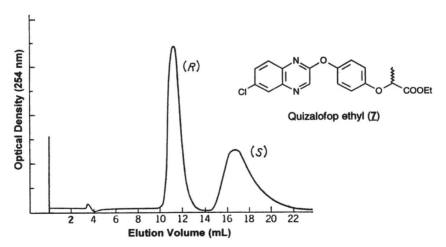

Figure 4. Liquid chromatographic resolution of the enantiomers of quizalofop ethyl (7) on Chiralcel OB (eluent: hexane–2-propanol 7:3). (Reproduced with permission from reference 35. Copyright 1986.)

enantiomerically pure *R* form, is a new postemergence herbicide used to control most temperate and tropical grasses in a variety of crops including cotton, peanuts, potatoes, soybeans, and vegetables. For the chiral analysis, the ionic DNBPG HPLC column was applied (*36*). However, for the determination of high optical excesses (>90% ee) with this CSP, it is crucial—because of some peak tailing—that the minor enantiomer be eluted first.

The 2-aryloxypropionic acid derivative (*RS*)-*N*,*N*-diethyl-2-(1-naphthyloxy)-propionamide (**9**) (napropamide) is used as soil-acting herbicide for preemergence control of annual grasses and sensitive broad-leaved weed species in crops such as rapeseed and tobacco and in horticulture. A chiral resolution method has been worked out on a commercial cellulose trisphenylcarbamate CSP (Chiralcel OC) (*37*). The resolution was good enough (α = 2.9) that the method could be applied for the preparative separation of the enantiomers to compare their herbicidal activity in a wheat germ test. A recent attempt to separate the enantiomers of **9** with a (3,5-dinitrophenyl)carbamated β-cyclodextrin-based CSP was unsuccessful (α = 1.03) (*38*).

One more published chromatographic procedure for the chiral resolution of 2-aryloxypropionates that should be mentioned is the patented method using as stationary phases glycerol-terminated silica or "diol" (Chromega) and a chiral mobile phase containing (*R*)-(–)- or (*S*)-(+)-2-pyrrolidinemethanol and an alkanoic acid (*39*). The examples listed in the patent include the free acids fluazifop (α = 1.16), haloxyfop (α = 1.14), diclofop (α = 1.09), and quizalofop (α = 1.11).

2-Propynyl(*R*)-2-[4-(5-chloro-3-fluoro-2-pyridyloxy)phenoxy]propionate (**10**) (clodinafop propargyl) is a 2-aryloxypropionate herbicide currently being introduced to the market in enantiomerically pure form for the selective control of annual grasses in cereals (*40*). Analytical determination of the optical purities of its enantiomers can easily be performed on cellulose triacetate (CTA-I), as shown by the large α value (5.70) (Figure 5). Other ester derivatives of (*R*)-2-(4-(5-chloro-3-fluoro-2-pyridyloxy)phenoxy)propionic acid were investigated and chromatographed on CTA-I as well, but the degree of resolution was strongly influenced by the ester moiety.

N-Arylalanines

A series of *N*-arylalanine esters have been introduced as postemergence herbicides for the selective control of wild oats in cereals. Their herbicidal activity generally originates from the (*R*)-enantiomers, while the (*S*)-enantiomers are virtually inactive. Furthermore, it has been found that the activity against wild oats is dependent on conversion of the applied esters to the corresponding biologically active acids (*41–43*). The herbicides are selective because the ester hydrolysis occurs much more slowly in wheat and barley than in wild oats. The stereoselectivity of the esterase catalyzing the hydrolysis could explain the superior herbicidal activity of the (*R*)-enantiomers (*44*).

N-Benzoyl-*N*-(3-chloro-4-fluorophenyl)-DL-alanine (**11**) (flamprop) is an eco-

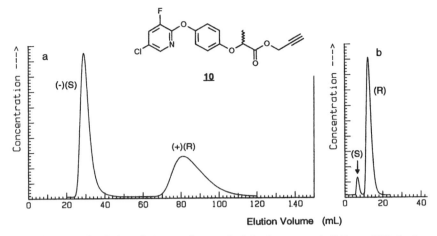

*Figure 5. Resolution of the enantiomers of clodinafop propargyl (**10**) on CTA-I; eluent: ethanol–water 95:5; (a) 2.5 mg of racemate on a 1.25 × 30 cm column; (b) enriched sample (+)-(R)-enantiomer with 84% ee on a 0.4 × 25 cm HPLC column.*

nomically important *N*-arylalaninate (Figure 6). Initially, its racemic methyl and isopropyl esters—flamprop methyl (**12**) and flamprop isopropyl (**13**)—were developed and marketed for the postemergence control of wild oats in wheat. But now the racemates have been almost entirely replaced by the (*R*)-enantiomers flamprop-M methyl and flamprop-M isopropyl.

Both flamprop methyl and flamprop isopropyl were resolved on an HPLC column containing *N*-formyl-(*S*)-isoleucine covalently bonded via aminopropyl groups to microparticulate silica (*45*), and this CSP produced quite a different enantiomer resolution from that produced by the apparently similar Pirkle stationary phases. The isopropyl ester gave almost baseline separation, while the resolution of the methyl ester remained relatively poor. Changing the mobile phase to methanol plus hexane and lowering the column temperature to 5 °C improved the separation. With this method, samples of (*R*)-**13** containing less than 1% of the (*S*)-enantiomer were analyzed. The same authors reported an even more efficient chiral resolution of flamprop isopropyl by capillary GC using

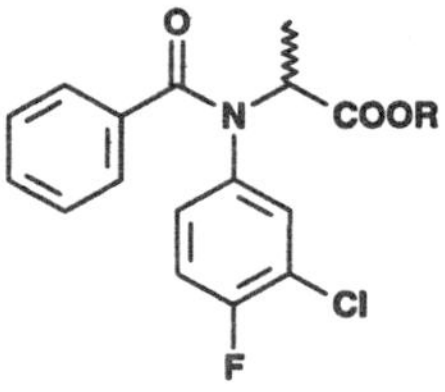

Flamprop (R=H) (**11**)
Flamprop methyl (R=Me) (**12**)
Flamprop isopropyl (R=i-Pr) (**13**)

Figure 6. Structure of flamprop and derivatives.

a Chirasil-L-Val column (*45*). This analytical method was later applied to flamprop methyl for metabolic studies (*46*).

Chloroacetanilides

The chloroacetanilides are the most important class of herbicides for the pre-emergence control of grass weeds in maize. A typical representative of the class is 2-chloro-*N*-(2-ethyl-6-methylphenyl)-*N*-(2-methoxy-1-methyl(ethyl)-acetamide (**14**) (metolachlor), which has been developed and marketed as a grass herbicide for use in maize, soybeans, and other broad-leaved crops and has become very important economically (*47*). The four stereoisomers (resulting from one asymmetric carbon and atropisomerism due to hindered rotation around the aryl carbon–nitrogen bond) of **14** exhibit very different herbicidal activities. The herbicidal activity is mainly influenced by the configuration at the chiral center, the (*S*)-isomers being the most active (*48, 49*). Two chiral HPLC methods for the resolution of the stereoisomers of metolachlor have been reported recently, but neither allowed baseline separation of all four stereoisomers (*50*) (Figure 7). According to the same article, on a Chiralcel OD column (cellulose tris(3,5-dimethylphenyl)carbamate coated on silica gel) the two (*S*)-isomers were fully resolved, while the (*R*)-isomers were only partially separated. On an amylose tris((*S*)-*N*-1-phenylethyl)carbamate column (Chiralpak-AS), however, the (*R*)-isomers were baseline-separated, while the (*S*)-isomers remained unresolved. These two meth-

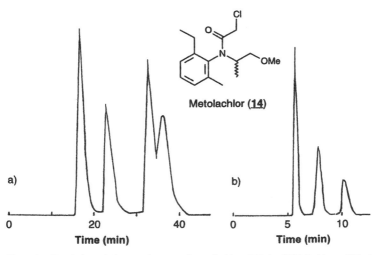

*Figure 7. Resolution of the stereoisomers of metolachlor (**14**) by HPLC (a) on Chiralcel OD (eluent: hexane–2-propanol 100:1) and (b) on Chiralpak AS (eluent: hexane–2-propanol 90:10). (Adapted from reference 50: chromatograms kindly provided by M. Ueji, National Institute of Agro-Environmental Sciences, Japan.)*

ods were used for the preparative separation of the stereoisomers of metolachlor (**14**) to investigate their herbicidal activity on rice and their degradation (*50*).

Sulfonylureas

The sulfonylureas, a relatively new class of herbicides, have become important because of their high activities at very low application rates and their selectivity. It has been demonstrated that the enzyme at which the highly specific action of these herbicides takes place is acetolactate synthease (*51–53*). Most sulfonylureas are achiral, but some chiral compounds belonging to this structural family have been investigated. One example from our laboratories is compound **15,** for which the enantiomers were needed in larger amounts for individual biological testing (Figure 8). This compound is not resolved on a CTA-I column, but its precursor (**16**) can be (α = 1.36) (Figure 8a). The optical purities of the isolated fractions were determined on an HPLC column (0.46 × 25 cm) filled with the same CSP (Figure 8b).

Miscellaneous Herbicides

Some compounds have herbicidal or plant-growth-controlling activity but do not belong to any of the large compound classes discussed earlier.

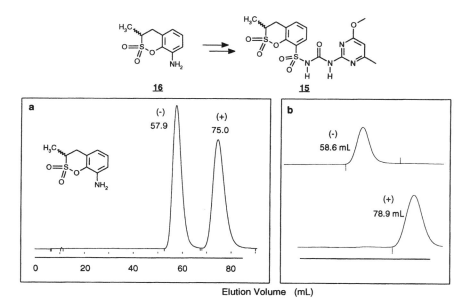

*Figure 8. (a) Resolution of racemic **16** and (b) determination of the optical purity of the preparatively isolated (–)- and (+)-enantiomers by HPLC on CTA-I; column: 0.46 × 25 cm; eluent: ethanol–water 95:5.*

(−)-2,3:4,6-Di-*O*-isopropylidene-2-keto-L-gulonic acid represents a special case. While its sodium salt (dikegulac sodium) has been developed and marketed as a systemic plant growth regulator, the acid can be applied as chiral counterion for the resolution of chiral amines by reverse-phase ion-pair chromatography (*54*).

In our laboratories, a wide range of benzoyl aryl epoxides of the structural type **18** have been investigated as experimental herbicides. Most of these chiral compounds could be easily resolved, on both analytical and preparative scales, by HPLC on cellulose triacetate. Figure 9 shows the analytical chromatographic resolution of **19** on CTA-I. The method has been applied to determining the optical purities of the fractions isolated from preparative chromatography.

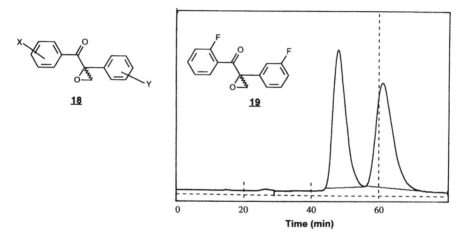

*Figure 9. Separation of the enantiomers of benzoyl aryl epoxide **19** by HPLC on CTA-I; column: 0.46 × 25 cm; eluent: ethanol–water 95:5.*

In the course of a standard herbicide screening program, unexpected activity was found in compound **20,** which belongs to a group of imidazole-5-carboxylic esters with hypnotic and antimycotic activity (*55*). After intensive synthetic work toward optimization of the herbicidal activity of this sterol biosynthesis inhibitor, the analogue **21** (with a five-membered ring) emerged as the most active in the series. Good separation of the enantiomers of this compound was achieved on a β-cyclodextrin-based CSP developed in our laboratories (*56*). Figure 10 shows the chromatographic resolution of **21** on this CSP under various conditions, clearly indicating the influence of the composition of the mobile phase and the content of cyclodextrin per gram of CSP on the chiral separation. Compound **20** and some other derivatives of **21** are also very well resolved on the same CSP.

The two 4,6-dimethoxypyrimidine derivatives **22** and **23** were developed in

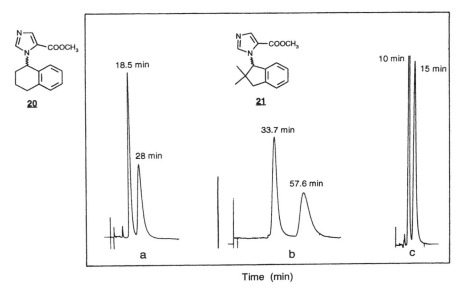

*Figure 10. Separation of the enantiomers of **21** by HPLC on experimental β-cyclodextrin CSPs (columns: 0.46 × 25 cm) (56) with (a) 0.116 mmol β-cyclodextrin per gram of CSP (eluent: methanol–water 64:36); (b) 0.143 mmol β-cyclodextrin per gram of CSP (eluent: methanol–water 64:36); and (c) 0.143 mmol β-cyclodextrin per gram of CSP (eluent: methanol–water 80:20).*

our laboratories as experimental rice herbicides (57). Racemic **22** could be well resolved on tribenzoyl cellulose (TBC) with pure methanol as the mobile phase, whereas **23** is baseline-resolved on Chiralcel OD (Figure 11).

Biorational design led to the synthesis of compound **24,** a herbicidally active inhibitor of histidine biosynthesis, which inhibits the enzyme imidazole glycerol phosphate dehydratase (58). The synthetic procedure provided the *trans*-isomer almost exclusively. The enantiomers of its synthetic precursor **25** can be base-line-separated by HPLC on a Chiralcel OD column (Figure 12). To provide enough of both enantiomers of **24** for individual biological testing, we carried out a preparative chiral resolution of the racemic precursor **25** using a CTA-I column. The optical purities of the isolated enantiomers of **25** (both greater than 99% ee) were determined on Chiralcel OD.

Safeners

Safeners are compounds that increase the selectivity of herbicides. The safener benoxacor (**26**), for example, increases the tolerance of maize to metolachlor. The enantiomers of this chiral compound were resolved by preparative HPLC (Figure 13). Up to 50 g of racemate could be separated in one run on a CTA-I column (20 × 100 cm; see Chapter 10). The optical purities of the isolated frac-

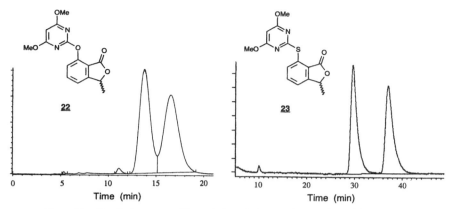

Figure 11. Liquid chromatographic resolution of the enantiomers of (a) 22 on TBC (eluent: methanol–water 95:5) and (b) 23 on Chiralcel OD (eluent: hexane–2-propanol 95:5).

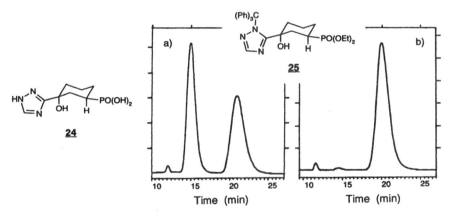

Figure 12. Resolution of the enantiomers of 25, the synthetic precursor of the histidine biosynthesis inhibitor 24, by HPLC on Chiralcel OD: (a) racemic sample, (b) sample with >99% ee (eluent: hexane–2-propanol 90:10).

tions (97% to 100%) were determined on an analytical HPLC column (0.46 × 25 cm) containing the same CSP. For the chloromethyl and the methyl derivatives **27** and **28,** the chiral selectivity obtained on CTA-I is inversely proportional to the number of chlorine atoms attached to the *exo*-methyl group (Figure 13). The enantiomers of **28** are also well resolved on *m*-methylbenzoyl cellulose beads (MMBCs) and on *p*-methylbenzoyl cellulose beads (PMBCs), CSPs (*59*) developed in our laboratories (*60*).

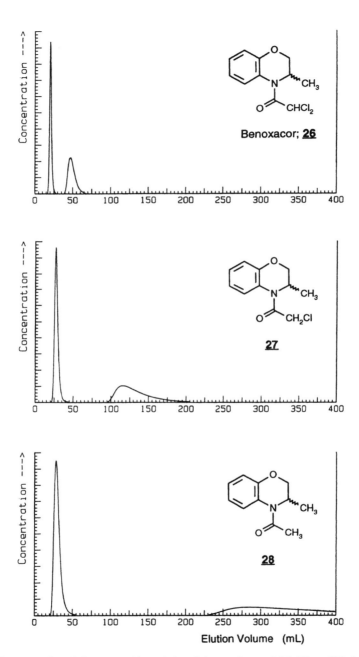

*Figure 13. Liquid chromatographic resolution of the enantiomers of **26–28** on CTA-I (column: 1.25 × 15 cm; eluent: ethanol–water 95:5).*

Disease Control

The implications of stereoisomerism for the biological activity of fungicides (*61*) are difficult to characterize because of the complex system involved. Receptors and metabolic pathways in both the pathogen and the host plant must be considered.

A century of research on chemical control of plant diseases has only recently led to the development and use of so-called systemic fungicides. Among these chemically diverse fungicides, the ergosterol biosynthesis inhibitors (EBIs) have rapidly gained importance. The EBIs include a number of chiral compounds. In fact, most of the important modern selective fungicides are chiral EBIs. Although the influence of stereoisomerism on their biological activity must have been generally recognized by now, astonishingly little work has been published about chromatographic resolution and analysis of chiral fungicides.

Ergosterol Biosynthesis Inhibitors

The large group of EBIs is chemically highly heterogeneous. As their name implies, they are characterized by a common mode of action, not a special structure or structural moiety.

Triazole Fungicides

The triazole fungicides (Figure 14) are the most important group among the EBIs (*61*). The primary mode of action of the 1,2,4-triazole fungicides is the inhibition of the cytochrome P-450–dependent C-14-demethylation of lanosterol or 24-methylene-24,25-dihydrolanosterol, which are intermediates in fungal sterol biosynthesis (*62*). Besides their fungicidal activity, some of these compounds also act as plant growth regulators. This activity can often be explained by their inhibition of gibberellin biosynthesis, and it has led to the development

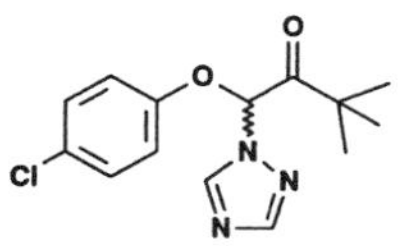

Triadimefon (**29**) 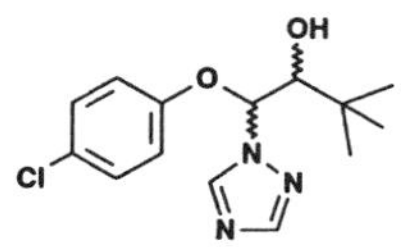   Triadimenol (**30**) 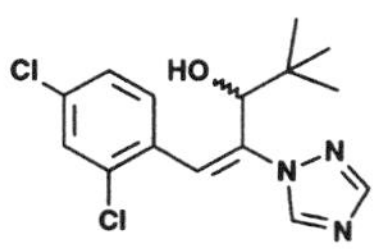   Diniconazole (**32**) 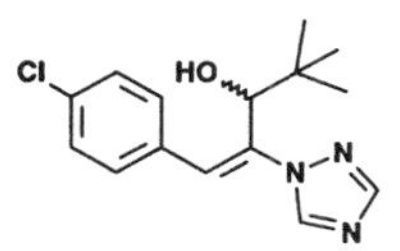   Uniconazole (**33**)

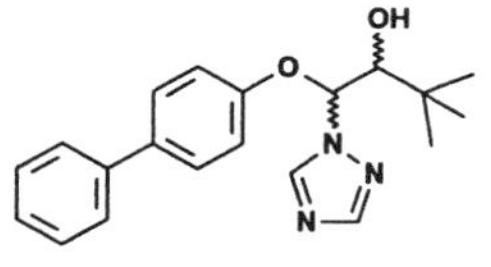

Bitertanol (**34**) 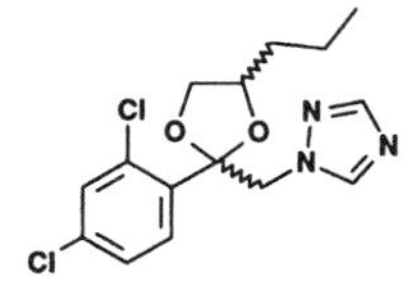   Propiconazole (**35**)

Figure 14. Structure of some commercial chiral 1,2,4-triazole fungicides.

of several plant growth regulators containing a triazole moiety (e.g., paclobutra-zol) (*62*).

The relationship between stereoisomerism and biological activity has been extensively investigated for the triazole fungicides triadimefon (**29**) and triadi-menol (**30**). As a consequence, quite a few papers have been published mention-ing analytical methods for determining the enantiomers of triadimefon and the four stereoisomers of triadimenol. A good chiral resolution of triadimefon was achieved on a CSP consisting of poly(triphenylmethyl methacrylate) (Chiralpak OT) (*63*). Column packing of *N*-formyl-(*S*)-isoleucine aminopropylated silica gel, however, gave only partial resolution (*64*). The enantiomers of triadimefon could be baseline-separated on silica gel coated with cellulose tris(*p*-methylbenzoate) (Chiralcel OJ) (*65*).

While for the enantiomers of the relatively weak fungicide triadimefon no significant activity differences have been found (*66*), the isomers of the much more potent fungicide triadimenol differ considerably in their activity, the (1*S*,2*R*)-isomer being clearly the most active (*63*). Furthermore, it is well docu-mented that many species of fungi reduce triadimefon to triadimenol (*67*). The four stereoisomers can be separated by GC on a Chirasil-L-Val CSP (Figure 15) (*63, 68–70*).

Two other chromatographic separations of the four stereoisomers of triadi-menol have been reported, both involving a derivatization. On an SE-52 fused-silica GC column, the diastereomeric (1*S*,3*S*)-*trans*-chrysanthemoyl esters were

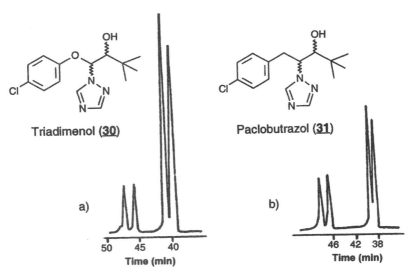

Triadimenol (**30**)

Paclobutrazol (**31**)

a)

b)

50   45   40
Time (min)

46   42   38
Time (min)

*Figure 15. Resolution of the stereoisomers of (a) triadimenol (**30**) and (b) paclobutrazol (**31**) by GC on a Chirasil-Val fused-silica capillary column (carrier: hydrogen). (Reproduced with permission from reference 60. Copyright 1985.)*

baseline-separated (*71*); on a Pirkle-type (*S*)-*N*-(3,5-dinitrobenzoyl leucine HPLC column (DNBLeu), the *N*-isopropylurethane derivatives could be resolved (*72*).

The structurally similar paclobutrazol (**31**) was developed and marketed as a plant growth regulator for pot-grown ornamentals and fruit trees. Like triadimenol, it possesses two asymmetric carbon atoms and therefore consists of four stereoisomers. These four isomers exhibit very different biological activity profiles. The (2*S*,3*S*)-isomer is responsible for plant growth regulation by the inhibition of gibberellin biosynthesis, while the (2*R*,3*R*)-isomer is a fungicide with low plant-growth-regulating activity (*73, 74*). The market product consists of these two stereoisomers. The four stereoisomers of paclobutrazol can be virtually baseline-separated by capillary GC with a Chirasil-L-Val column (Figure 15) (*69, 70*). The (1*S*,3*S*)-*trans*-chrysanthemoyl ester derivatives of the enantiomers of the biologically relevant (2*RS*,3*RS*)-diastereomer are also resolved on an achiral SE-52 fused-silica column (*71*).

The 1-vinyltriazoles diniconazole (**32**) and uniconazole (**33**) (Figure 14) are also well documented, as far as the analysis of stereoisomers is concerned. The 1-vinyltriazoles have been developed as fungicides or as plant growth regulators. Diniconazole was developed and marketed as fungicide with excellent activity against powdery mildews and other leaf and ear diseases (e.g., *Septoria, Fusarium*, rusts) in cereals, against powdery mildew in vines, and against powdery mildew, rust, and black spot in roses. Its monochloro analogue, uniconazole, is an effective plant growth regulator. The biological activities of both compounds are concentrated in the racemates with the (*E*)-configuration at the double bond. The first reported chromatographic resolutions of **32** and **33** were achieved on a CSP consisting of (1*R*,3*R*)-*N-trans*-chrysanthemoyl-(*R*)-phenylglycine covalently bound to γ-aminopropyl silanized silica (Sumichiral OA-2200) (α = 1.50 for **32** and α = 1.31 for **33**) (*75*). The similar CSP (*R*)-*N*-1-(α-naphthyl)ethylaminocarbonyl-(*S*)-valine covalently bound to γ-aminopropyl silanized silica (Sumichiral OA-4100) also gave good separations of the enantiomers of **32** (α = 1.20) and **33** (α = 1.15) (*76*). The other three diastereoisomers of this CSP were tested as well, and the α values obtained were similar (*77*). Chromatograms with good resolutions (α = 1.34 for **32** and α = 1.29 for **33**) were also achieved on a structurally related CSP (with (*S*)-*tert*-leucine instead of (*S*)-valine; Chirex 3020) (Figure 16) (*78*). In comparison, a Pirkle-type column consisting of (*R*)-*N*-(3,5-dinitrobenzoyl)-1-naphthylglycine ionically bound to γ-aminopropyl silanized silica (Sumichiral OA-2500-I) gave a somewhat less effective chiral resolution (*79, 80*). A Cyclobond I CSP (β-clycodextrin chemically bound to spherical silical gel) proved to be well suited for the liquid chromatographic resolution of the enantiomers of **32** (α = 1.19) but not for **33** (*81*). This method was applied for the analysis of diniconazole-M, a formulation for enriched (*R*)-**32** (*81*). A Cyclobond II column (γ-cyclodextrin chemically bound to spherical silica gel) gave almost baseline separation of the enantiomers of **33** (*82*).

Bitertanol (**34**) (Figure 14) is another commercially successful triazole fungicide whose chiral analysis is reported in the literature. It is mainly used to control

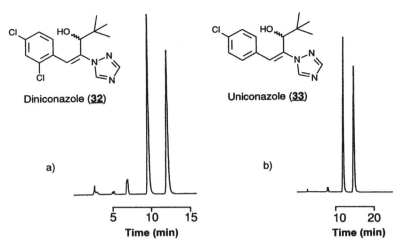

Figure 16. Liquid chromatographic resolutions of (a) diniconazole (32) and (b) uniconazole (33) on Chirex 3020 (eluent: hexane–1,2-dichloroethane–ethanol 100:20:1). (Reproduced with permission from reference 78.)

fruit diseases caused by *Venturia* and *Sclerotinia*. Direct chromatographic resolution was attempted by capillary GC on a Chirasil-L-Val column (*69*), but the resolution remained poor. The diastereomeric (1*S*,3*S*)-*trans*-chrysanthemoyl ester derivatives of the enantiomers, however, were baseline-resolved on an SE-52 fused-silica column (*71*).

One of the most important market products in this class of compounds is propiconazole (**35**) (Figure 15), a systemic foliar fungicide with a broad range of activity. Some years ago, remarkable differences in in vitro activity against *Ustilago maydis* were reported for the four stereoisomers (*83*). Therefore, a novel stereoselective synthesis of all four stereoisomers was developed (*3, 84, 85*) to provide the large quantities needed for extensive greenhouse and field testing. The tests showed that the different pathogens discriminate between the stereoisomers, so each has a different activity profile. As a consequence, a mixture of all four was required for strong activity against a broad spectrum of pathogens and thus for the extraordinary commercial success of propiconazole. On Chiralcel OD, all four stereoisomers can be detected, but their resolution remains incomplete. Baseline separation is obtained, however, on the high-resolution CSP Chiralcel OD-H (Figure 17).

The triazole **36** is an experimental fungicide developed in our laboratories that has good fungicidal activity, particularly against *Piricularia oryzae*. It consists of four stereoisomers, which were separable through tedious crystallization procedures (*86*). Individual biological testing revealed great differences in their biological profiles. Chiral analysis proved quite difficult. In fact, baseline separation of all four stereoisomers could never be achieved, but one CSP, Chiralcel OC,

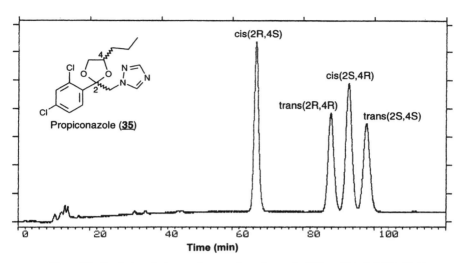

Figure 17. Resolution of the stereoisomers of propiconazole (35) on Chiralcel OD-H (eluent: hexane–2-propanol 98:2).

proved to be best suited for this case (Figure 18). While the so-called racemate B (consisting of the (*S,R*)- and (*R,S*)-stereoisomers) was completely resolved, only partial resolution was obtained for racemate A (the (*S,S*)- and (*R,R*)-stereoisomers). However, the enantiomers of the synthetic precursor of **36,** the ketone **37,** were baseline-separated on Chiralcel OD (Figure 18).

A number of chiral experimental triazole fungicides from our laboratories

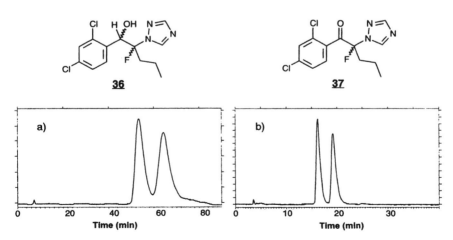

*Figure 18. Resolution (a) of racemate B of **36** on Chiralcel OC (eluent: hexane–2-propanol 95:5) and (b) of the enantiomers of **37** on Chiralcel OD (eluent: hexane–2-propanol 95:5).*

were investigated to determine the biological activities of their enantiomers (*87*). To obtain these enantiomers in sufficient quantities, we attempted preparative chiral chromatographic resolutions of their epoxide precursors **38–42** with cellulose triacetate as a CSP. Interestingly, the ester groups exert a strong influence on the selectivity: While the chiral resolution of the methyl ester **38** is exceptionally good (α = 10.19), it is much worse for the corresponding ethyl ester **39** (α = 1.48), and the isopropyl ester **40** is not resolved at all (Figure 19). In contrast, the sterically more hindered *tert*-butyl ester **41** is again reasonably well resolved (α = 1.84). In the case of the methyl ester **38,** the corresponding enantiomers of the (triazolylmethyl)mandelic acid were obtained from the optically pure epoxides by ring opening with triazole (*87*). The enantiomers of this triazole arylester derivative can also be separated on the MMBC CSP (*60*).

Morpholines

Morpholine fungicides inhibit sterol biosynthesis at two different sites of the same pathway: the Δ^{14}-reductase and the Δ^{8}-Δ^{7}-isomerase (*88–90*). Because of their high activity against powdery mildew and rusts, the morpholine fungicides

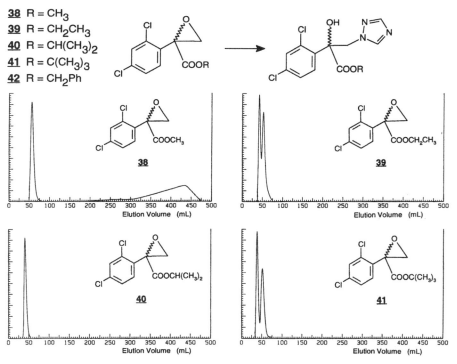

*Figure 19. Liquid chromatographic resolution of the enantiomers of the epoxide intermediates **38–42** on CTA-I (eluent: ethanol–water 95:5).*

were developed mainly for use in cereals. In spite of their importance, no chiral analysis of the main morpholine market products can be found in the literature.

We have recently designed morpholine compounds based on the postulated principle of action of EBIs on the Δ^8-Δ^7 isomerization (*91*). The main representative of the new class of EBIs is compound **43,** which contains two asymmetric centers in the five-membered ring. We attempted to prepare the four stereoisomers to evaluate their biological activities. The syntheses started from the optically pure enantiomers of 3-phenylcyclopentanone, obtained by preparative chromatographic resolution on CTA-I (*91*). Introduction of the *tert*-butyl group into the *para* position of the phenyl ring led to the corresponding enantiomers of 3-(4-*tert*-butylphenyl) cyclopentanone, whose optical purity was determined by chiral HPLC on MMBC (Figure 20) (*91*). Treating these optically active intermediates with *cis*-2,6-dimethylmorpholine produced two pairs of diastereoisomers that could be separated by chromatography on silica gel. Both diastereomeric enantiomer pairs of **43** could be partially resolved by HPLC on MMBC or PMBC (*56*). Biological tests showed that the (1*R*,3*S*)-isomer was the most potent (*91*).

Pyrimidines

The systemic pyrimidine fungicide fenarimol (**44**), also an inhibitor of the C-14 demethylation of lanosterol, is used against *Venturia* spp., and *Erysiphales* on pome fruit and *Erysiphales* on cherries, eggplants, grapes, peppers, and tomatoes (Figure 21). The chiral chromatographic resolution of its enantiomers was attempted on

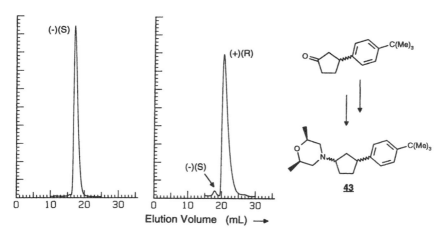

*Figure 20. Structure of the experimental morpholine fungicide **43** and analytical determination of the optical purities of the enantiomers (a) (–)-(S) and (b) (+)-(R) of its precursor 3-(4-tert-butylphenyl)cyclopentanone on MMBC (eluent: hexane–2-propanol 90:10).*

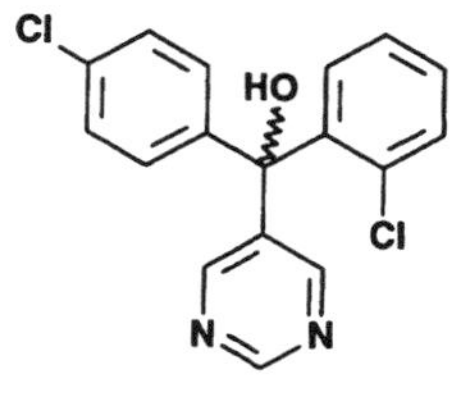

Figure 21. Structure of fenarimol.   **Fenarimol (44)**

two CSPs. A chiral urea ion-pair column gave only poor resolution, and the application of a β-cyclodextrin column was reported without any details (*92*).

Acylalanines and Related Acylanilides

The origin of the acylaniline fungicides is closely linked with herbicide research on chloroacetanilides such as metolachlor (**14**) (Figure 7). Replacing the alkoxy-alkyl substituent in metolachlor with an ester moiety led to a prototype compound **45,** which exhibited curative and systemic activity against *Phytophthora infestans* (*93*) (Figure 22). Resolution and biological testing of the enantiomers revealed that (*R*)-**45** was an excellent fungicide with low herbicidal activity, while its (*S*)-enantiomer showed high herbicidal but no fungicidal activity (*93*). An intensive optimization program produces the biologically more powerful metalaxyl (**46**) (*94*). In this case, the (*R*)-enantiomer had 3 to 10 times the fungicidal activity of (*S*)-metalaxyl, depending on the pathogens tested (*92, 95, 96*). Because racemic metalaxyl shows no phytotoxic effects, separating its enantiomers was not necessary and the compound was developed as a racemate. The enantiomers of metalaxyl can be resolved by HPLC on a Chiralcel OJ column (Figure 23).

A later result of structural optimization is the experimental fungicide clozylacon (**47**) (*97–99*), which, because of its greater stability against hydrolysis, is especially suited for soil application against oomycetes, for which long-term protection is required (*99, 100*). Because of the stereogenic center at C-3 and the atropisomerism due to hindered rotation around the aryl carbon–nitrogen bond, clozylacon consists of four stereoisomers, which have been separated by several

(*R*)-**45**   Metalaxyl (**46**)   Clozylacon (**47**)

Figure 22. Structure of some chiral acylanilide fungicides.

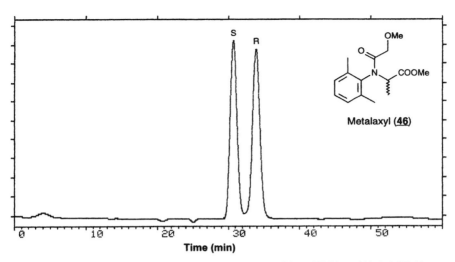

*Figure 23. Separation of the enantiomers of metalaxyl (**46**) by HPLC on Chiralcel OJ (eluent: hexane–2-propanol 99:1).*

methods (*99, 101, 102*). The individual biological tests showed that the fungicidal activity arises mainly from the (αS,3R)-stereoisomer (*102*). Large amounts of the most active isomer have been obtained by chromatographic resolution on cellulose triacetate (Chapter 10). The four stereoisomers can be baseline-separated either on a ChiraSpher column or on MMBC beads (Figure 24).

In the course of work toward an economically feasible stereoselective synthesis of the biologically most active (αS,3R)-isomer of clozylacon (*103–105*),

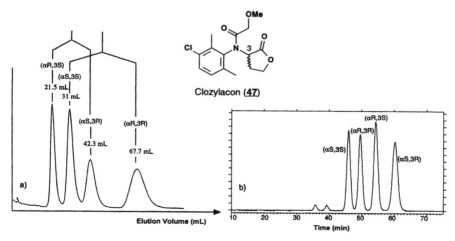

*Figure 24. Liquid chromatographic resolution of the stereoisomers of clozylacon (**47**) (a) on ChiraSpher (eluent: hexane–2-propanol 95:5) and (b) on MMBC (eluent: hexane–2-propanol 6:4).*

methods for the liquid chromatographic resolution of the enantiomers of the unchlorinated intermediate have also been developed. Baseline separation was achieved on both the ChiraSpher (*105*) and CTA-I CSPs. ChiraSpher has been the standard method for determining the optical purities of all the enantioselectively synthesized samples of this unchlorinated precursor (ee values of up to 99% measured).

The structurally related carboximide vinclozolin (**48**) is a selective contact fungicide effective against *Botrytis cinerea*, *Sclerotinia sclerotiorum*, and *Monilinia* spp. in grapes, oilseed rape, hops, strawberries, stone fruit, and vegetables (Figure 25). Its chiral resolution, with a poor resolution factor of 1.03, was carried out by capillary GC on an octakis(3-*O*-butyryl-2,6-di-*O*-pentyl)-γ-cyclodextrin column (Lipodex E) (*106*).

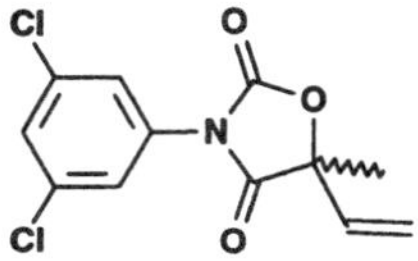

Figure 25. Structure of vinclozolin.   Vinclozolin (**48**)

Insect Control

There has been much research on the interrelation of stereochemistry and biological activity of synthetic and natural insecticides. The topic has recently been reviewed for the most important classes of insecticides containing chiral compounds, such as organochlorines, carbamates, pyrethroids, and organophosphates (*107*). Most of the commercially important insect control agents are organophosphates or pyrethroids, which hold an estimated 50% to 60% of the world's insecticide market (*108*). The number of publications dealing with chiral analysis of organophosphates and pyrethroids reflects the great interest in the implications of stereoisomerism in these two classes of compounds.

Organophosphates

Organophosphates that act as cholinesterase inhibitors began to be introduced as insecticides in the 1950s, and some of them are chiral (Figure 26). The biochemical and toxicological properties of the enantiomers of chiral organophosphates differ widely (*109*). In fact, enantiomers of chiral organophosphates having the phosphorus atom as a stereocenter often exhibit widely different anticholinesterase activity, whereas only slight activity differences are found for enantiomers of organophosphates with stereogenic centers other than phosphorus (*109*).

Malathion (**49**) is one of the most widely used organophosphorus insecticides. It can be thermally isomerized to isomalathion (**50**), which has a second

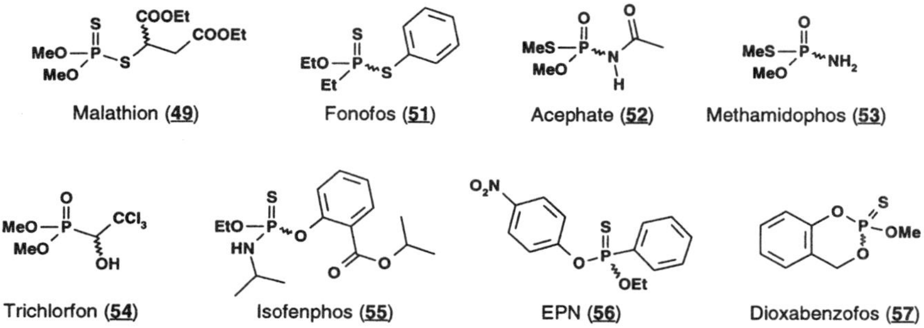

Malathion (**49**)   Fonofos (**51**)   Acephate (**52**)   Methamidophos (**53**)

Trichlorfon (**54**)   Isofenphos (**55**)   EPN (**56**)   Dioxabenzofos (**57**)

Figure 26. Structure of some commercial chiral organophosphate insecticides.

stereogenic center at the phosphorus atom and therefore consists of four stereoisomers (*110*). Whereas malathion is virtually nontoxic, isomalathion is a potent inhibitor of acetylcholinesterase (*111*). The four stereoisomers of isomalathion were baseline-separated by HPLC on an amylose tris(3,5-dimethylphenyl)carbamate column (Chiralpak AD) (Figure 27) (*110*). With this analytical method, the stereoisomeric purities of all four stereoselectively synthesized stereoisomers were determined (values of up to 98% measured).

Fonofos (**51**) is another commercially important organophosphorus insecticide used to control *Gryllotalpa* spp., and soil pests (e.g., *Diabrotica, Symphyla, Agriotes* spp.). Two methods for the chiral resolution of its enantiomers were reported in the literature. Baseline separation was achieved by capillary GC employing a column coated with octakis(6-*O*-methyl-2,3-di-*O*-pentyl)-γ-cyclodextrin (Figure 28)

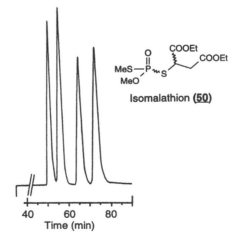

Isomalathion (**50**)

*Figure 27. Liquid chromatographic resolution of the stereoisomers of isomalathion (**50**) on Chiralpak AD (eluent: hexane–2-propanol 90:10). (Reproduced with permission from reference 110. Copyright 1993.)*

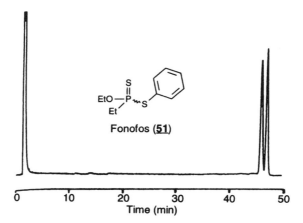

*Figure 28. Resolution of the enantiomers of fonofos (**51**) by GC on a pyrex glass capillary column coated with octakis(6-O-methyl-2,3-di-O-pentyl)-γ-cyclodextrin (carrier gas: hydrogen, 0.5 bar). (Reproduced with permission from reference 24. Copyright 1991.)*

(*24*). On an HPLC column filled with *N*-(naphthylethyl)carbamate-substituted β-cyclodextrin (Cyclobond I RN), no effective resolution was obtained (α = 1.02) (*38*).

Acephate (**52**) and methamidophos (**53**) are systemic broad-spectrum insecticides (Figure 26). Acephate does not inhibit acetylcholinesterase directly, but it is converted in insects to methamidophos, which produces the inhibition (*112*). While the enantiomers of acephate can be separated by liquid chromatography on Chiralcel OC (*65, 113*), methamidophos was resolved on a fused-silica capillary GC column coated with heptakis(3-O-acetyl-2,6-di-O-pentyl)-β-cyclodextrin (*24*).

The organophosphorus compound trichlorfon (**54**) is a contact and ingestive insecticide used to control Diptera and Lepidoptera on beets, cereals, cotton, fruits, and vegetables. It is also used against household pests and ectoparasites on domestic animals. The same GC CSP used for methamidophos proved to be well suited for the chiral resolution of its enantiomers: heptakis(3-O-acetyl-2,6-di-O-pentyl)-β-cyclodextrin (Figure 29) (*24*). Furthermore, with a Chiralcel OB HPLC column, a very high α value (1.62) was reported (*114*).

For isofenphos (*55*), a contact and ingestive insecticide effective against soil-dwelling insects (Coleoptera) and leaf-eating pests (Diptera and Homoptera), two HPLC methods for the enantiomer resolution were published. On a chiral Sumipax OA-4000 column (structure of CSP not reported), close to baseline separation was achieved (*115*), while the use of a Chiralcel OC column was reported without any details about the quality of the resolution (*116*).

O-Ethyl *O*-*p*-nitrophenyl phenylphosphonothioate (**56**) (EPN) is a nonsystemic insecticide effective against a wide range of lepidopterous larvae, especially bollworms and *Alabama argillacea* in cotton, *Chilo* spp. in rice, and other leaf-eat-

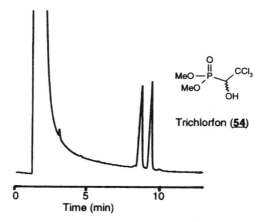

Trichlorfon (**54**)

*Figure 29. Separation of the enantiomers of trichlorfon (**54**) by GC on a pyrex glass capillary column coated with heptakis(3-O-acetyl-2,6-di-O-pentyl)-β-cyclodextrin (carrier gas: hydrogen, 0.5 bar). (Reproduced with permission from reference 24. Copyright 1991.)*

ing larvae in fruit and vegetables. The first reported separation of its enantiomers (α = 1.14) was done on a Chiralpak OT column (*117, 118*). The same research group used EPN as one of a number of chiral organophosphorus compounds to test new CSPs. While Chiralcel OA and OB columns gave good resolution factors (1.40 and 1.22, respectively), Chiralcel OC and OK did not resolve the EPN enantiomers at all (*114*). On a Chiralcel OJ column, however, baseline separation was obtained (*65*).

Dioxabenzofos (**57**) is an organophosphorus insecticide with vapor action against a wide range of boring, chewing, and sucking insects in fruit, rice, tea, tobacco, and vegetables. Recent studies with its enantiomers illustrated the complex interplay of stereochemical, toxicological, and metabolic factors (*119–121*). The first liquid chromatographic resolution of the enantiomers of dioxabenzofos was developed on Chiralcel OK (α = 1.09) (*114*). A much more efficient separation (α = 1.45) was later achieved on Chiralpak OT (*117, 118*).

Pyrethroids

The pyrethroid insecticides were derived from natural products that occur in the flower heads of various *Chrysanthemum* species. In fact, the insecticidal properties of pyrethrum extracts isolated from the flower heads of *Chrysanthemum cinerariae-folium* have been well documented for more than 100 years (*122*). The natural product's rapid knockdown of insects, high level of activity, and low toxicity to mammals caused intense interest in synthesizing similar compounds. Early synthetic pyrethroids had significant success as household insecticides in spite of their inherent instability to light. In the early 1970s, the combination of phenoxybenzyl alcohols with chrysanthemic acid derivatives, in which the dimethylvinyl substituent was replaced by dihalovinyl moieties, led to a series of highly active and light-stable synthetic pyrethroids (*123*). After that breakthrough the pyrethroids quickly gained commercial importance, and today they account for more than 20% of the world's insecticide market (*108*).

Pyrethroids constitute a chemically diverse class of neuroactive insecticides. In general, they contain one to three chiral centers and therefore consist of two to eight stereoisomers. Often the individual stereoisomers exhibit large differences in toxicity to insects. Not surprisingly, much research effort has been put into the resolution and analysis of pyrethroid stereoisomers to investigate their individual biological activities. Most HPLC analysis of pyrethroid stereoisomers has been done on Pirkle-type CSPs.

The synthetic pyrethroids are often divided into three groups: the α-cyano pyrethroids, the noncyano pyrethroids, and the pyrethroids lacking the cyclopropane moiety.

α-Cyano and Noncyano Pyrethroids with a Cyclopropane Moiety

Cypermethrin (**58**), an ingestive and contact insecticide effective against a wide range of insect pests, particularly Lepidoptera in cereals, citrus, fruits, tobacco, vegetables, and other crops, is one of the economically important pyrethroids marketed by several companies (Figure 30). The eight stereoisomers of cypermethrin were resolved by HPLC on Pirkle DNBPG columns, covalently (*92, 124*) or ionically bonded (*125–127*), and on Cyclobond I (*124*). However, baseline separation was achieved only by using two Pirkle-type columns in series, the first containing *N*-(*S*)-1-(α-naphthyl)ethylaminocarbonyl-(*S*)-*tert*-leucine (Sumichiral OA-4600) and the second *N*-(*R*)-1-(α-naphthyl)ethylaminocarbonyl-(*S*)-*tert*-leucine (Sumichiral OA-4700) covalently bonded to γ-aminopropyl silanized silica (Figure 31) (*78, 128*).

Cyfluthrin (**59**), a fluoro derivative of cypermethrin, is an insecticide used to control many pests, especially Homoptera, Hemiptera, and Lepidoptera on cereals, cotton, fruits, and vegetables. Its eight stereoisomers have been separated by chiral HPLC on DNBPG: In the ionic case, five baseline-separated peaks were observed (*125*); in the covalent case, all eight stereoisomers could be detected, although not all of them were baseline-separated (*127*).

The economically very important cyhalothrin (**60**) is used to control animal ectoparasites, especially *Boophilus micropus* and *Haematobia irritans* on cattle and *Bovicola ovis*, *Linognathus* spp., and *Melophagus ovinus* on sheep. Chiral analytics of cyhalothrin are mentioned in three papers. While apparently all four stereoisomers were detectable on DNBPG (no details given) (*126, 127*), only the two diastereomers could be resolved on a column containing *N*-formyl-(*S*)-phenylalanine aminopropylated silica gel (*64*).

Deltamethrin (**61**), developed and marketed as a single isomer, the (α*S*,1*R*)-*cis*-configuration, is widely acknowledged as the leading product in the

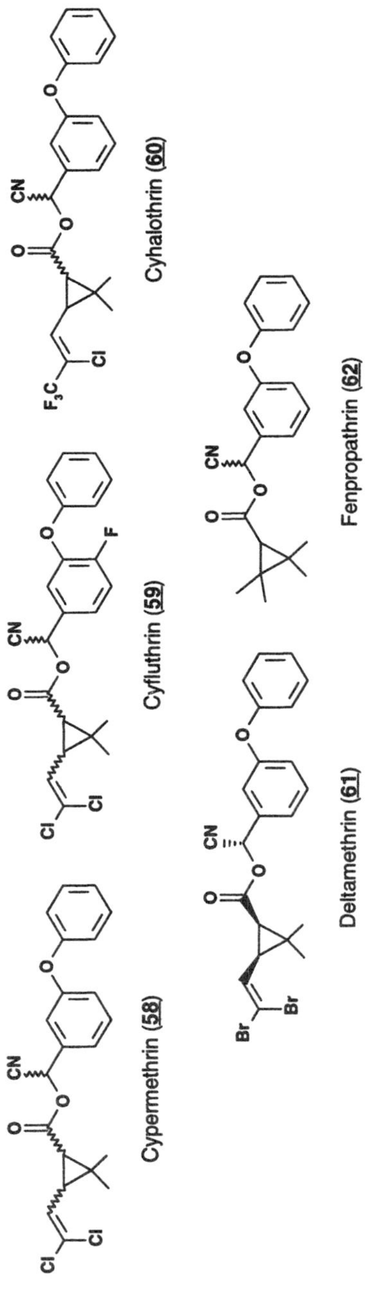

Figure 30. Structure of some commercial α-cyano pyrethroids.

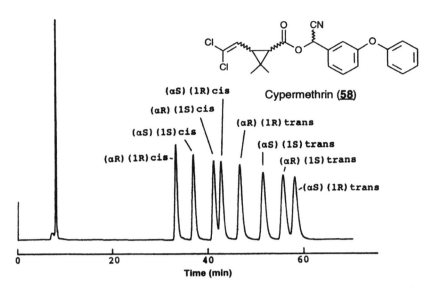

*Figure 31. Resolution of the stereoisomers of cypermethrin (**58**) by HPLC on two columns, Sumichiral OA-4600 and OA-4700, in series (eluent: hexane–1,2-dichloroethane–ethanol 500:10:0.05). (Reproduced with permission from reference 128. Copyright 1990.)*

pyrethroid market (*108*). It is a potent insecticide effective by contact and ingestion against a wide range of pests, such as Homoptera, Lepidoptera, and Thysanoptera in cereals, citrus, cotton, grapes, soybeans, vegetables, and other crops. The marketed stereoisomer and its enantiomer were separated by HPLC on Chiralpak OT(+) (*129*). The separation of the marketed stereoisomer from its (αR)-epimer has been reported twice; in both cases DNBPG columns were applied (*125, 126*). Finally, the separation of all eight stereoisomers of deltamethrin was attempted on a covalent DNBPG column. Quite a good resolution was obtained, even though not all eight stereoisomers were baseline-separated (*130*).

Fenpropathrin (**62**) is effective at relatively high application rates against various species of Acari, Aleyrodidae, Aphididae, and Lepidoptera in cotton, grapes, ornamentals, vegetables, and other field crops. Unlike the pyrethroids discussed earlier, it has only one asymmetric carbon. Its enantiomers were readily resolved on CSPs consisting of *N*-(*S*)-(Sumichiral OA-4000) and *N*-(*R*)-1-(1-naphthyl)ethylaminocarbonyl-(*S*)-valine (Sumichiral OA-4100) covalently bonded to γ-aminopropyl silanized silica (α values of 1.22 and 1.12, respectively) (*76*). The same type of CSP with varied terminal (e.g., *N-tert*-butyl) and amino acid groups (e.g., (*R*)-valine, (*S*)-*tert*-leucine) gave similar α values (between 1.04 and 1.26) (*77, 78, 128*), the best suited being Sumichiral OA-4600 (Figure 32) (*128*). On Sumichiral OA-2500-I, α values of up to 1.09 were obtained (*79, 80, 127*). Not quite baseline separation was achieved on a covalent DNBPG CSP (*126*).

Permethrin (**63**), one of the so-called noncyano pyrethroids (Figure 33), is a

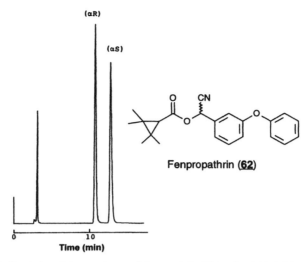

*Figure 32. Resolution of the enantiomers of fenpropathrin (**62**) on Sumichiral OA-4600 (eluent: hexane–1,2-dichloroethane–ethanol 500:10:0.05). (Reproduced with permission from reference 128. Copyright 1990.)*

contact insecticide for the control of a broad range of pests, such as leaf- and fruit-eating lepidopterous and coleopterous pests in cotton, fruit, tobacco, vines, and other crops. It is also effective against a wide range of animal ectoparasites. The chiral analysis of its four stereoisomers was first attempted on DNBPG. While in an early paper only the *cis-* and *trans-*stereoisomers could be separated (*125*), later the detection of four peaks was reported (*126, 127*). However, no baseline resolution of the *cis-* and *trans-*racemates could be achieved. Somewhat better results, with α values of 1.15 and 1.05 for the enantiomers of the *cis-* and the *trans-*racemate, respectively, were obtained on Sumichiral OA-2500-I (*79, 128*).

Allethrin (**64**) is a contact insecticide used against household insect pests such as Blattodea, Cimicidae, Culicidae, Muscidae, Siphonaptera, and Acari. On DNBPG HPLC columns, the resolution of the eight stereoisomers was not

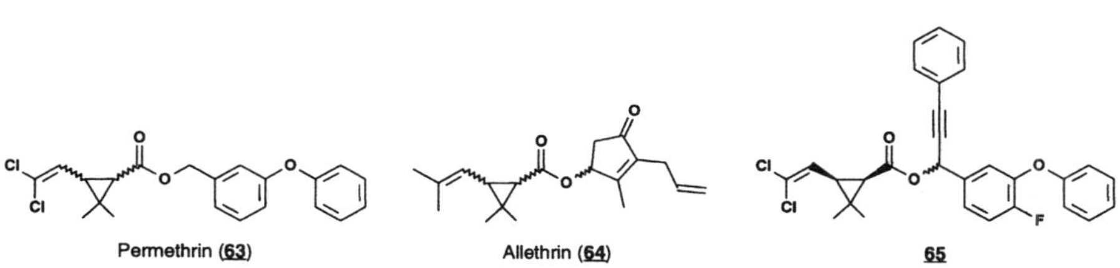

Permethrin (**63**)   Allethrin (**64**)   **65**

Figure 33. Structure of some noncyano pyrethroids.

very efficient (*126, 127*). On Cyclobond I, the four *trans*-stereoisomers were almost baseline-separated, while the *cis*-stereoisomers were not resolved at all (*124*). A stunning baseline resolution of all eight stereoisomers was achieved on Sumichiral OA-4600 (Figure 34) (*78, 128*).

The experimental pyrethroid **65**, a mixture of the epimers with the configurations (α*R*,1*R*)-*cis* and (α*S*,1*R*)-*cis* (*131*), contains a phenylethynyl substituent in place of the cyano group. It is a fish-safe pyrethroid insecticide with good activity against rice hoppers and rice water weevils. In the course of synthetic efforts in our laboratories toward a stereoselective preparation of the alcohol moiety of **65**, a chiral analytical method for the accurate and efficient determination of optical purities was needed. The enantiomers of the alcohol **66** could be resolved without derivatization by HPLC on a Chiralcel OC CSP (Figure 35). With this method, samples of **66** with optical purities up to 95% ee were analyzed.

Replacing the phenoxybenzyl alcohol moiety in permethrin (**63**) with a phenoxypyridyl alcohol led to a series of potent pyrethroid insecticides of structure **68** (Figure 36) (*132*). For the investigation of the biological activities of the individual isomers of these pyrethroids, the optically pure enantiomers of the corresponding alcohols **67** had to be prepared. They were accessible via chiral resolution of their acetate derivatives by preparative liquid chromatography on cellulose triacetate (*132*).

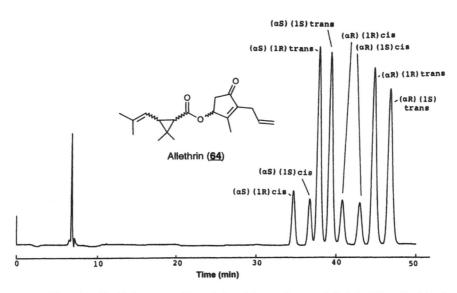

*Figure 34. Liquid chromatographic resolution of the stereoisomers of allethrin (**64**) on Sumichiral OA-4600 (eluent hexane–1,2-dichloroethane–ethanol 500:30:0.15). (Reproduced with permission from reference 128. Copyright 1990.)*

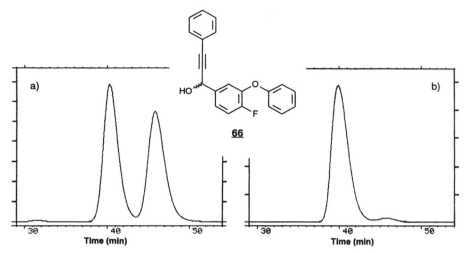

*Figure 35. Separation of the enantiomers of the alcohol moiety **66** of the experimental pyrethroid **65** by HPLC on Chiralcel OC: (a) racemic sample, (b) enriched sample with 93% ee (eluent: hexane–2-propanol 95:5).*

Pyrethroids Lacking the Cyclopropane Moiety

Fenvalerate (**69**), an economically important pyrethroid lacking the cyclopropane moiety (Figure 37), is a highly active contact insecticide used to control a wide range of pests, including Aleyrodidae, Aphididae, and Lepidoptera in cereals, cotton, grapes, maize, potatoes, and vegetables. Several methods for the chiral resolution of its four stereoisomers have been published. The first reported chiral analyses were done on DNBPG CSPs, both ionic and covalent (*125–127, 133, 134*). In one case, baseline resolution of all stereoisomers was achieved (*133, 134*). The most efficient resolution, however, was obtained on Sumichiral OA-2500-I (Figure 38) (*78, 128*).

τ-Fluvalinate (**70**) is a potent, broad-spectrum, foliar-applied insecticide and

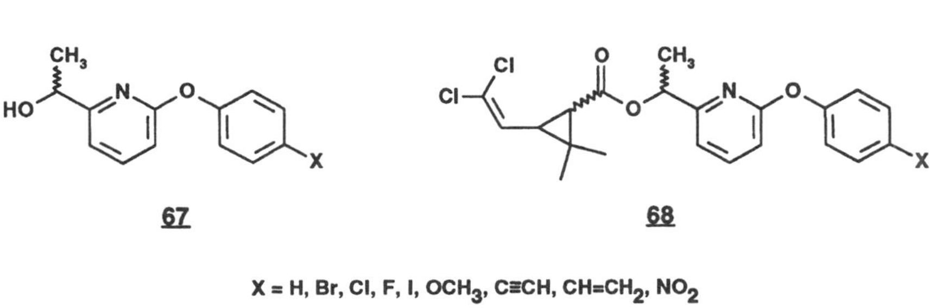

X = H, Br, Cl, F, I, OCH₃, C≡CH, CH=CH₂, NO₂

Figure 36. Structure of phenoxypyridyl pyrethroid derivatives and their alcohol precursors.

Fenvalerate (69)

τ-Fluvalinate (70)

Flufenprox (71)

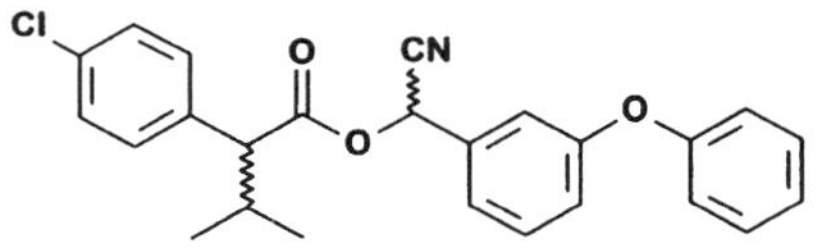

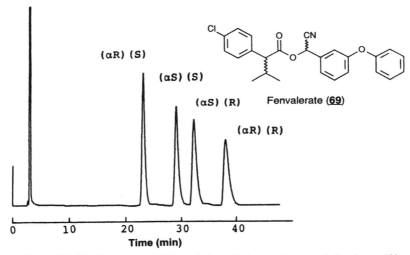

Figure 37. Structure of some pyrethroids lacking the cyclopropane moiety.

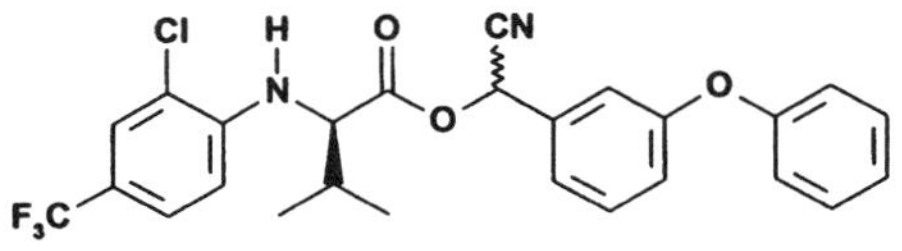

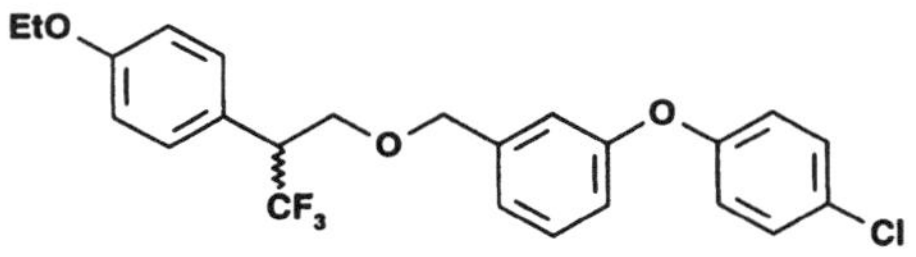

*Figure 38. Liquid chromatographic resolution of the stereoisomers of fenvalerate (**69**) on Sumichiral OA-2500-I (eluent: hexane–1,2-dichloroethane–ethanol 500:30:0.15). (Reproduced with permission from reference 128. Copyright 1990.)*

acaricide used against Aphididae, Cicadellidae, Lepidoptera, Thysanoptera, and other pests in cotton, fruit trees, vegetables, trees, and vines. It is normally sold as the epimer mixture synthesized from the resolved acid based on (*R*)-valine and the racemic α-cyano-3-phenoxybenzyl alcohol. The only reported separation of these epimers was done by HPLC on an ionic DNBPG column (*126*).

Flufenprox (**71**) is a very new bis(aralkyl)ether rice insecticide that combines broad-spectrum insect control (against Homoptera, Heteroptera, Coleoptera, and Lepidoptera) with low toxicity to fish and mammals (*135*). Its enantiomers were resolved by HPLC on a Chira-Chrom-2 column containing the 3,5-dini-troanilide derivative of (*R,R*)-tartaric acid covalently bonded to γ-aminopropyl silanized silica (Figure 39) (*28*).

Benzoylureas

Benzoylureas are insect growth regulators that act by inhibiting chitin synthesis and so interfering with the formation of the cuticle (*136*).

Lufenuron (**72**) is an interesting chiral example of this class of compounds. It is a potent insecticide (*137, 138*) and is also used to kill ticks on animals. The enantiomers of lufenuron can be separated on Chiralpak AD with a separation factor α of 1.33 (Figure 40). The racemic intermediate **73** has also been re-solved, on both analytical and preparative scales, by liquid chromatography us-ing cellulose tribenzoate beads (TBCs) as CSP (*138–140*).

Pheromones

Pheromones, which are often blends of different components, are ubiquitous semiochemicals (from Greek semeion, a mark, signal) that act as messengers among members of the same species. They have been studied mainly in insects, probably because pheromones play a dominant role in insect communication and therefore have great potential as insect control agents. In the exploration of alternative methods for pest control, much attention has been given to such com-

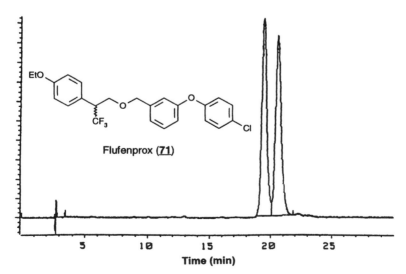

*Figure 39. Separation of the enantiomers of flufenprox (**71**) by HPLC on Chira-Chrom-2 (elu-ent: hexane–ethanol 99.9:0.1). (Reproduced with permission from reference 28. Copyright 1994.)*

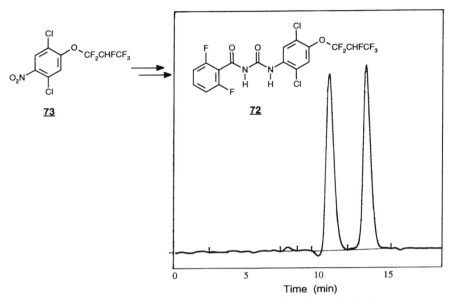

Figure 40. Separation of the enantiomers of **72** *by HPLC on Chiralpak AD (eluent: hexane–2-propanol 95:5).*

pounds, and outstanding synthetic work has been performed to obtain them in optically pure form. An excellent review of this topic was published recently (*141*). Whenever pheromones are produced synthetically, it is essential to determine their enantiomeric purity because this often is crucial to their biological activity.

Aggregation pheromones secreted by insects are a promising method for insect control. (*S*)-(+)-Ipsdienol (**74**) and (*S*)-(–)-ipsenol (**75**), for example, are two chiral components of the aggregation pheromone of the bark beetle *Ips paraconfusus* (*142*). The optical purity of these two pheromone components has been determined by HPLC and GC, respectively. Racemic ipsdienol was resolved as its benzoate derivate on a Chiralpak OT(+) column (*143*); the enantiomers of ipsenol were separated, after derivatization with isopropyl isocyanate, by GC on Chirasil-L-Val (*144*). In our laboratories, we found that racemic ipsenol can also be resolved as its *p*-nitrobenzoate derivative by HPLC on three different CSPs: TBC, MMBC, and PMBC, with separation factors of 1.98, 1.50, and 1.77, respectively (Figure 41).

Juvenile Hormone Mimics

Juvenile hormones are compounds secreted by insects that act in conjunction with the molting hormones to suppress adult differentiation and to maintain the

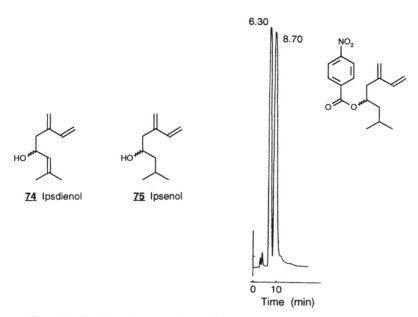

Figure 41. Resolution of the enantiomers of ipsenol p-nitrobenzoate by HPLC on PMBC (eluent: hexane–2-propanol 9:1).

immature nature of the developing stages. Since their discovery in the late 1960s, these hormones have attracted the attention of many research laboratories. Besides intensive work on the synthesis and the biological activity of the natural juvenile hormones, research has been focused on the development of mimics with greater chemical and metabolic stability, biological activity, and selectivity.

In the course of these investigations, new potent 4-phenoxyphenoxy juvenile hormone mimics were discovered, among them 2-ethyl-4-(4-phenoxyphenoxy)-methyl-1,3-dioxolane (**76**) (diofenolan) (*145*) and its fluoro derivative 2-ethyl-4-[4-(3-fluorophenoxy)phenoxy]methyl-1,3-dioxolane (**77**) (Figure 42) (*146*).

Various stereoselective syntheses were developed to prepare the stereoisomers of these compounds in sufficient amounts for individual biological testing (*147–149*). The optical purities of the synthetic diol intermediates **78** and **79** were determined by HPLC on Chiralcel OD (*149*). However, because of peak tailing, only high enantiomeric excesses of the last eluted (*S*)-enantiomers could be accurately measured with this method (*148*). On the same CSP, baseline separation of all four stereoisomers of **77** was obtained (Figure 43), while the stereoisomers of **76** were only partially resolved.

Another phenoxyphenyl ether with good juvenile hormone mimic activity is pyriproxyfen (**80**), a highly effective fly, mosquito, and cockroach control agent. Its enantiomers have been resolved by HPLC on Chiralcel OJ (Figure 44) (*150, 151*).

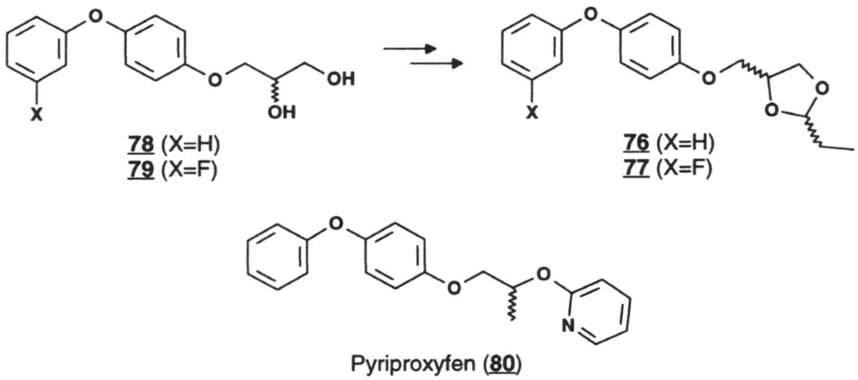

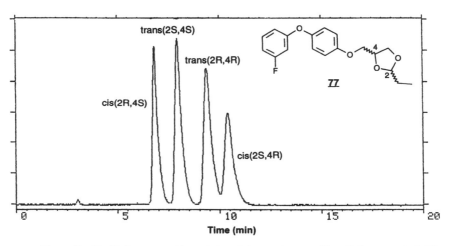

78 (X=H)
79 (X=F)

76 (X=H)
77 (X=F)

Pyriproxyfen (**80**)

*Figure 42. 4-Phenoxyphenoxy juvenile hormone mimics **76** and **77**, their synthetic diol precursors **78** and **79**, and the commercial juvenile hormone mimic pyriproxyfen (**80**).*

*Figure 43. Liquid chromatographic resolution of the stereoisomers of juvenile hormone mimic **77** on Chiralcel OD (eluent: hexane–2-propanol 95:5).*

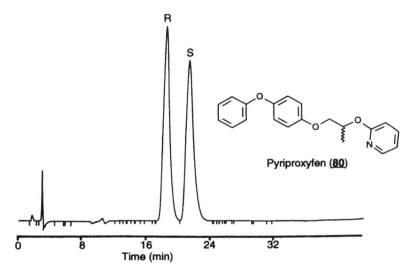

*Figure 44. Resolution of the enantiomers of pyriproxyfen (**80**) by HPLC on Chiralcel OJ (eluent: hexane–2-propanol 9:1). (Reproduced with permission from reference 150. Copyright 1991.)*

Concluding Remarks

This review illustrates how the powerful analytical techniques of HPLC and capillary GC have become indispensable tools in the search for and investigation of chiral biologically active compounds for weed control, disease control, and insect control. This development must be even more dramatic than the literature suggests, since doubtless many chiral analytics developed in industry have never been made public.

To date, only a limited number of the many available CSPs have been extensively studied, as far as the chiral recognition process is concerned, and the search for an appropriate chiral analytical method for a given compound is still very empirical. Nevertheless, there is a solution nowadays to almost every separation problem. Further rationalization of the mechanisms of enantiomer resolution will make the development of chiral analytical methods even more effective.

Acknowledgments

The authors heartily thank their many colleagues in Ciba-Geigy's Crop Protection research units for their great support and John Dingwall, Saleem Farooq, Urs Müller, and David Wadsworth for critically reading this chapter.

Appendix

Chiral Stationary Phases Mentioned (alphabetical order)

Abbreviation/ Trade Name	Chiral Selector of the CSP	Method	Supplier/Ref
Acetylquinine-II-silica	acetylquinine covalently bound to silica	HPLC	32
Chira-Chrom-2	3,5-dinitroanilide derivative of (R,R)-tartaric acid bound to silica	HPLC	28
Chiral-2	tartaric acid–phenylethylamine–modified silica	HPLC	21
Chiral AGP	α_1-acid glycoprotein immobilized on silica	HPLC	Chrom Tech
Chiralcel OA	cellulose triacetate coated on silica	HPLC	Daicel
Chiralcel OB	cellulose tribenzoate coated on silica	HPLC	Daicel
Chiralcel OC	cellulose tris(phenylcarbamate) coated on silica	HPLC	Daicel
Chiralcel OD	cellulose tris(3,5-dimethylphenyl-carbamate) coated on silica	HPLC	Daicel
Chiralcel OD-H	cellulose tris(3,5-dimethylphenyl-carbamate) coated on silica	HPLC	Daicel
Chiralcel OJ	cellulose tris(p-methylbenzoate) coated on silica	HPLC	Daicel
Chiralcel OK	cellulose tricinnamate coated on silica	HPLC	Daicel
Chiralpak AD	amylose tris(3,5-dimethylphenyl-carbamate) coated on silica	HPLC	Daicel
Chiralpak AS	amylose tris[(S)-phenylethylcarbamate] coated on silica	HPLC	Daicel
Chiralpak OT(+)	poly(triphenylmethyl methacrylate) coated on silica	HPLC	Daicel
Chirasil-L-Val	L-valine immobilized on polysiloxane	GC	Alltech
ChiraSpher	poly-[(S)-N-acryloylphenylalanine ethyl ester] bound to silica	HPLC	Merck
Chirex 3020	(R)-N-1-(α-naphthyl)ethylamino-carbonyl-(S)-*tert*-leucine bound to silica	HPLC	Phenomenex
CTA-I	cellulose triacetate (pure polymer)	HPLC	Daicel, Merck
Cyclobond I	β-cyclodextrin bound to silica	HPLC	Astec
Cyclobond I RN	β-cyclodextrin (R)-N-naphthylethyl-carbamate bound to silica	HPLC	Astec
Cyclobond II	γ-cyclodextrin bound to silica	HPLC	Astec
γ-Cyclodextrin derivative	octakis(6-O-methyl-2,3-di-O-pentyl)-γ-cyclodextrin	GC	24
β-Cyclodextrin derivative	heptakis(3-O-acetyl-2,6-di-O-pentyl)-β-cyclodextrin	GC	24
DACH-DNB	(R,R)-(N,N'-3-5-dinitrobenzoyl)-*trans*-1,2-diaminocyclohexane	HPLC	22
DNBLeu	(S)-N-(3,5-dinitrobenzoyl)leucine covalently bound to silica	HPLC	Baker, Regis
DNBPG-covalent	N-(3,5-dinitrobenzoyl)phenylglycine covalently bound to silica	HPLC	Baker, Regis

Abbreviation / Trade Name	Chiral Selector of the CSP	Method	Supplier/Ref
DNBPG ionic	*N*-(3,5-dinitrobenzoyl)phenylglycine ionically bound to silica	HPLC	Baker, Regis
EnantioPac	α_1-acid glycoprotein bound to silica	HPLC	LKB
Lipodex E	octakis(3-*O*-butyryl-2,6-di-*O*-pentyl-γ-cyclodextrin)	GC	Macherey-Nagel
MMBC	*m*-methylbenzoyl cellulose beads	HPLC	59, 60
N-formyl-(*S*)-isoleucine	*N*-formyl-(*S*)-isoleucine covalently bound to silica	HPLC	45, 64
N-formyl-(*S*)-phenylalanine	*N*-formyl-(*S*)-phenylalanine covalently bound to silica	HPLC	64
PMBC	*p*-methylbenzoyl cellulose beads	HPLC	59, 60
Sumichiral OA-2200	(1*R*,3*R*)-*N*-*trans*-chrysanthemoyl-(*R*)-phenylglycine bound to silica	HPLC	SCAS
Sumichiral OA-2500-I	(*R*)-*N*-(3,5-dinitrobenzoyl)-1-naphthylglycine ionically bound to silica	HPLC	SCAS
Sumichiral OA-4000	(*S*)-*N*-1-(α-naphthyl)ethylaminocarbonyl-(*S*)-valine bound to silica	HPLC	SCAS
Sumichiral OA-4100	(*R*)-*N*-1-(α-naphthyl)ethylaminocarbonyl-(*S*)-valine bound to silica	HPLC	SCAS
Sumichiral OA-4600	(*S*)-*N*-1-(α-naphthyl)ethylaminocarbonyl-(*S*)-*tert*-leucine bound to silica	HPLC	SCAS
Sumichiral OA-4700	(*R*)-*N*-1-(α-naphthyl)ethylaminocarbonyl-(*S*)-*tert*-leucine bound to silica	HPLC	SCAS
TBC	tribenzoyl cellulose beads	HPLC	59

NOTES: Astec: Advanced Separation Technologies (Whippany, NJ, USA). Baker: J. T. Baker (Phillipsburg, NJ, USA). ChromTech: ChromTech AB (Hägersten, Sweden). Daicel: Daicel Chemical Industries (Tokyo, Japan). LKB: LKB AB (Bromma, Sweden). Merck: E. Merck (Darmstadt, Germany). Macherey-Nagel: Macherey-Nagel GmbH & Co (Düren, Germany). Phenomenex: Phenomenex (Torrance, CA, USA). Regis: Regis Chemical Company (Morton Grove, IL, USA). SCAS: Sumika Chemical Analysis Service (Osaka, Japan).

References

1. *Drug Stereochemistry: Analytical Methods and Pharmacology;* Wainer, I. W.; Drayer, D. E., Eds.; Marcel Dekker: New York, 1988.
2. *Stereoselectivity of Pesticides;* Ariens, E. J.; van Rensen, J. J. S.; Welling, W., Eds.; Elsevier: Amsterdam, Netherlands, 1988.
3. Ramos Tombo, G. M.; Bellus, D. *Angew. Chem. Int. Ed. Engl.* **1991,** *30*, 1193.
4. Crossley, R. *Tetrahedron* **1992,** *48*, 8155.
5. Asymmetric Synthesis; Morrison, J. D., Ed.; Academic: Orlando, FL, 1983-1985; Vols. 1-5.
6. Testa, B.; Trager, W. F. *Chirality* **1990,** *2*, 129.
7. Cayen, M. N. *Chirality* **1991,** *3*, 94.
8. Kagan, H. B. *Bull. Soc. Chim. Fr.* **1988,** 846.
9. Sheldon, R. *Chem. Ind.* **1990,** 212.
10. Federsel, H.-J. *CHEMTECH,* 1993, December, 24.

11. Vermeulen, N. P.; Testa, B. In *Stereoselectivity of Pesticides;* Ariens, E. J.; van Rensen, J. J. S.; Welling, W., Eds.; Elsevier: Amsterdam, Netherlands 1988; p 375.
12. *The Biochemistry and Uses of Pesticides;* Hassall, K. A., Ed.; VCH Verlagsgesellschaft mbH: Weinheim, Germany, 1990.
13. Luers, H. *The Agronomist;* BASF: Ipswich, England, 1988; No. 1, p 6.
14. Dicks, J. W.; Slater, J. W.; Bewick, D. W. *Proceedings of the British Crop Protection Conference on Weeds;* BCPC Publications: Croyden, England, 1985, Vol. 1, p 271.
15. Wegler, R.; Eue, L. In *Chemie der Pflanzenschutz und Schädlingsbekämpfungsmittel;* Wegler, R., Ed.; Springer; Berlin, Germany, 1970; Vol. 2, p 278.
16. Naber, J. D.; van Rensen, J. J. S. In *Stereoselectivity of Pesticides;* Ariens, E. J.; van Rensen, J. J. S.; Welling, W.; Eds.; Elsevier: Amsterdam, Netherlands, 1988; p 263.
17. Blessington, B.; Crabb, N.; O'Sullivan, J. *J. Chromatogr.* **1987,** *396,* 177.
18. Blessington, B.; Crabb, N. *J. Chromatogr.* **1988,** *454,* 450.
19. Blessington, B.; Crabb, N. *J. Chromatogr.* **1989,** *483,* 349.
20. Ludwig, P.; Gunkel, W.; Hühnerfuss, H. *Chemosphere* **1992,** *24,* 1423.
21. Müller, M. D.; Bosshardt, H.-P. *J. Assoc. Off. Anal. Chem.* **1988,** *71,* 614.
22. Tambuté, A.; Siret, L.; Caude, M.; Rosset, R. *J. Chromatogr.* **1991,** *541,* 349.
23. Bettoni, G.; Ferorelli, S.; Loiodice, F.; Tangari, N.; Tortorella, V.; Gasparrini, F., Misiti, D.; Villani, C. *Chirality* **1992,** *4,* 193.
24. König, W. A.; Icheln, D.; Runge, T.; Pfaffenberger, B.; Ludwig, P.; Hühnerfuss, H. *J. High Resol. Chromatogr.* **1991,** *14,* 530.
25. König, W. A.; Gehrke, B. *J. High Resol. Chromatogr.* **1993,** *16,* 175.
26. König, W. A. *J. High Resol. Chromatogr.* **1993,** *16,* 338.
27. Bewick, D. W. *Pestic. Sci.* **1986,** *17,* 349.
28. Massey, P. R.; Tandy, M. J. *Chirality* **1994,** *6,* 63.
29. Wu, Z.; Goodall, D. M.; Lloyd, D. K.; Massey, P. R.; Sandy, K. C. *J. Chromatogr.* **1990,** *513,* 209.
30. Davy, G. S.; Francis, P. D. *J. Chromatogr.* **1987,** *394,* 323.
31. Mutsaers, J. H. G. M.; Kooreman, H. J. *Recl. Trav. Chim. Pays-Bas* **1991,** *110,* 185.
32. Pettersson, C.; Gioeli, C. *J. Chromatogr.* **1987,** *398,* 247.
33. Karlsson, A.; Pettersson, C. *Chirality* **1992,** *4,* 323.
34. Wink, O.; Luley, U. *Pestic. Sci.* **1988,** *22,* 31.
35. Uchiyama, M.; Washio, N.; Ikai, T.; Igarashi, H.; Suzuki, K. *Nippon Noyaku Gakkaishi* **1986,** *11,* 459.
36. Meduna, V.; Hoffmann-LaRoche and Company AG, Basel, Switzerland, personal communication.
37. Müller, M. D.; Wacker, R.; Bosshardt, H.-P. *Chimia* **1991,** *45,* 160.
38. Armstrong, D. W.; Reid III, G. L.; Hilton, M. L.; Chang, C.-D. *Envir. Pollut.* **1993,** *79,* 51.
39. Russel, J. W. U.S. Patent 4 786 732, 1988.
40. Amrein, J.; Nyffeler, A.; Rufener, J. *Proceedings of the British Crop Protection Conference on Weeds;* BCPC Publications: Farnham, England, 1989, Vol. 1, p 71.
41. Jeffcoat, B.; Harries, W. N. *Pestic. Sci.* **1973,** *4,* 891.
42. Jeffcoat, B.; Harries, W. N. *Pestic. Sci.* **1975,** *6,* 283.
43. Jeffcoat, B.; Harries, W. N.; Thomas, D. B. *Pestic. Sci.* **1977,** *8,* 1.
44. Venis, M. A. *Pestic. Sci.* **1982,** *13,* 309.
45. Gray, A.; Mathews, B. L.; Self, C. *Pesticide Science Biotechnology: Proceedings of the 6th International Congress on Pesticide Chemistry, 1986;* International Union of Pure and Applied Chemistry: Ottawa, Canada, 1987; p 249.
46. Hutson, D. H.; Logan, C. J.; Taylor, B. *Pestic. Sci.* **1991,** *31,* 151.
47. Gerber, H. R.; Müller, G.; Ebner, L. *Proceedings of the British Crop Protection Conference on Weeds;* BCPC Publications: London, 1974, p 787.
48. Moser, H.; Rihs, G.; Sauter, H. *Z. Naturforscher* **1982,** *37b,* 451.

49. Moser, H.; Rihs, G.; Sauter, H. P.; Böhner, B. In *Proceedings of the 5th International Congress of Pesticide Chemstry, Kyoto, 1982;* Miyamoto, J.; Karney, P. C., Eds.; Pergamon: Oxford, England, 1983; Vol. 1, p 315.
50. Ueji, M.; Ishizaka, M.; Kobayashi, K.; Sugiyama, H. *Abstracts of Papers,* 8th International Congress on Pesticide Chemistry, 1994; Poster No. 251.
51. LaRossa, R. A.; Schloss, J. V. *J. Biol. Chem.* **1984,** *259,* 8753.
52. Chaleff, R. S.; Mauvais, C. J. *Science (Washington, D.C.)* **1984,** *224,* 1443.
53. Ray, T. B. *Proceedings of the British Crop Protection Conference on Weeds;* BCPC Publications: Croyden, England, 1985; p 131.
54. Pettersson, C.; Gioeli, C. *Chirality* **1993,** *5,* 241.
55. van Lommen, G.; Lutz, W.; Sipido, V.; Huxley, P.; van Gestel, J.; Fischer, H. In *Abstracts of Papers,* 7th International Congress on Pesticide Chemistry, Hamburg, Germany, 1990; Frehse, H.; Kesseler-Schmitz, E.; Conway, S., Eds.; International Union of Pure and Applied Chemistry: Hamburg, Germany, 1990; Vol. 3.
56. Francotte, E., CIBA, Basel, unpublished results.
57. Lüthy, C., CIBA, Basel, unpublished results.
58. Mori, I.; Iwasaki, G.; Kimura, Y.; Matsunaga, S.; Ogawa, A.; Nakano, T.; Buser, H.-P.; Hatano, M.; Tada, S.; Hayakawa, K.; *J. Am. Chem. Soc.* **1995,** *117,* 4411.
59. Francotte, E.; Wolf, R. M. *J. Chromatogr.* **1992,** *595,* 63.
60. Zhang, T.; Francotte, E., *Chirality* **1995,** *7,* 425.
61. Fuchs, A. In *Stereoselectivity of Pesticides;* Ariens, E. J.; van Rensen, J. J. S.; Welling, W., Eds.; Elsevier: Amsterdam, Netherlands, 1988; p 203.
62. Köller, W. *Pestic. Sci.* **1987,** *18,* 129.
63. Deas, A. H. B.; Carter, G. A. *Proceedings of the British Crop Protection Conference on Pests and Diseases;* BCPC Publications: Croyden, England, 1986, p 835.
64. Akanya, J. N.; Taylor, D. R. *J. Liq. Chromatogr.* **1987,** *10,* 805.
65. *Application Guide for Chiral Column Selection,* 2nd ed.; Daicel Chemical Industries: Tokyo, Japan.
66. Krämer, W.; Büchel, K. H.; Draber, W. In *Proceedings of the 5th International Congress of Pesticide Chemistry, Kyoto, 1982;* Miyamoto, J.; Karney, P. C., Eds.; Pergamon: Oxford, England, 1983; Vol. 1, p 223.
67. Clark, T.; Clifford, D. R.; Deas, A. H. B.; Gendle, P.; Watkins, D. A. M. *Pestic. Sci.* **1978,** *9,* 497.
68. Deas, A. H. B.; Clark, T.; Carter, G. A. *Pestic. Sci.* **1984,** *15,* 63.
69. Clark, T.; Deas, A. H. B. *J. Chromatogr.* **1985,** *329,* 181.
70. Burden, R. S.; Carter, G. A.; Clark, T.; Cooke, D. T.; Croker, S. J.; Deas, A. H. B.; Hedden, P.; James; C. S.; Lenton, J. R. *Pestic. Sci.* **1987,** *21,* 253.
71. Burden, R. S.; Deas, A. H. B.; Clark, T. *J. Chromatogr.* **1987,** *391,* 273.
72. Clark, T.; Wong, W.-C.; Vogeler, K. *Pestic. Sci.* **1991,** *33,* 447.
73. Sugavanam, B. *Pestic. Sci.* **1984,** *15,* 296.
74. Hedden, P.; Graebe, J. E. *J. Plant Growth Regul.* **1985,** *4,* 111.
75. Oî, N.; Nagase, M.; Inda, Y.; Doi, T. *J. Chromatogr.* **1983,** *259,* 487.
76. Oî, N.; Kitahara, H. *J. Liq. Chromatogr.* **1986,** *9,* 511.
77. Oî, N.; Kitahara, H.; Kira, R. *J. Chromatogr.* **1990,** *535,* 213.
78. *Chirex Application Guide;* Phenomenex: Torrance, CA.
79. Oî, N.; Kitahara, H.; Doi, T. Eur. Patent 299 793, 1988.
80. Oî, N.; Kitahara, H.; Matsumoto, Y.; Nakajima, H,; Horikawa, Y. *J. Chromatogr.* **1989,** *462,* 382.
81. Furuta, R.; Nakazawa, H. *J. Chromatogr.* **1992,** *625,* 231.
82. Furuta, R.; Nakazawa, H. *Chromatographia* **1993,** *35,* 555.
83. Ebert, E.; Eckhardt, W.; Jäkel, K.; Moser, P.; Sozzi, D.; Vogel, C. Z. *Naturforsch. C:* **1989,** *44,* 85.
84. Ramos Tombo, G. M.; Nyfeler, R.; Speich, J. *Abstracts of Papers,* 7th International

Congress on Pesticide Chemistry, Hamburg, Germany; International Union of Pure and Applied Chemistry: Hamburg, Germany, 1990; Poster No. 01B-38.

85. Buser, H. P.; Ramos Tombo, G. M. *AGRO-INDUSTRY Hi-Tech* **1993,** *November–December*, 10.

86. Th. Früh, CIBA, Basel, unpublished results.

87. Kunz, W.; Eckhardt, W. *Abstracts of Papers*, 6th International Congress on Pesticide Chemistry, Ottawa, Canada; International Union of Pure and Applied Chemistry: Ottawa, Canada, 1986; Poster No. 1B-03.

88. Baloch, R. I.; Mercer, E. I.; Wiggins, T. E.; Baldwin, B. C. *Phytochemistry* **1984,** *23,* 2219.

89. *Fungicide Resistance in North America;* Brentin, J. K.; Delp, C. J., Eds.; APS Press: St. Paul, MN, 1988; p 19.

90. Himmele, W.; Pommer, E.-H. *Angew. Chem. Int. Ed. Engl.* **1980,** *19,* 184.

91. Huxley-Tencer, A.; Francotte, E.; Bladocha-Moreau, M. *Pestic. Sci.* **1992,** *34,* 65.

92. Baker, P. G. *Pesticide Science Biotechnology: Proceedings of the 6th International Congress on Pesticide Chemistry, 1986;* International Union of Pure and Applied Chemistry: Ottawa, Canada, 1987; p 329.

93. Hubele, A.; Kunz, W.; Eckhardt, W.; Sturm, E. In *Proceedings of the 5th International Congress of Pesticide Chemistry, Kyoto, 1982*; Miyamoto, J.; Karney, P. C., Eds.; Pergamon: Oxford, England, 1983, p 233.

94. Moser, H.; Vogel, C. *Abstracts of Papers, 4th International Congress on Pesticide Chemistry, Zürich, Switzerland;* International Union of Pure and Applied Chemistry: Zurich, Switzerland, 1978; Abstract II-310.

95. Nyfeler, R.; Huxley, P. In *Fungicides for Crop Protection, 100 Years of Progress;* Smith, I. M., Ed.; Monograph 31; BCPC Publication: Croydon, England, 1985; p 45.

96. Nyfeler, R.; Huxley, P. In *Proceedings of the British Crop Protection Conference on Pests and Diseases, 1986;* BCPC Publications: Thornton Heath, England, 1986; p 207.

97. Kunz, W.; Eckhardt, W. Br. Patent 1 577 702, 1977.

98. Suess, H.; Eckhardt, W. U.S. Patent 4 721 797, 1986.

99. Eckhardt, W.; Francotte, E.; Herzog, J.; Margot, P.; Rihs, G.; Kunz, W. *Pestic. Sci.* **1992,** *36,* 223.

100. Margot, P.; Eckhardt, W.; Dahmen, H. In *Proceedings of the British Crop Protection Conference on Pests and Diseases, 1988;* BCPC Publications: Thornton Heath, England, 1988; p 527.

101. Eckhardt, W.; Francotte, E.; Kunz, W.; Hubele, A. Eur. Patent 275 523, 1988.

102. Eckhardt, W.; Francotte, E.; Margot, P. *Abstracts of Papers,* 7th International Congress on Pesticicide Chemistry, Hamburg, Germany; International Union of Pure and Applied Chemistry: Hamburg, Germany, 1990; Poster No. 01B-11.

103. Sutter, M.; Buser, H. P.; Pugin, B.; Eckhardt, W.; Spindler, F. Eur. Patent 348 351 A, 1988.

104. Buser, H. P.; Sutter, M.; Spindler, F.; Pugin, B. *Abstracts of Papers,* 7th International Congress on Pesticide Chemistry, Hamburg, Gemany; International Union of Pure and Applied Chemistry: Hamburg, Germany, 1990; Poster No. 01B-10.

105. Buser, H. P.; Pugin, B.; Spindler, F.; Sutter, M. *Tetrahedron* **1991,** *47,* 5709.

106. König, W. *J. High Resol. Chromatogr.* **1993,** *16,* 569.

107. Naber, J. D.; van Rensen, J. J. S. In *Stereoselectivity of Pesticides;* Ariens, E. J.; van Rensen, J. J. S.; Welling, W., Eds.; Elsevier: Amsterdam, Netherlands, 1988; p 263.

108. Williams, A. In *Agrochemical Chirality, Agrow Reports DS70;* PJB Publications: Richmond, United Kingdom, 1992; p 41.

109. De Jong, L. P. A.; Benschop, H. P. In *Stereoselectivity of Pesticides;* Ariens, E. J.; van Rensen, J. J. S.; Welling, W., Eds.; Elsevier: Amsterdam, 1988; p 103.

110. Berkman, C. E.; Thompson, C. M.; Perrin, S. R. *Chem. Res. Toxicol.* **1993,** *6,* 718.

111. Toia, R. F.; March, R. B.; Umetsu, N.; Mallipudi, N. M.; Allahyari, R.; Fukuto, T. R. *J. Agric. Food Chem.* **1980,** *28,* 599.

112. Magee, P. S. In *Insecticide Mode of Action;* Coats, J. R., Ed.; Academic: Orlando, FL, 1982; p 101.
113. Miyazaki, A.; Nakamura, T.; Kawaradani, M.; Marumo, S. *J. Agric. Food Chem.* **1988,** *36,* 835.
114. Ichida, A.; Shibata, T.; Okamoto, I.; Yuki, Y.; Namikoshi, H.; Toga, Y. *Chromatographia* **1984,** *19,* 280.
115. Gorder, G. W.; Kirino, O.; Hirashima, A.; Casida, J. E. *J. Agric. Food Chem.* **1986,** *34,* 941.
116. Shibata, T. Poster Presentation at the 2nd International Symposium on Chiral Separations, September 1989, Guildford, England.
117. Okamoto, Y.; Honda, S.; Hatada, K.; Okamoto, I.; Toga, Y.; Kobayashi, S. *Bull. Chem. Soc. Jpn.* **1984,** *57,* 1681.
118. Okamoto, Y.; Hatada, K. *J. Liq. Chromatogr.* **1986,** *9,* 369.
119. Wu, S. Y.; Hirashima, A.; Eto, M.; Yanagi, K.; Nishioka, E.; Moriguchi, K. *Agric. Biol. Chem.* **1989,** *53,* 157.
120. Hirashima, A.; Ishaaya, I.; Ueno, R.; Ichiyama, Y.; Wu, S. Y.; Eto, M. *Agric. Biol. Chem.* **1989,** *53,* 175.
121. Hirashima, A.; Ueno, R.; Oyama, K.; Koga, H.; Eto, M. *Agric. Biol. Chem.* **1990,** *54,* 1013.
122. Naumann, K. In *Chemistry of Plant Protection;* Bowers, W. S.; Ebing, W.; Martin, D.; Wegler, R., Eds.; Springer: Berlin, Germany, 1990; Vol. 4.
123. Elliott, M. *Pestic. Sci.* **1989,** *27,* 337.
124. Kutter, J. P.; Class, T. J. *Chromatographia* **1992,** *33,* 103.
125. Chapman, R. A. *J. Chromatogr.* **1983,** *258,* 175.
126. Cayley, G. R.; Simpson, B. W. *J. Chromatogr.* **1986,** *356,* 123.
127. Lisseter, S. G.; Hambling, S. G. *J. Chromatogr.* **1991,** *539,* 207.
128. Oî, N.; Kitahara, H.; Kira, R. *J. Chromatogr.* **1990,** *515,* 441.
129. Meinard, C.; Bruneau, P. *J. Chromatogr.* **1988,** *450,* 169.
130. Maguire, R. J. *J. Agric. Food Chem.* **1990,** *38,* 1613.
131. Greuter, H.; Dingwall, J.; Martin, P.; Bellus, D. *Helv. Chim. Acta* **1981,** *64,* 2812.
132. Francotte, E.; Ackermann, P. In *Chirality and Biological Activity;* Holmstedt, B.; Frank, H.; Testa, B., Eds.; Alan R. Liss: New York, 1990; p 63.
133. Papadopoulou-Mourkidou, E. *Chromatographia* **1985,** *20,* 376.
134. Papadopoulou-Mourkidou, E. *Anal. Chem.* **1989,** *61,* 1149.
135. Gordon, R. F. S.; Bushell, M. J.; Pascoe, R.; Enoyoshi, T. *Proceedings of the British Crop Protection Conference on Pests and Diseases;* BCPC Publishers: Farnham, England, 1992; p 81.
136. Maas, W.; van Hes, R.; Groscurt, A. C.; Deul, D. H. In *Chemie der Pflanzenschutz- und Schädlingsbekämpfungsmittel;* Wegler, R., Ed.; Springer Verlag: Berlin, Germany, 1980; Vol. 6, p 424.
137. Drabek, J.; Francotte, E.; Lohmann, D.; Bourgeois, F. Ger. Patent DE-B 4000 355, 1990, Ciba-Geigy.
138. Drabek, J.; Bourgeois, F.; Francotte, E.; Lohmann, D. Poster presentation at the Symposium on Advances in the Chemistry of Insect Control, Oxford, England, 1989.
139. Francotte, E.; Junker-Buchheit, A. *J. Chromatogr.* **1992,** *576,* 1.
140. Francotte, E.; Wolf, R. M. Poster presentation at the 2nd International Symposium on Chiral Discrimination, Rome, Italy, 1991.
141. Mori, K. *Tetrahedron* **1989,** *45,* 3233.
142. Silverstein, R. M.; Rodin, J. O.; Wood, D. L. *Science (Washington, D.C.)* **1966,** *36,* 509.
143. Kubo, I.; Komatsu, S.; Iwagawa, T.; Wood, D. L. *J. Chromatogr.* **1986,** *363,* 309.
144. Oertle, K.; Beyeler, H.; Duthaler, R. O.; Lottenbach, W.; Riediker, M.; Steiner, E. *Helv. Chim. Acta* **1990,** *73,* 352.
145. Karrer, F.; Farooq, S. U.S. Patent 4 097 581, 1975.

146. Karrer, F. U.S. Patent 4 971 981, 1987.
147. Karrer, F.; Buser, H. P.; Ramos, G.; Rindlisbacher, A.; Venanzi, L. M.; Ward, T. R. Eur. Patent EP 501 912 A1, 1991.
148. Buser, H.-P.; Spindler, F. *Tetrahedron: Asym.* **1993,** *4,* 2451.
149. Karrer, F.; Kayser, H.; Buser, H.-P.; Ramos Tombo, G. M. *Chimia* **1993,** *47,* 302.
150. Okamoto, M.; Nakazawa, H. *J. Chromatogr.* **1991,** *588,* 177.
151. Okamoto, M.; Nakazawa, H. *Anal. Sci. Suppl.* **1991,** *7,* 147.

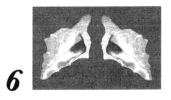

6

Chiral Separation Methods

Satinder Ahuja

Development of chiral separation methods requires an understanding of stereochemistry. Biological activity is intimately related to the stereochemistry of a molecule, the difference in spatial arrangements of atoms. Pronounced differences in pharmacologic activity have been observed in molecules that are mirror images of each other. This explains the regulatory requirements for detailed investigations on chiral molecules. Basic information on stereochemistry is provided in this chapter to prepare the reader for better understanding of the mechanisms of chiral separation methods. The choice of a method may be influenced by molecular structure or the individual chemist's preference for an achiral or chiral stationary phase. Logical reasons for selecting a method are discussed.

A molecule that—like a hand—cannot be superimposed on its mirror image is called *chiral* (from Greek word *cheir,* meaning hand). Molecules that are mirror images of each other are called *enantiomers.* Molecules that are isomeric but have a different spatial arrangement are called *stereoisomers.* Symmetry classifies stereoisomers as either enantiomers or *diastereomers.* Here are some important points:

- Any molecule containing a single chiral carbon atom (one with four different groups attached) can have two enantiomers.

3407–8/96/0139$15.00/0

- Diastereomers (sometimes called *diastereoisomers*) are stereoisomers that are not enantiomers of each other.
- A chiral molecule can have only one enantiomer, but it can have several diastereomers or none.
- Two stereoisomers cannot be both enantiomers and diastereomers of each other simultaneously.

Stereoisomerism

Stereoisomerism can result from a variety of sources besides a single chiral carbon (stereogenic or chiral center). A molecule need not have a chiral carbon to exist in enantiomeric forms, but it is necessary that the molecule, as a whole, be chiral. There are two simple molecular sources of chirality: molecules having a stereogenic center and those having a stereogenic axis. Detailed discussion on these topics may be found in several books and review articles (*1–8*); a short summary follows.

Stereoisomerism is possible in molecules with one or more of the following properties:

- one or more centers of chirality
- helicity (e.g., helical nature of tertiary structures of proteins, polysaccharides, and nucleic acids)
- planar chirality (e.g., paracyclophanes)
- axial chirality (e.g., spiranes with cyclic skeleton)
- torsional chirality (e.g., torsion about double or single bonds as in *cis*- and *trans*-isomers and rotamers)
- topological asymmetry (e.g., catenanes)

Stereoisomers may be either configurational or conformational. Configuration relates to a particular spatial arrangement of the atoms in molecules of defined constitution without regard to arrangement, differing only in torsion around single bonds. Conformation, on the other hand, deals with arrangement of the atoms in molecules of defined configuration resulting from torsion around *one or more* single bonds. A full definition of molecular structure and shape includes constitution, configuration, and conformation.

As mentioned earlier, the symmetry factor classifies stereoisomers as either enantiomers or diastereomers. The amounts of energy necessary to convert stereoisomers into their isomeric forms may be used for their classification. Stereoisomers with low-energy barriers to this conversion are termed conformational isomers, whereas those with high-energy barriers are configurational isomers. Diastereomers differ in energy content, and thus in every physical and chemical property; however, the differences may be so minute as to be nearly indistinguishable.

Biological Activity

It is necessary to determine the stereoisomeric composition of chemical compounds, especially those of pharmaceutical importance (*9–11*). About half the 2050 drugs listed in the *USP Dictionary of USAN and International Drug Names* contain at least one asymmetric center, and almost 25% of them have been used in racemic or diastereomeric forms (*12*). The differences in the pharmacologic properties between enantiomers of these racemic drugs have not yet been examined in many cases, probably because of difficulties in obtaining both enantiomers in optically pure forms. Some enantiomers have different pharmacologic activities, and a patient may be taking a useless or even undesirable enantiomer when ingesting a racemic mixture. To ensure the safety and effectiveness of drugs, it is important to isolate and examine both enantiomers separately. Furthermore, it is necessary to measure and control the stereochemical composition of drugs in at least three situations: (1) during manufacture, when problems of preparative-scale separations may be involved; (2) during quality control (or regulatory analysis), when analytical questions of purity and stability predominate; and (3) during pharmacologic studies of plasma disposition and drug efficacy, when ultratrace methods may be required (*2*).

Accurate assessment of the isomeric purity of substances is critical because isomeric impurities may have unwanted toxicologic, pharmacologic, or other effects. Such impurities may be carried through a synthesis, preferentially react at one or more steps, and yield an undesirable level of another impurity. Frequently one isomer of a series produces a desired effect, while another is inactive or even produces an undesired effect. Some examples of activity differences are given in Table I. These differences in therapeutic activity frequently exist between enantiomers, the most difficult stereoisomers to separate.

Table I. Activities of Some Chiral Compounds

Compound	Activity
Amphetamine	*d*-Isomer is a potent central nervous system stimulant; *l*-isomer has little effect.
Ascorbic acid	*d*-Isomer is antiscorbutic; *l*-isomer is not active.
Epinephrine	*l*-Isomer has 10 times the vasoconstrictive activity of *d*-isomer.
Synephrine	*l*-Isomer has 60 times the pressor activity of *d*-isomer.
Propoxyphene	α-*l*-Isomer is antitussive; α-*d*-isomer is analgesic.
Propranolol	Only (*S*)-(–)-isomer has β-adrenergic blocking activity.
Warfarin	(*S*)-(–)-Isomer is 5 times as potent an anticoagulant as (*R*)-(+)-isomer.

Regulatory Requirements

In 1987 the U.S. Food and Drug Administration issued a set of initial guidelines on the submission of new drug applications. The question of stereochemistry was approached directly in the guidelines on the manufacture of drug substances (*13, 14*). The finalized guidelines require a full description of manufacturing methods, including tests to demonstrate the drug's identity, strength, quality, and purity. Submissions should show the applicant's knowledge of the molecular structure of the drug substance. For chiral compounds, this includes identification of all chiral centers. The enantiomer ratio is 50:50 by definition for a racemate, but should be determined for any nonracemic mixture of stereoisomers. The proof of structure should consider stereochemistry and provide appropriate descriptions of the molecular structure. An unwanted enantiomer can be considered an impurity, and therefore it is desirable to explore potential in vivo differences between enantiomers.

Stereoisomeric Interactions

Enantiomers have identical physical properties except for the sign of optical rotation. A chemical reaction carried out in an achiral environment produces a *racemate*, a mixture consisting of equal amounts of two enantiomers. Therefore, the separation of an enantiomeric mixture (optical resolution) is necessary to obtain optically pure species.

Optical resolution of racemates is one way to obtain pure isomers. Since Pasteur reported the first example of optical resolution in 1848, more than a thousand compounds have been resolved, mainly by fractional recrystallization of diastereomeric salts. Resolution by entrainment or with enzymes or bacteria has also been applied to large-scale separation of some racemates—for example, amino acids. Of the various methods, more chiral compounds are being resolved today by chromatographic methods than crystallization.

Many factors influence interactions of stereoisomeric molecules in any environment. These factors can affect chromatography of stereoisomers and must be carefully reviewed before developing a separation method (*8*): dipole–dipole interactions, electrostatic forces, hydrogen bonding, hydrophobic bonding, inductive effects, ion–dipole interactions, ligand formation, partition coefficient differences, pk differences (extent of ionization), resonance interactions–stabilization, solubilities, steric interference (size, orientation, and spacing of groups), structural rigidity–conformational flexibility, temperature, and van der Waals forces.

Diastereomers are easier to separate than enantiomers because they differ in physical properties, but recent advances allow the chromatographic resolution of enantiomers. There are two basic approaches. In the indirect approach, the enantiomers are converted into covalent, diastereomeric compounds by a reaction with a chiral reagent, and these diastereomers are typically separated on a routine, achiral stationary phase. In the direct approach, either (1) the enantiomers or their derivatives are passed through a column containing a chiral stationary phase (CSP), or (2) the derivatives are passed through an achiral column using a chiral solvent or, more commonly, a mobile phase that contains a chiral additive. Both variants of the indirect approach depend on differential, transient diastereomer formation between the solutes and on the selector to effect the observed separation.

Separation Methods

Chromatographic methods such as thin-layer chromatography, gas–liquid chromatography, and high-pressure liquid chromatography (HPLC) offer distinct advantages over classical techniques in the separation and analysis of stereoisomers, especially enantiomers, which are generally difficult to separate (*4, 15–19*). These methods show promise for moderate-scale separations of synthetic intermediates as well as for final products. For large-scale separations and in consideration of the cost of plant-scale resolution processes, the sorption methods offer substantial increases in efficiency over recrystallization. Most of the discussion in this chapter is on HPLC, since it offers the greatest promise.

Treating an enantiomeric mixture with a chiral reagent to produce a pair of diastereomers (the indirect method) allows separation of samples by chromatography. On the other hand, using a chiral stationary or mobile phase (the direct method) is a useful alternative. The direct approach has been examined extensively by many research scientists. Early successful results did not attract much interest, and little was done to develop it into a generally applicable method. About 20 years ago, systematic research was initiated for the design of CSPs for separating enantiomers by gas chromatography. Molecular design and preparation of the CSPs for liquid chromatography have been examined since then. More recently, efforts have been directed toward finding new types of chiral stationary and mobile phases on the basis of the stereochemical viewpoint and the technical evolution of modern liquid chromatography.

Since HPLC is now one of the most powerful separation techniques, resolution of enantiomers by HPLC is expected to move rapidly with the availability of efficient chiral stationary phases. Large-scale, preparative liquid chromatographic systems have already been put on the market as process units for isolating and purifying chemicals and natural products. Chiral HPLC is ideally suited for large-scale preparation of optical isomers.

Mechanism

To develop optimum methods, it is important to understand the mechanism of chiral separation. The chromatographic process is considered a series of equilibria, and the equilibrium constant describes the distribution of a compound between the stationary and mobile phases. The separation occurs because of differences in retention of various compounds on the stationary phase. It is customary to distinguish between partition chromatography, in which the stationary phase is a liquid, and adsorption chromatography, in which the stationary phase is a solid. The introduction of bonded organic phases for HPLC has blurred this distinction. This section focuses on the types of molecular interactions with the stationary phase that lead to retention.

Our understanding of chiral separations with some systems is reasonably good, but it remains poor for protein and cellulose stationary phases. A number of chiral recognition models have been proposed for optical resolutions by HPLC. The models are often based on the three-point interaction rule advanced by Dalgleish (20) in 1952. He arrived at his conclusions from paper chromatography studies of certain aromatic amino acids. He assumed that the hydroxyl groups of the cellulose were hydrogen-bonded to the amino and carboxyl groups of the amino acid. A third interaction was caused, according to Dalgliesh, by the aromatic ring substituents. He postulated that three simultaneous interactions between an enantiomer and the stationary phase are needed for chiral discrimination. However, this is not always necessary because steric interactions can also lead to discrimination.

In adsorption chromatography, some kind of bonding interaction with the sorbent must occur. This interaction may arise through a noncovalent attachment depending on conditions. For example, hydrogen bonding and ionic or dipole attraction are enhanced by nonpolar solvents, whereas hydrophobic interactions are more important in aqueous media.

Chiral separations are also possible through reversible diastereomeric association between a chiral environment, introduced into a column, and an enantiomeric solute. Since chromatographic resolutions are possible under a variety of conditions, the necessary difference in association can be produced by many types of molecular interactions. The association, which may be expressed quantitatively as an equilibrium constant, is a function of the magnitudes of the binding and repulsive interactions involved. Repulsions are usually steric, although dipole–dipole repulsions may also occur. Binding interactions include hydrogen bonding, electrostatic and dipole–dipole attractions, charge-transfer interactions, and hydrophobic interactions (in aqueous systems). At times, a single type of binding interaction may be sufficient to differentiate enantiomers. For example, hydrogen bonding alone is sufficient for optical resolution in some HPLC separations. The fact that enantiomeric solutes bearing only one hydrogen-bonding substituent can be separated under such conditions implies that only one attractive force is necessary for chiral discrimination in this type of chromatography. With a one-point binding interaction as a model, the equilibrium constants of the two enantiomers at the chiral binding site may differ because effects from the site force one of the enantiomers to assume an unfavorable conformation. This resembles the mechanism accounting for specificity in enzyme–substrate interactions.

It may be possible to construct chiral cavities that preferentially include only one enantiomer. Molecular imprinting techniques are very interesting in this respect (*21*). The idea is to create rigid chiral cavities in a polymer network such that only one of two enantiomers will find the environment acceptable. Other CSPs for which steric fit is of primary importance include those based on cyclodextrins and crown ethers.

Modes of Separation in HPLC

Various methods can be used for chromatographic separation of enantiomers but some kind of chiral discriminator or selector is always necessary (*22, 23*). Two types of selectors can be distinguished: a chiral additive in the mobile phase and a chiral stationary phase. Another possibility is precolumn derivatization of the sample with a chiral reagent to produce diastereomeric molecules that can be separated by achiral chromatographic methods.

Although the details have not been worked out, the mechanism is discussed for each mode of separation. The proposals made by certain scientists appear attractive; however, vigorous differences prevail, so an attempt has been made not to highlight a single proposal.

Chromatography of Diastereomeric Derivatives

Diastereomeric derivatives are the oldest and most widely used chromatographic approach to the resolution of enantiomers (*7*). The precolumn derivatization of an optically active solute with another optically active molecule depends on the ability to derivatize the target molecule. A large number of functional groups and derivatives have been investigated, including amino groups (derivatized to amides, carbamates, ureas, thioureas, and sulfonamides), hydroxyl groups (esters, carbonates, and carbamates), carboxyl groups (esters and amides), epoxides (isothiocyanates), olefins (chiral platinum complexes), and thiols (thioesters).

This method has been used with a wide variety of HPLC columns and mobile phases, including normal- and reversed-phase approaches. At present there is no definitive way of determining which chromatographic approach will work. Derivatization has several advantages:

- The method has been extensively studied, making the application relatively easy and accessible.
- It uses standard HPLC supports and mobile phases.
- Detectability can be improved by selecting a derivatizing agent with a strong chromophore or fluorophore.

The method also has limitations:

- Synthesizing the diastereomeric derivatives requires isolating the compounds of interest first. This hinders the development of an automated procedure for large numbers of samples.
- The method often cannot be applied to routine assays because of enantiomeric contamination of the derivatizing agent, which can lead to inaccurate determinations. Such contamination has been encountered in a number of studies. Silber and Riegelman (24), for example, used (−)-*N*-trifluoroacetyl-L-prolyl chloride (TPC) in the determination of the enantiomeric composition of propranolol in biological samples. They found that commercial TPC was contaminated with 4% to 15% of the (+)-enantiomer and that the reagent rapidly racemized during storage.
- Enantiomers can have different rates of reaction or different equilibrium constants when they react with another chiral molecule. As a result, two diastereomeric products may be generated in proportions different from the starting enantiomeric composition (25).

Enantiomeric Resolution Using Chiral Mobile-Phase Additives

Enantiomeric compounds have been resolved through the formation of diastereomeric complexes with chiral molecules added to the mobile phase. The chiral resolution is due to differences in the stabilities of the diastereomeric complexes, solvation in the mobile phase, or binding of the complexes to the solid support. Lindner and Pettersson (26) have published an overview of this method.

Three major approaches to the formation of diastereomeric complexes are transition metal ion complexes (ligand exchange), ion pairs, and inclusion complexes. Each method is based on the formation of reversible complexes and uses an achiral chromatographic packing.

The solutes resolved by chiral ion-pair chromatography have included amino alcohols such as alprenolol, carboxylic acids such as tropic acid and naproxen, and amino acids such as tryptophan.

The composition of the mobile phase depends on the chiral agent used. When a chiral counterion is used, a mobile phase of low polarity, such as methylene chloride, is used to promote ion-pair formation. The retention of the solute can be decreased by increasing the concentration of the counterion or by adding a polar modifier, such as 1-pentanol (27). The latter approach usually reduces stereoselectivity. The water content of the mobile phase also appears to be important, and a water content of 80 to 90 ppm has been recommended.

With serum albumin as the chiral agent, aqueous mobile phases containing phosphate buffers are used. The retention and stereoselectivity can be altered by changing the pH. Aqueous or nonaqueous mobile phases can be used when (+)-dibutyl tartrate is the chiral modifier.

In some cases it appears that the modifier is retained by the stationary phase when the column is equilibrated with an aqueous mobile phase. The system then can be used with an organic mobile phase (*27*).

The chiral ion-pair systems are not stable. The chromatography can be affected by the water content of the mobile phase, temperature, pH, and other variables. This makes routine applications difficult. In addition, the counterions often absorb in the UV region, reducing the sensitivity of the system; indirect photometric detection (*28*) or other detection methods must be used.

Inclusion

Cyclodextrins are cyclic oligosaccharides of α-D-glucose units linked through the 1,4 position. The three most common forms are α-, β-, and γ-cyclodextrin, which contain six, seven, and eight glucose units, respectively. Cyclodextrin has a stereospecific, doughnut-shaped structure. The interior cavity is relatively hydrophobic, and a variety of water-soluble and insoluble compounds can fit into it to form inclusion complexes. For example, a naphthyl group can fit into the β-cyclodextrin cavity. The cavity is smaller for α- and larger for γ-cyclodextrin. If the included compounds are chiral, the inclusion complexes are diastereoisomeric.

Sybilska et al. (*29*) have used β-cyclodextrin as a chiral mobile-phase additive in the resolution of mephenytoin, methylphenobarbital, and hexobarbital. They attribute the observed resolution to two different mechanisms. The resolution of mephenytoin is due to a difference in the adsorption of the diastereoisomeric complexes on the achiral C_{18} support. For methylphenobarbital and hexobarbital, the relative stabilities of the diastereoisomeric complexes are responsible for the resolution.

Mobile phases modified with β-cyclodextrin can also resolve mandelic acid and some of its derivatives (*30, 31*). Aqueous mobile phases modified with a buffer such as sodium acetate are commonly used. Alcoholic modifiers, such as ethanol, can be added to the mobile phase to reduce retention.

Automation is possible for the direct measurement of biological samples. Sybilska et al. (*29*) have automated the preparative separations of mephenytoin, methylphenobarbital, and hexobarbital.

The applications of this approach seem limited. For example, unlike hexobarbital and methylphenobarbital, the chiral barbiturates secobarbital, pentobarbital, and thiopental are not resolved when chromatographed with a β-cyclodextrin-containing mobile phase (*27*). Cyclodextrin inclusion complexes are discussed further in the next section.

Enantiomeric Resolution Using Chiral Stationary Phases

Enantiomers can be resolved by the formation of diastereomeric complexes between the solute and a chiral molecule that is bound to the stationary phase. The stationary phase is called a CSP, and the use of these phases is the fastest growing area of chiral separations. The first commercially available CSP for HPLC was introduced by Pirkle in 1981 (*32*), and today many chiral phases are commercially available.

The separation of enantiomeric compounds on a CSP is due to differences in energy between temporary diastereomeric complexes formed between the solute isomers and the CSP; the larger the difference, the greater the separation. The observed retention and efficiency of a CSP results from all the interactions between the solutes and the CSP, including achiral interactions.

Selecting the Stationary Phase

The traditional classification of CSPs is shown in Table II, and a flowchart for the selection procedure is given in Figure 1. The chromatographic analysis of enantiomeric compounds can also be broadly divided into normal-phase and reversed-phase methods (*33*).

Normal-Phase Analyses. Type 1 or 2 CSPs are preferable for normal-phase analyses. Derivatization by an achiral reagent can be used to increase analyte–CSP interactions and improve the resolution. Derivatization can also bring other benefits, such as increased detectability by UV or fluorescence detectors or enhanced solubility in normal-phase eluents. If derivatization is possible, a Type 1 CSP (donor–acceptor) is favored. Otherwise, a Type 2 CSP (e.g., silica-supported cellulose triesters) should be selected except when a π-acid or a π-base is already located close to a chiral center in the analyte. For example, an arylsulfoxide is better separated on a Type 1 CSP.

The analytes identified as candidates for separation with a Type 1 CSP can

Table II. Classification of Chiral Stationary Phases

Type	Description	Examples	Mode (modifiers)
1	Pirkle-type (donor–acceptors)	DNB-phenylglycine, DNB-leucine, naphthylalanine	Normal phase (polar)
2	Cellulose triesters or carbamates on silica	Chiralcel OA, OB, OF, OJ	Normal phase (polar)
3	Inclusion (cyclodextrins, polyacrylates, polyacrylamides, crown ethers)	Cyclobond I, II, III; Chiralpak OP, OT; Chiralcel CR	Reversed phase (aqueous aceto-nitrile or methanol)
4	Ligand exchange	Proline, hydroxyproline	Reversed phase (aqueous buffers)
5	Proteins	Albumin, glycoprotein	Reversed phase, (aqueous buffers)

SOURCE: Data taken from reference 33.

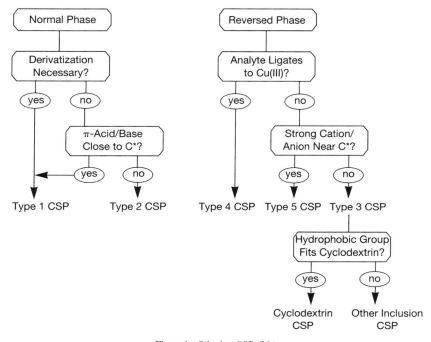

Figure 1. Selecting CSP (33).

be further divided into those with π-donor groups and those with π-acceptor groups. When there is a π-donor group (e.g., phenyl or naphthyl) close to the chiral center, a π-acceptor Type 1 CSP (e.g., *N*-(3,5-dinitrobenzoyl)dinitrobenzoyl-D-phenylglycine) may be initially selected. Otherwise, a π-donor Type 1 CSP (e.g., naphthylvaline) would be indicated if a π-acceptor group was similarly located in the analyte or its derivative. These requirements are not overriding, however, since Type 1 CSPs also operate by dipole–dipole, hydrogen-bonding, and steric interactions.

Reversed-Phase Analyses. If bidentate ligation to a metal ion such as Cu(II) is possible, Type 4 CSPs (ligand exchangers) are indicated. Otherwise, the choice is between Types 3 and 5.

For nonligating reversed-phase analytes, if a strongly cationic or anionic site is close to the chiral center, the analyte may be resolvable on a protein-based CSP (Type 5). Alternatively, inclusion CSPs (Type 3) should be tried first.

Inclusion Analyses. If a nonpolar includable group (ideally 5 to 9 Å across) is close to the chiral center, one of the Cyclobond range of cyclodextrin CSPs should be selected. Otherwise, other types of inclusion CSPs, such as the chiral acrylate polymers marketed by Merck and Daicel, should be used.

Macrocyclic antibiotics have been investigated as a new class of chiral selectors for liquid chromatography (34). These stationary phases can be used both in normal- and reversed-phase modes.

Because many CSPs are available for HPLC, it is difficult to determine which is most suitable for a particular problem. This difficulty can be partially overcome by grouping the CSPs for chiral separations according to a common characteristic. The first step—that is, the formation of the solute–CSP complexes—can be used as the basis for a classification system that divides the commercially available CSPs into five categories (33).

Types of Stationary Phases

Pirkle-Type and Related CSPs (Type 1). Type 1 CSPs are often based on an amino acid derivative. Both π-acid and π-base columns are available (with an amide, urea, or ester moiety in the CSP). The solute binds to the stationary phase through dipole–dipole, hydrogen-bonding, and π–π interactions.

The mobile phase usually contains a nonpolar organic solvent with various amounts of a polar modifier (e.g., hexane–2-propanol).

Polar functional groups (e.g., amino, carboxyl, and some hydroxyl) must be made less polar by derivatization (Table III).

Pirkle (35) has recently developed a hybrid column with both π-acid (3,5-dinitrobenzoyl) and π-base (naphthyl) groups. It resolves a variety of compounds, including underivatized carboxylic acids.

Derivatized Cellulose (Type 2). Cellulose is a crystalline polymer composed of β-D-glucose units. Crystalline cellulose is unable to withstand HPLC pressures and therefore has to be modified. Several approaches have been reported, which are based on derivatizing the available hydroxyl groups: cellulose triacetate (CTA), and cellulose tribenzoate (CTB).

Alcohols are needed as the mobile phase for CTA, whereas hexane–2-propanol is commonly used for CTB.

Table III. Derivatizations with Achiral Reagents

Group	Reagent	Product
NH	COOH deriv.	amide
	isocyanate	urea
	chloroformate	carbamate
COOH	alcohol	ester
	aniline	anilide
	amine	amide
OH	COOH deriv.	ester
	isocyanate	carbamate

A variety of analytes can be resolved. CTA prefers the phenyl group, which is also well suited for preparative work. Multiple interactions are involved in chiral recognition, which includes inclusion into channels and cavities.

Cellulose can also be depolymerized to shorter units, then derivatized to ester or carbamate derivative, followed by coating onto silica, i.e., noncovalent attachment (Table IV).

The mobile phase is usually hexane–2-propanol with base or acid added, depending on the analyte.

The chiral recognition mechanism is not well known; multiple mechanisms, including fit into cavities, are possible and successful in many applications. Many compounds resolve without derivatization; some require derivatization.

Cyclodextrins and Other Inclusion-Complex-Based CSPs (Type 3). Cyclodextrins, described earlier as chiral mobile-phase additives, can also be incorporated into CSPs. Chiral recognition is based on entry of a solute or a portion of the solute molecule into the cyclodextrin cavity to form an inclusion complex stabilized by hydrogen bonds and other forces.

Mobile phases are generally composed of aqueous (often buffered) methanol, ethanol, or acetonitrile mixtures. Polar organic acetonitrile mobile phase (with triethylamine–acetic acid) has been used recently.

Several CSPs are obtained by derivatizing hydroxyl groups to produce acetylates or other esters and carbamate derivatives—for example, (*R*)- and (*S*)-*N*-[1-(1-naphthyl)ethyl]carbamate derivatives. The chiral recognition mechanism does not appear to be an inclusion complex.

Cyclodextrin CSPs can be used with organic mobile phases (e.g., hexane–2-propanol).

Table IV. Derivatized Cellulose Columns

Name	*Separation*	*Mobile Phase*
Chiralpak OT(+) Chiralpak OP(+)	Compounds with aromatic group	methanol or hexane–2-propanol
Chiralcel OA Chiralcel OB Chiralcel OC Chiralcel OK Chiralcel OD	Compounds with aromatic group, carbonyl group, nitro group, sulfinyl group, cyano group, hydroxyl group	hexane–2-propanol or ethanol + water
Chiralcel CA-1		ethanol + water
Chiralpak WH Chiralpak WM Chiralpak WE	DL-Amino acid or its derivative	aqueous $CuSO_4$

SOURCE: Daicel Chemical Industries.

Ligand Exchangers (Type 4). The ligand exchange technique is based on complexation to a metal ion such as Cu(II). The chiral ligand immobilized on column is usually an amino acid (e.g., proline). Solutes must be able to form coordination complexes with the metal ion; examples are amino acids and derivatives, hydantoins, and amino alcohols.

The mobile phase contains the metal ion, typically added as copper sulfate. Retention and resolution can be manipulated by adjusting the concentration of metal ion, pH, modifiers, and temperature.

Protein CSPs (Type 5). Proteins are polymers composed of chiral units (L-amino acids) and are known to be able to bind small organic molecules. The protein can be immobilized on solid support by a variety of bonding chemistries; the choice of bonding method affects the selectivity of the CSP obtained. Protein CSPs include α_1 acid glycoprotein, ovomucoid (a glycoprotein from egg white), human serum albumin, and bovine serum albumin.

The retention and chiral recognition mechanisms may be unrelated to drug binding by free protein (in vivo or in vitro). The retention is most likely based on hydrogen bonding, ionic interactions, or other mechanisms. Recent investigations with commercial chicken ovomucoid indicate that chiral recognition may be due to an ovoglycoprotein that is present as an impurity (*36*).

The mobile phases are generally aqueous buffers that are used in limited pH ranges with organic modifiers. Retention and stereoselectivity can be manipulated by varying the mobile phase composition, pH, and temperature. Although the capacity of protein CSPs is limited, they offer broad applicability.

Computerized Methods

Computerized methods are based on evaluation of the potential energy contours obtained in molecular docking processes. By computerized matching of the contours for docking two enantiomeric molecules with the same chiral entity, an energy difference between the two complexes at their potential energy minima can be calculated. This calculation offers, at least in principle, a means of predicting the elution order in a chromatographic system. However, solvent effects cannot be taken specifically into account by such methods, and so far the predictive value is limited. Furthermore, even in well-defined chromatographic systems containing a simple chiral selector, different retention mechanisms may compete, making the situation more complex than assumed in the theoretical approach. Nevertheless, data from computational work on chiral discrimination will continue to add to our understanding of the molecular interactions (*37, 38*). Chirule, a computer program for developing chiral separations, has been described by Stauffer and Dessy (*39*). It constructs a multidimensional information space from a large number of known chiral separations by fragmenting the molecules at their chiral centers.

References

1. *Drug Stereochemistry: Analytical Methods and Pharmacology*, 2nd ed.; Wainer, I. W.; Drayer, D., Eds.; Marcel Dekker: New York, 1993.
2. *Chiral Separations by Liquid Chromatography;* Ahuja, S., Ed.; ACS Symposium Series 471; American Chemical Society: Washington, DC, 1991.
3. Allenmark, S. *Chromatographic Enantioseparation;* Ellis Harwood: New York, 1991.
4. Ahuja, S. *Selectivity and Detectability Optimizations in HPLC;* Wiley: New York, 1989.
5. Lough, W. J. *Chiral Liquid Chromatography;* Blackie and Son: Glasgow, Scotland, 1989.
6. Gal, J. *LC-GC* **1987,** *5,* 106.
7. Hara, S.; Cazes, J. *Liq. Chromatogr.* **1986,** *9,* 2, 3.
8. Souter, R.W. *Chromatographic Separations of Stereoisomers;* CRC: Boca Raton, FL, 1985.
9. Ahuja, S. Presented at the 1st International Symposium on Separation of Chiral Molecules, Paris, France, 1988.
10. Blaschke, G.; Kraft, H. P.; Fickentscher, K.; Koehler, F. *Arzneim. Forsch.* **1979,** *29,* 1640.
11. *Pharm. Times* **1982,** April, 25.
12. *USP Dictionary of USAN and International Drug Names;* U.S. Pharmacopeial Convention: Rockville, MD, 1995.
13. DeCamp, W H. *Chirality* **1989,** *1,* 2.
14. *Guidelines for Submitting Supporting Documentation in Drug Applications for the Manufacture of Drug Substances;* Office of Drug Evaluation and Research, U.S. Food and Drug Administration: Rockville, MD, 1987.
15. *Chromatography of Pharmaceuticals: Natural, Synthetic, and Recombinant Products;* Ahuja, S., Ed.; ACS Symposium Series 512; American Chemical Society: Washington, DC, 1992.
16. Palamareva, M. S.; Snyder, L. R. *Chromatographia* **1984,** *19,* 352.
17. Bitterova, J.; Soltes, L.; Fanovee, T. R. *Pharmazie* **1990,** *45,* H.6.
18. Gil-Av, E. In *Chiral Separations by Liquid Chromatography;* Ahuja, S., Ed.; ACS Symposium Series 471; American Chemical Society, Washington, DC, 1991.
19. Schurig, V.; Jung, M.; Schmalzing, D.; Schliemer, M.; Duvekot, J.; Buyten, J. C.; Peene, J. A.; Musschee, P. *J. High Resol. Chromatogr. Chromatogr. Commun.* **1990,** *13,* 470.
20. Dalgliesh, C. *J. Chem. Soc.* **1952,** 137.
21. Wulff, G. *Polymeric Reagents and Catalysts;* Ford, W. T., Ed.; ACS Symposium Series 308; American Chemical Society: Washington, DC, 1986; pp 186–230.
22. *Drug Stereochemistry: Analytical Methods and Pharmacology;* Wainer, I. W.; Drayer, D. E., Eds.; Marcel Dekker: New York, 1988.
23. Dappen, R.; Arm, H.; Mayer, V.R. *J. Chromatogr.* **1986,** *373,* 1.
24. Silber, B; Riegelman, S. *J. Pharmacol. Exp. Ther.* **1980,** *215,* 643.
25. Krull, I. S. *Advances in Chromatography;* Giddings, J. C.; Grushka, E.; Cazes, J.; Brown, P. R., Eds.; Lippincott: Philadelphia, PA, 1977; Vol. 16, p 146.
26. Lindner, W. H.; Pettersson, C. *Liquid Chromatography in Pharmaceutical Development;* Wainer, I. W., Ed.; Aster: Springfield, [State], 1985; p 63.
27. Pettersson, C.; Schill, G. *J. Liq. Chromatogr.* **1986,** *9,* 269.
28. Allenmark, S. *J. Liq. Chromatogr.* **1986,** *9,* 425.
29. Sybilska, D.; Zukowski, J.; Bojarski, J *J. Liq. Chromatogr. 9,* 591.
30. Debowski, J.; Sybilska, D.; Jirczak, J. *J. Chromatogr.* **1982,** *237,* 303.
31. Debowski, J.; Sybilska, D.; Jirczak, J. *J. Chromatogr.* **1983,** *282,* 83.
32. Pirkle, W. H.; Finn, J. M.; Schreiner, J. L.; Hamper, B. C. *J. Am. Chem. Soc.* **1981,** *103,* 3964.
33. Taylor, D. R. *Recent Advances in Chiral Separations;* Stevenson, D.; Wilson, I. D., Eds.; Plenum: New York, 1990.

34. Armstrong, D.W.; Tang, Y.; Chen, S.; Zhou, Y.; Bagwill C.; Chen, J.-R. *Anal. Chem.* **1994,** *66,* 1473.
35. Pirkle, W. J. *Org. Chem.* **1992,** *57,* 3854.
36. Haginaka, J.; Seyama, C.; Kanasugi, N. *Anal. Chem.* **1995,** *67,* 2539.
37. Lipkowitz K. B.; Baker, B. *Anal. Chem.* **1990,** *62,* 774.
38. Topol, S. *Chirality* **1989,** *1,* 69.
39. Stauffer, S. T.; Dessy, R. E. *J. Chromatogr. Sci.* **1994,** *32,* 228.

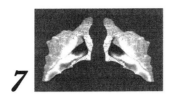

7

Development
of Chiral Methods

Thomas D. Doyle, Charlotte A. Brunner,
and Robert L. Hunt

Advances in techniques for chiral chromatographic analysis of enantiomeric substances such as drugs have been paralleled by the establishment of regulatory policies that address stereochemical methods. Analytical criteria unique to chiral substances include those relating to stereochemical resolution, identity, purity, and stability. We explain the rationale behind establishing such criteria and how they are met in practice in an analytical drug laboratory, using the analyses of racemic ibuprofen and of enantiomeric naproxen as examples. A fundamental conclusion is that current technology both allows and requires laboratory objectives to be essentially the same for both chiral and achiral analytes.

The 1992 notice in the *Federal Register* regarding the release of the "Policy Statement for the Development of New Stereoisomeric Drugs" (*1*) by the U.S. Food and Drug Administration (FDA) represented the culmination of more than a decade of rapid technological advances in the synthesis, resolution, and quantitative analysis of chiral molecules. The key concept in the statement is given in the opening paragraph of the General Policy section:

> The stereoisomeric composition of a drug with a chiral center should be known, and the quantitative isomeric composition of the material used in pharmacologic, toxicologic, and clinical studies known. Specifications for the final product should assure identity, strength, quality, and purity from a stereochemical viewpoint.

Because the human body presents a chiral environment, which can and usually does—often dramatically—distinguish between enantiomers and between other stereoisomeric forms, the need for such a policy is self-evident. As examples of drugs for which stereochemistry is important, the FDA document cites propranolol, sotalol, levodopa, levamisole, and carnitine. The scientific basis for this need is well established in the pharmaceutical community. Ariens (2) has cautioned that to disregard the chirality of drug molecules is to risk generating "scientific nonsense."

Yet chirality has often been disregarded. In investigating the stereochemistry of the diuretic cyclothiazide, Nusser et al. (3) developed methods for measuring its stereochemical composition. The molecular structure of cyclothiazide permits the existence of eight stereoisomers (four diastereomers, each an enantiomeric pair). Yet the 1990 edition of the *United States Pharmacopeia* (*USP*)—the last in which this substance appears—lists cyclothiazide as a single entity (4). Nusser et al. demonstrated that the commercial drug contains all eight isomers, in markedly unequal proportions, and they found significant variations between batches in the relative amounts of the isomers. Yet the *USP* compendium was silent on even the possibility of isomerism, and we have found no published studies on whether the isomers differ in their disposition, metabolism, and effects in the body.

Obviously, scientific policy must be based on the availability of adequate methods. Before 1980, there were essentially only two established and generally applicable approaches to quantitative enantiomeric analysis: optical rotation measurement and achiral chromatography of diastereomers produced by reaction with chiral reagents. Both methods, when properly applied, provide useful and valid information, but both have limitations, as discussed later. Other, less generally used methods of chiral analysis, such as NMR (employing chiral solvents, chiral shift reagents, or diastereomeric derivatization), are useful approaches but are not discussed in this chapter.

Chromatographic resolution of enantiomers with either chiral stationary phases (CSPs) or chiral mobile-phase additives (including recent, rapid developments in chiral electrophoresis) is now the dominant technique for chiral analysis. It is largely the emergence, success, and acceptance of these direct methods of resolution, as well as concomitant advances in the synthesis of enantiomerically pure molecules, that have made the FDA policy statement possible.

This chapter describes work in our laboratory on the design and validation of analytical methods for chiral drugs, with emphasis on procedures for assessing identity, strength, quality, and purity. In particular, we discuss our work on racemic ibuprofen and enantiomeric (S)-naproxen, since our experiences with these drugs well illustrate many of the issues addressed by the FDA policy statement. This chapter is not intended to be a comprehensive review of recent work, but rather an account of progress in the development of methods for analysis of chiral drugs in our research laboratory at the FDA and of the principles that guided this work.

Differences Between Chiral and Achiral Methods

Insofar as laboratory objectives and analytical criteria are concerned, and particularly with respect to the rigor of the required specifications, there are today essentially no differences between chiral and achiral methods. Several excellent critical articles have discussed method validation and the associated criteria for achiral drug samples in pharmaceutical analysis and in the bioanalytical laboratory (*5–7*). We contend that such specifications as those for recovery, accuracy, precision, reproducibility, linearity, specificity, limits of detection and quantitation, and ruggedness are the same whether or not the analyzed substances are stereoisomeric.

Criteria for testing the suitability of chromatographic systems for achiral samples have also been well discussed (*8*). While some commercially available chiral stationary phases for high-pressure liquid chromatography (HPLC) do not afford notably high efficiency or peak symmetry, many others are similar in performance to achiral phases. Thus, the criteria for system suitability of chiral analyses are in most cases the same as for their achiral counterparts, especially with regard to the most important parameter, the resolution factor.

The only analytical factors we consider, therefore, are those that are unique to or arise specifically from the stereoisomeric nature of chiral samples:

- stereoisomeric resolution
- stereoisomeric identity
- stereoisomeric purity
- stereoisomeric stability

Stereoisomeric Resolution

Exactly as with achiral mixtures, the analytical objective for the separation of enantiomers is to resolve all components of the mixture, including the enantiomers, with a resolution factor (R_s) of at least 1.5 (widely accepted in the industry as a minimum for baseline resolution). This criterion is indeed a minimum; it is certainly not ideal. Resolution factors of 2.0 or greater provide a degree of ruggedness, especially when peaks are of unequal areas, as in purity or stability determinations.

In pursuit of this objective, a choice must be made among three chromatographic methods:

- direct resolution of underivatized enantiomers on a CSP (clearly the preferred approach when feasible and when other factors such as detection characteristics and requirements allow)

- direct resolution, again on a CSP, after derivatization with achiral reagents, where the solutes to be resolved and analyzed are still enantiomeric
- indirect resolution on an achiral stationary phase by formation of diastereomeric derivatives with a chiral agent

In development work, the choice is dictated not only by the resulting chromatographic separation but also by such practical factors as availability and stability of stationary phases, availability of reagents, ease of derivatization (when required), and cost.

Chiral Analysis of Racemic (RS)-Ibuprofen

Studies in our laboratory on the chiral analysis of the analgesic ibuprofen illustrate these approaches, as well as practical considerations in developing a satisfactory method.

Ibuprofen (Figure 1) belongs to the class of nonsteroidal anti-inflammatory drugs (NSAIDs). It is well known that most, if not all, NSAIDs undergo in vivo a unidirectional conversion of the (R)- to the (S)-enantiomer. Ibuprofen is currently manufactured for over-the-counter sale in the United States exclusively as the racemate.

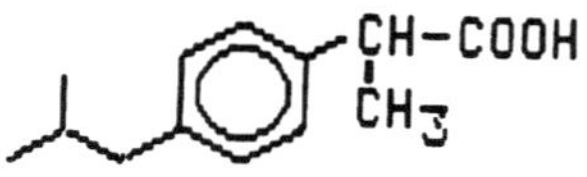

Figure 1. Structure of ibuprofen.

Most of the work in our laboratory on the chiral analysis of ibuprofen has used direct resolution following achiral derivatization. As early as 1984, we had resolved (RS)-ibuprofen as a number of aromatic and aliphatic amide derivatives (9) on the (R)-N-(3,5-dinitrobenzoyl)phenylglycine (DNBPG) CSP of Pirkle et al. (10) (Figure 2). Maximum separation ($\alpha = 1.12$) and resolution ($R_s = 1.75$) were obtained with the 1-(1-naphthyl)methylamide under normal-phase conditions (Figure 3a). This original method was procedurally inconvenient, however, requiring conversion of the carboxylic acid to the intermediate acid chloride, by reflux in thionyl chloride, followed by reaction with 1-(1-naphthyl)methylamine.

Other workers have successfully adapted and modified this approach. Nicoll-Griffith and co-workers (11–13) simplified carboxyl activation by using ethyl chloroformate; investigated the use of other amide derivatives, such as the 1-naphthyl and 4-methoxyphenyl analogues; applied the method to chiral analysis of ibuprofen in urine; and automated the derivatization.

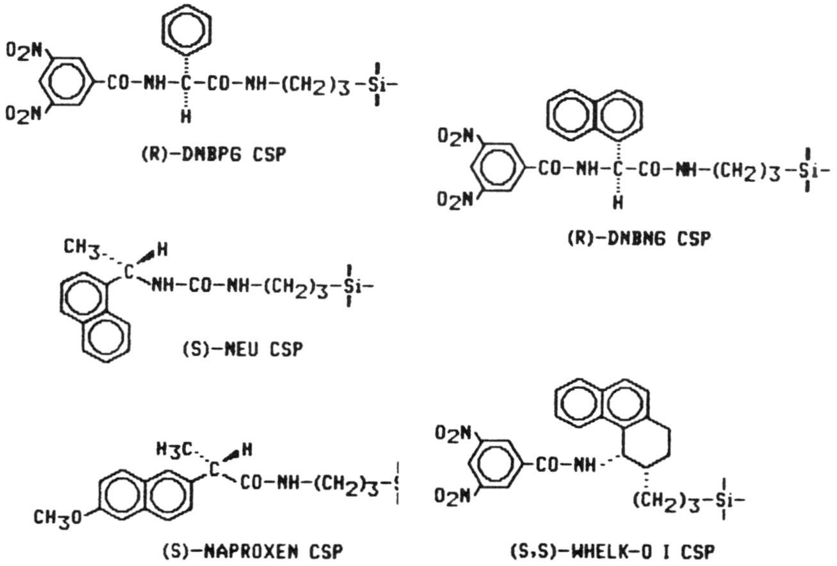

Figure 2. Chiral stationary phases (CSPs) used in this study.

Method Optimization

Subsequent investigations in our laboratory showed that ibuprofen could be resolved as amide derivatives with selectivities (α) as high as 3.81 by suitable choice of derivative and CSP. Chromatograms exhibiting extremely high performance can be obtained by suitable selection of the mobile phase. The chromatogram in Figure 3b was obtained on a column 25 cm long in less than 3 min at a flow rate of 2 mL/min. Some of our findings are shown in Table I, which shows previously unpublished results for 24 ibuprofen amides on two CSPs. The highest α value was obtained for the 3,5-dinitroanilide on the (S)-1-(1-naphthyl)ethylurea (NEU) CSP of Oi et al. (14) (see Figure 2 for the structure). The derivatives affording notably high selectivities on NEU CSP were principally those containing π-acidic (electron-withdrawing) aromatic groups. Unfortunately, the corresponding reagent amines invariably gave slow, incomplete reaction during derivatization.

Eventually, we solved these problems and developed a satisfactory method. In pursuit of an optimized procedure, we changed the identity of the derivative, its method of formation, the identity of the CSP, and even the physical construction of the CSP.

To activate the ibuprofen carboxyl, we used 2-ethoxy-1-ethoxycarbonyl-1,2-dihydroquinoline (EEDQ), which reacts rapidly at room temperature. EEDQ was later, in collaboration with Ahn et al. (15), found to produce no observable racemization of the chiral analyte, though measurable racemization occurred with such alternative activators as ethyl chloroformate. This important issue of

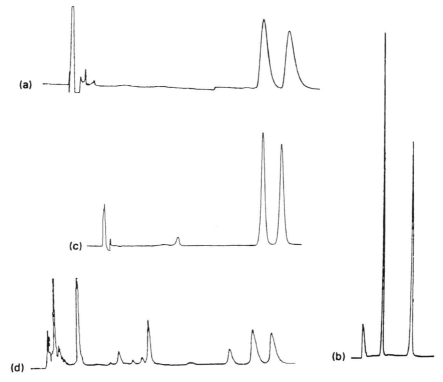

Figure 3. Enantiomeric resolution of ibuprofen after achiral derivatization: (a) 1-(1-naphth-yl)methylamide derivative on the (R)-*DNBPG; (b) 3,5-dinitrophenylanilide derivative on the* (S)-*NEU CSP; (c) optimized procedure and resolution for the 4-nitrobenzylamide derivative on two 5-cm* (S)-*NEU CSPs in series (3-μm particles); (d) optimized analysis of ibuprofen in canine plasma under conditions similar to (c). All mobile phases consisted of varying proportions of hexane (or heptane), 2-propanol, and acetonitrile. The* (R)-*enantiomer eluted first in all cases.*

stereochemical stability during the analytical procedure is discussed in more detail later.

As the derivatizing amine, we eventually selected 4-nitrobenzylamine (PNBA). This reagent contains an electron-withdrawing aromatic group, but the group's effect on the reaction rate is diminished because it is separated from the amine function by a methylene unit. PNBA reacts essentially quantitatively with EEDQ-activated ibuprofen during a 10-min reflux in 1,2-dichloroethane (Figure 4).

To complement the π-acidic nature of the derivative, we selected the π-basic NEU CSP. However, the resolution (R_s = 1.47) obtained on a commercial 25-cm NEU column with a particle size of 5 μm, under analytical conditions suitable for the intended plasma samples, was less than optimum.

Table I. Resolution of Amide Derivatives of Racemic Ibuprofen on the (R)-DNBPG CSP and on the (S)-NEU CSP

Ibuprofen Derivative	(R)-DNBPG[a]			(S)-NEU[b]		
	k_1'	k_2'	α	k_1'	k_2'	α
1 Anilide	3.7	4.5	1.21	1.4	1.6	1.13
2 2-Nitroanilide	5.6	5.6	1.00	2.2	2.2	1.00
3 3-Nitroanilide	9.4	10.6	1.14	2.1	3.7	1.75
4 4-Nitroanilide	10.0	11.6	1.15	2.8	5.3	1.88
5 3,5-Dinitroanilide	10.3	10.7	1.04	7.4	28.2	3.81
6 2-Methoxyanilide	3.0	3.2	1.10	0.6	0.6	1.00
7 3-Methoxyanilide	6.4	7.9	1.24	2.0	2.2	1.13
8 4-Methoxyanilide	9.7	11.9	1.16	3.0	3.5	1.16
9 3,5-Dimethoxyanilide	11.2	14.0	1.25	2.4	2.8	1.17
10 3,4-Dimethoxyanilide	35.1	41.6	1.18	8.7	10.6	1.21
11 Pentafluoroanilide	1.5	1.6	1.10	0.7	1.1	1.48
12 3,5-Bis(trifluoromethyl)anilide	0.9	1.0	1.04	0.4	0.7	1.73
13 4-Acetylanilide	19.5	23.0	1.18	4.1	5.5	1.36
14 4-Benzoylanilide	15.1	17.7	1.17	2.2	3.9	1.54
15 Benzylamide	6.0	6.5	1.09	1.3	1.3	1.00
16 4-Nitrobenzylamide	18.6	20.3	1.09	3.7	4.2	1.13
17 4-Methoxybenzylamide	12.1	13.3	1.10	2.2	2.3	1.07
18 1-(1-Naphthyl)methylamide	13.1	14.4	1.10	1.4	1.4	1.00
19 1-Naphthylamide	11.4	13.9	1.23	2.0	2.0	1.00
20 2-Naphthylamide	9.1	11.6	1.27	2.4	2.4	1.25
21 3-Quinolinoylamide	6.5	7.5	1.15	1.7	3.0	1.77
22 2-Fluorenylamide	9.9	11.8	1.20	1.4	1.8	1.32
23 9-Oxo-2-fluorenylamide	10.6	11.9	1.12	1.8	2.5	1.41
24 3-Fluoranthenylamide	27.2	32.1	1.18	1.8	2.1	1.77

NOTE: The (S)-enantiomer eluted first on both stationary phases.
[a](R)-N-(3,5-Dinitrobenzoyl)phenylglycine stationary phase with mobile phase of hexane-2-propanol (97:3).
[b](S)-1-(1-Naphthyl)ethylurea stationary phase with mobile phase of hexane-2-propanol (90:10).

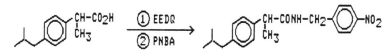

Figure 4. Derivatization of ibuprofen with PNBA.

This last obstacle to method development was solved by constructing NEU columns with improved efficiency. The only available commercial columns of this type had 5-μm silica particles, but we had previously shown (*16*) that NEU CSPs can be generated in situ by a simple and economical laboratory procedure, starting with achiral aminopropyl columns and a chiral isocyanate. Columns of any desired dimensions and, more important, with the improved efficiency of 3-μm particles, can thus be prepared.

In our laboratory, we find it convenient to employ such columns in lengths of 5 or 10 cm and to join them in series as required. We have established, as chromatographic theory readily predicts, that selectivity (α) and retentivity (k') are unaffected by these modifications and that efficiency (N) is proportional to total column length. Resolution (R_s) should thus increase with the square root of the length; systematic studies in our laboratory have confirmed that this relationship is very closely followed.

Figure 3c shows an optimized resolution for (RS)-ibuprofen, incorporating all the modifications discussed. This chromatogram is for the 4-nitrobenzyl amide and was obtained with three 5-cm (R)-NEU columns with 3-μm particles in series. The efficiency was greater than 10,000 plates for the 15-cm effective length, affording a resolution factor (R_s) of 2.54.

Ahn et al. (*15*) successfully applied this procedure to studies of the disposition of ibuprofen in canine plasma, using two 10-cm NEU columns in series (again with 3-μm particles). A chromatogram of a plasma sample taken from that study is shown in Figure 3d.

Comparison with Diastereomeric Methods

Ahn et al. also compared this method, involving resolution of enantiomeric derivatives, with alternative procedures employing achiral chromatography of diastereomeric derivatives obtained by reacting ibuprofen with chiral (S)-1-(1-naphthyl)ethylamine. The primary disadvantage of such diastereomeric methods is uncertainty about the enantiomeric purity of the reagent. Thus it is necessary to perform a preliminary chiral assay of the reagent (in which these same considerations are of course a factor) and, if necessary, correct for isomeric contamination of the reagent to obtain the final analytical results. The one advantage of diastereomeric methods is that an achiral stationary phase can be used.

In comparing these approaches, there are other factors to consider as well; these are summarized in Table II for diastereomeric methods and for enantiomeric methods with and without achiral derivatization. Some of these factors are often overlooked during method development. For example, diastereomeric derivatization may yield ratios of products differing from the starting composition, and the source of this inequality may be either kinetic or thermodynamic. This problem does not exist for achiral derivatization, except perhaps if other chiral substances are present in the mixture and participate in the reaction or if stereospecific self-association of the analyte produces species that react at significantly different rates.

Another factor is that diastereomers may have different properties in electronic absorption or emission (and hence detection). This is rarely a problem for UV absorption, though the potential for error is greater for fluorescence emission.

The stereoisomeric stability of the analyte to the analytical conditions must

Table II. Comparison of Diastereomeric and Enantiomeric Methods of Chromatographic Chiral Analysis

Factor	Diastereomeric Methods	Enantiomeric Methods With Derivatization	Enantiomeric Methods Without Derivatization
Chiral HPLC system	no	yes	yes
Derivatization	yes	yes	no
Chiral purity of reagent	yes	no	NA
Racemization of reagent	yes	no	NA
Racemization of analyte	yes	yes	(yes)
Differing reaction rates	yes	(no)	NA
Differing detection	(yes)	no	NA

NOTE: NA is not applicable. Entries in parentheses are subject to exception or qualification.

also be taken into account. Such stability is always of greatest concern for diastereomeric methods, if only because of the complication of an additional chiral center (or centers), but all reactions, diastereomeric or not, increase the risk of racemization or inversion. Even for direct methods, when no derivatization is performed, it is necessary to consider the stereoisomeric stability of the analyte to the conditions of analysis, including perhaps exposure to the (often complex) chiral environment of the chromatographic system itself.

Direct Resolution of Ibuprofen

Advances in analytical technology, not only the design of new and more powerful CSPs but also refinements in the selection of mobile phases or of other analytical conditions, make it increasingly possible to resolve chiral molecules directly, without derivatization. Recently, direct resolution has become possible for ibuprofen. Figure 5a shows the reversed-phase resolution of (*RS*)-ibuprofen on an (*R*)-*N*-(3,5-dinitrobenzoyl)naphthylglycine (DNBNG) CSP developed by Oi et al. (*17*); the CSP is shown in Figure 2; the reference is to the ionic modification of this CSP; the covalently bound analogue is commercially available. An atypical mobile phase of 2-propanol and aqueous acetate buffer (50:50) provided the key to the success of this resolution. It is well known that 2-propanol tends to promote enantioselectivity in normal-phase mode, but only recently have studies in our laboratory shown that this is true in reversed-phase mode as well.

An example of the great strides made in chiral separation is the direct, normal-phase resolution of (*RS*)-ibuprofen on the (*S,S*)-Whelk-O I CSP of Pirkle and Welch (*18*) (*see* structure in Figure 2). The chromatogram, shown in Figure 5b, can fairly be called an ideal result: a complete baseline separation with resolution factor (R_s) of 2.85, highly symmetrical peaks, and an efficiency (N) of 26,000 plates per meter. This separation is an optimization developed in our laboratory by minor modification of a resolution reported by Pirkle and Welch (*18*).

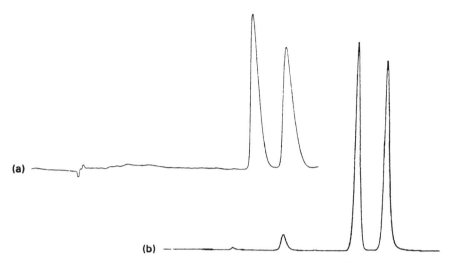

Figure 5. Enantiomeric resolution of ibuprofen without derivatization: (a) direct reversed-phase (2-propanol–aqueous acetate) resolution on the (R)-DNBNG CSP; (b) direct normal-phase (hexane–2-propanol–trace acetic acid) resolution on the (S,S)-Whelk-O I CSP. The (R)-enantiomer eluted first in both cases.

Chiral Analysis of Enantiomeric (*S*)-Naproxen

Like ibuprofen, naproxen is an NSAID that is converted in vivo from the (*R*)- to the (*S*)-isomer. Naproxen is dispensed in the United States only as the (*S*)-enantiomer (Figure 6) and has recently been made available for over-the-counter sale.

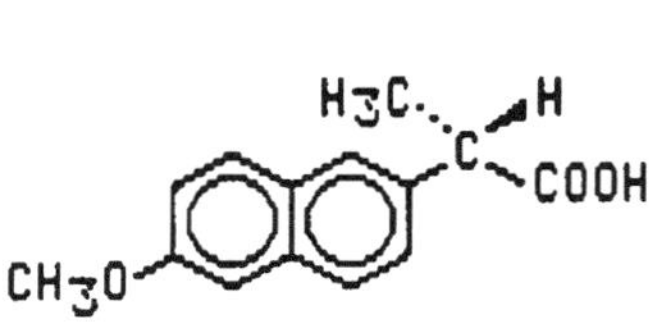

$$H_3C\cdots C\diagdown{}_H$$
$$C-COOH$$
$$CH_3O$$

Figure 6. Structure of (S)-naproxen.

From a chemical standpoint, naproxen is a relatively simple enantiomeric substance to resolve. The molecule contains a strongly electron-donating naphthyl ring system, attached directly to the chiral center, that can interact stereospecifically with complementary electron-withdrawing groups such as the dinitrophenyl rings of several of the CSPs shown in Figure 2. In addition, the carboxyl group, also directly attached to the asymmetric carbon, can contribute hydrogen-bonding or other polar interactions in either free or derivatized form.

We first reported the resolution of amide derivatives of naproxen on the DNBPG CSP in 1984 (*9*).

Indeed, naproxen well illustrates the principle of reciprocity first enunciated by Pirkle et al. (*19*), which can be restated as "Chiral recognition is a two-way street." The structural features of naproxen that make it a good candidate for chromatographic enantioresolution also make it a good candidate for binding to silica to serve as a CSP for resolving other chiral molecules.

Accordingly, we bound naproxen to silica through a propylamide linkage by well-known procedures to produce the (*S*)-naproxen CSP (Figure 2) (*20*). Among the many chiral substances we have resolved in this CSP, the most intriguing is racemic naproxen itself; it was resolved by chromatographing the (3,5-dinitro-phenyl)anilide (Figure 7a).

Given these favorable structural features, it is not surprising that (*R*)- and (*S*)-naproxen can be separated by direct, underivatized enantiomeric resolution on a

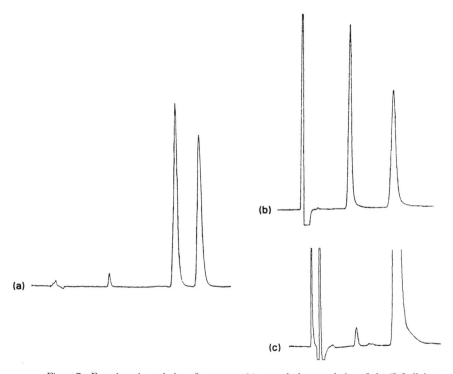

Figure 7. Enantiomeric resolution of naproxen: (a) normal-phase resolution of the (3,5-dinitro-phenyl)anilide of racemic naproxen on the (S)-naproxen CSP; (b) direct normal-phase resolution of racemic naproxen on the (S,S)-Whelk-O I CSP; (c) analysis of a commercial tablet of (S)-naproxen under the conditions of (b), showing 0.8% (R)-naproxen. The (R)-enantiomer eluted first in all cases.

suitable CSP. In fact, Pirkle and Welch (*18*) designed the Whelk-O I CSP (mentioned earlier in connection with the resolution of ibuprofen) specifically to resolve naproxen. A chromatogram from our laboratory of the direct resolution of (*RS*)-naproxen on the Whelk-O I CSP is shown in Figure 7b. With a mobile phase consisting of heptane–2-propanol–acetic acid (80:20:1), an impressive separation was obtained ($\alpha = 1.93$; $R_s = 5.25$).

With resolution so readily attainable, attention can be given to other analytical concerns. (Our work on naproxen has so far concerned only dosage-form analysis; the discussion below focuses on this aspect.)

Stereoisomeric Identity

It should be axiomatic that a drug substance with one or more chiral centers (and thus capable of existing in stereoisomeric forms) is not adequately described, nor are samples of the substance properly identified, unless the absolute configuration of the molecule is specified and then correlated with appropriate identification tests. Nevertheless, the official compendiums do not always include information on stereochemistry.

The situation with respect to naproxen is especially confusing. The 1995 *USP* (*21*) contains monographs for both naproxen (as the free acid) and naproxen sodium. Both refer, by inference but not explicitly, to the approved drug substance, the (*S*)-isomer [the absolute configuration of naproxen has been established (*22*)]. The *USP* gives a structure for free naproxen that is drawn, incorrectly, as the racemate. The official name specifies that the substance is dextrorotatory but does not include the "(*S*)" descriptor. The monograph includes a test for specific rotation, measured in chloroform, that specifies a range between +63.0° and +68.5°. Only from this requirement can the stereoisomeric identity of the drug substance naproxen be inferred.

While the assignment of dextrorotation to the (*S*)-enantiomer of free naproxen is correct, the situation is confused by the accompanying monograph for naproxen sodium. In it, the approved name (again without the Cahn–Ingold–Prelog descriptor, and with no structure given) specifies that the substance is levorotatory, as indeed (*S*)-naproxen is when measured in aqueous alkali (the range given for specific rotation is −15.3° to −17.0°).

The identification tests for naproxen drug substance and for several dosage forms specify only infrared, ultraviolet, and achiral chromatographic determinations. None of these will in principle distinguish between (*R*)- and (*S*)-enantiomers, although infrared determinations will in some cases distinguish racemic from enantiomeric forms.

The existence of powerful methods of chromatographic resolution provides an obvious remedy for such ambiguities. In our laboratory, we routinely and unambiguously establish stereochemical identity by a simple three-step procedure, which we term the elution order identification (EOID) test:

1. Using a racemic sample, develop (or confirm) an enantiomeric separation via a chiral chromatographic method—preferably, but not necessarily, by direct, underivatized resolution. Baseline resolution is not absolutely required for identification.

2. Using an authentic reference sample of one of the individual enantiomers, establish which enantiomer elutes first on the chromatographic system chosen. The retention time of a pure enantiomer should not be relied on (especially when resolution is not baseline), since it can vary with experimental conditions. The elution order can and always should be established with absolute certainty from a chromatograph of a mixture of one enantiomer and the racemate. The mixture should give peaks of clearly unequal areas, and it must be known unambiguously which of the enantiomers is in excess and to which peak it corresponds.

3. From the established elution order, determine the stereochemical identity of the sample in the same manner, first chromatographing the sample alone and then confirming the result with a mixture of the sample and the racemate

The EOID test is simple and quite powerful as an enantiomeric identification tool. It presupposes only the availability of an authentic enantiomeric standard and of the racemate (or the opposite enantiomer, or even a partially racemized sample). The only precaution is that Step 2 (the determination of the elution order) must always be performed. Many CSPs are commercially available in both configurations; using a column with the opposite configuration would reverse the elution order, resulting in the wrong enantiomeric assignment if previously recorded results are relied on. Step 2 eliminates this possibility (and indeed can be used to confirm the configuration of a chiral column).

Chromatographic enantiomeric identification is meant not to replace optical rotation measurement but to supplement it, providing an independent test. Chromatography does have many advantages, however, especially its suitability for microsamples. It also is superior for determining isomeric purity, as discussed in the following section.

Stereoisomeric Purity

Polarimetry

Specifications for drug substances and drug products have often been inadequate with respect to the isomeric purity of enantiomeric substances. Most compendial methods for assaying enantiomeric purity have depended on measurement of optical rotation by polarimetry. However, determination of enantiomeric purity is in essence equivalent to trace determination of the opposite enantiomer; polarimetric methods are inherently inferior for this purpose, for several reasons:

- A single numerical value (optical rotation) is measured and compared with that of a reference standard, and then the composition of the two components is calculated. The individual isomers are neither observed discretely nor measured independently. This aspect essentially precludes determination of trace isomeric composition.
- The chiral purity of the reference standard, and hence its specific rotation, is inherently uncertain, unless determined by nonpolarimetric methods. (Specific rotations have historically been revised only to higher absolute values, as the purity of standards improves.)
- Any chemical impurities, chiral or not, in either standard or sample, produce errors.
- Specific rotations are often low, especially at the sodium D-line.
- Rotations are strongly dependent on the solvent (even a change of sign is possible, as with naproxen) and thus also on the purity of the solvent.

Existing polarimetric methods for determination of isomeric purity often use extreme, sometimes ill-conceived, stratagems to overcome these limitations. An example may be found in the *USP* monographs for dextroamphetamine sulfate (as the elixir or as tablets) (*21*). In addition to requiring the equivalent of 130 mg of dextroamphetamine sulfate for the determination, the procedure requires derivatization to acetylamphetamine, followed by recrystallization of the product. (In this step, the isomeric composition of the sample, which is the property being tested, is subject to change by fractional crystallization!) The requirement for the analytical product is that the specific rotation be between $-37.5°$ and $-44.0°$. This $\pm8\%$ specification effectively allows for an enantiomeric contamination of 4%.

By contrast, we have previously shown (*23*) that the presence of as little as 0.1% (*R*)- in (*S*)-amphetamine can be determined chromatographically with good accuracy and precision; the limit of detection was established at below 0.01%. It is noteworthy that the *USP* reference standard for dextroamphetamine shows none of the levo-isomer at this 0.01% level, as determined by chiral HPLC.

Enantiomeric Purity of Naproxen Tablets

For (*S*)-naproxen, the *USP* monographs contain no specifications for isomeric purity for the drug substances or for any of the dosage forms, except that implied by the specific rotation test. But with the recent approval of this anti-inflammatory agent for over-the-counter sale, appropriate methods are certainly needed. Resolution on the Whelk-O I CSP, already mentioned, is clearly suitable for this purpose, especially when the (*S,S*)-configuration of the CSP is employed so that the minor (*R*)-component elutes first, enhancing detectability.

A chromatogram for the determination of the isomeric composition of a commercial tablet of (*S*)-naproxen, obtained by simply dissolving the tablet and resolving on the (*S,S*)-Whelk-O I CSP, under normal-phase conditions, is shown in Figure 7c. The result corresponds to an (*R*)-naproxen content of 0.8% relative to total naproxen. Although this work is still in progress at this writing, this result is clearly well within the analytical capabilities of the method. The chromatogram shown corresponds to an amount of naproxen on column of approximately 10 ng, whereas the limit of detection (signal-to-noise ratio of 4) was shown to be 120 pg by ultraviolet detection at 235 nm. The linearity of response was excellent for both (*R*)- and (*S*)-naproxen in the range 2.5 to 100 ng.

The major component, (*S*)-naproxen, has an ultraviolet spectrum that is perforce identical to that of the enantiomeric (*R*)-contaminant. This naturally simplifies calculations, but the observed detector response is not necessarily identical for the two isomers, since the concentration of the major component may fall outside the linear range and since the differing retention times (and hence peak heights) will ordinarily result in different limits for this linear range for each isomer. This difference is shown in the concentration response curve for the two naproxen isomers in Figure 8. A variety of methods can be used to correct for

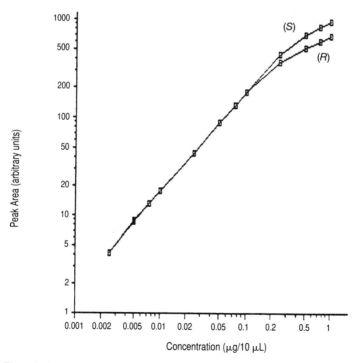

Figure 8. Linearity plot for (R)- *and* (S)-*naproxen. A 500-fold range is shown, wherein the responses of the two enantiomers are identical until the highest concentrations.*

these factors. At levels below 1% contamination, however, the analysis essentially becomes a limit test, so the need for absolute accuracy is diminished and calculations are correspondingly simplified. It is not certain that the observed minor peak is due to (R)-naproxen, only that it has a retention time identical to that for this isomer.

Stereoisomeric Stability

Aspects of Stability

The final analytical factor unique to chiral substances, stereoisomeric stability, is perhaps the most complex. There are many aspects to this issue:

- *Stability toward racemization of an enantiomerically pure substance with a single chiral center in normal storage and under stressed conditions.* Isomeric stability of the racemate will not normally be of concern, since inversion should be identical in both directions for a racemic substance stored in an achiral environment. However, the question of kinetic stability may be of interest. (But unidirectional inversion of, for example, ibuprofen in vivo, as mentioned previously, is a fascinating question of drug metabolism, not of stability.)
- *Stability of a substance with more than one chiral center toward inversion of one or more of those centers to form diastereomers.* This is of concern whether the original substance is a racemate or a pure enantiomer.
- *Stability of chiral substances when stored in the presence of other chiral materials.* For example, drug tablets can contain chiral excipients such as cellulose.
- *Stability of chiral substances, both enantiomers and racemates, to analytical conditions.* This is important whether or not a particular analytical procedure is designed to examine the stereochemistry of the analyte, but it is especially important in that case, since the analytical probe will then often be chiral, increasing the risk of differential instability.
- *Stability of the chiral analytical probes to storage, to analytical conditions, and even to the presence of the chiral analyte.* These probes include CSPs, chiral mobile-phase additives, and chiral derivatizing reagents.

Clearly, the subject of stereoisomeric stability is vast, and although much is known, much more remains to be determined. Few generalizations can be made; a case-by-case approach is prudent. Some observations from recent work in our laboratory may be of interest.

Concerning the first two aspects of stability, we have yet to observe significant isomeric instability for any simple amine or amino alcohol, either during storage or during analysis (including during and after formation of such derivatives as amides, ureides, and carbamates). We have generally not subjected these substances to thermal or chemical extremes, however, nor have we performed systematic long-term studies.

Amino acids, on the other hand, must be handled with caution, especially during and after derivatization. Significant inversion (1% or more) has sometimes been observed. Conditions can usually be found to eliminate or minimize this problem.

Stereoisomeric Stability of Naproxen

We have had limited experience with enantiomerically pure carboxylic acids; however, our work with (S)-naproxen is instructive. To determine the identity and purity of this drug, we needed a sample of racemic naproxen, which we did not have and were unable to acquire from any outside source. We assumed that the proton on the chiral carbon α to the carboxyl group would be readily exchangeable, with inversion, and so we refluxed (S)-naproxen in a variety of strongly basic media, including aqueous alkali and a solution of triethylamine in 1,2-dichloroethane.

Even on prolonged reflux under these conditions, inversion ranged from negligible to undetectable. The (S)-naproxen was recovered unchanged. Eventually, we obtained complete racemization by refluxing naproxen in a solution of sodium methoxide in methanol. Under these conditions, the racemic methyl ester was the recovered product, which could then be readily hydrolyzed to the racemic free acid. The lesson was twofold: First, the carboxyl proton of naproxen is of course preferentially abstracted in basic conditions; a second proton is then apparently extremely resistant to exchange. Second, simple chiral carboxylic acids should be added to the list of compounds with considerable stereochemical stability.

Concerning stability of the chiral analytical probes, all the CSPs discussed in this chapter and shown in Figure 2 appear to be stable in a stereochemical sense to long-term storage (in some cases more than 10 years) and are also stable to frequent use. We base this observation on unchanged selectivities (α values) of chiral test materials. However, these CSPs have to date been employed primarily in normal-phase mode; we have little data on stability to reversed-phase conditions. Additionally, for almost all CSPs (unlike chiral mobile-phase additives), there is no easy nondestructive way to determine the enantiomeric purity of the chiral molecules bound to the support.

Conclusion

Chromatographic chiral analysis is now well established as a mature technology. Chromatographic resolution of virtually any chiral molecule is a reasonable objective. Direct resolution of enantiomers, without derivatization, is increasingly feasible.

When intelligently selected and designed, methods for chiral analysis can now routinely afford the same accuracy, precision, and ruggedness as other, achiral procedures, without sacrifice of procedural simplicity or convenience.

In spite of these advances, some aspects of chiral analysis are unique, applying especially to the objectives of achieving acceptable stereochemical resolution and determining stereochemical identity, purity, and stability. Many questions about stereoisomeric stability remain to be answered.

But on the whole, as a result of the continuation of the technical advances that began in the early 1980s, chiral analysis has reached a point where analytical criteria for stereochemical parameters are the same as for any other analytical determination. Laboratory objectives, and commercial and regulatory decisions, should reflect this fact.

References

1. "FDA's Policy Statement for the Development of New Stereoisomeric Drugs," *Fed. Reg.* **1992,** *57(102),* 22249.
2. Ariens, E. J. *Eur. J. Clin. Pharmacol.* **1984,** *26,* 663–668.
3. Nusser, E.; Banerjee, A.; Gal, J. *Chirality* **1991,** *3,* 2–13.
4. *The United States Pharmacopeia XXII and National Formulary XVII;* The United States Pharmacopeial Convention: Rockville, MD, 1990; Suppl. 1.
5. Edwardson, P. A. D.; Bhaskar, G.; Fairbrother, J. E. *J. Pharm. Biomed. Anal.* **1990,** *8,* 929–933.
6. Carr, G. P.; Wahlich, J. C. *J. Pharm. Biomed. Anal.* **1990,** *8,* 613–618.
7. Buick, A. R.; Doig, M. V.; Jeal, S. C.; Land, G. S.; McDowall, R. D. *J. Pharm. Biomed. Anal.* **1990,** *8,* 629–637.
8. Wahlich, J. C.; Carr, G. P. *J. Pharm. Biomed. Anal.* **1990,** *8,* 619–623.
9. Wainer, I. W.; Doyle, T. D. *J. Chromatogr.* **1984,** *284,* 117–124.
10. Pirkle, W. H.; Finn, J. L.; Schreiner, J. L.; Hamper, B. C. *J. Am. Chem. Soc.* **1981,** *103,* 3964–3966.
11. Nicoll-Griffith, D. A. *J. Chromatogr.* **1987,** *402,* 179–187.
12. Nicoll-Griffith, D. A.; Inaba, T.; Tang, B. K.; Kalow, W. *J. Chromatogr.* **1988,** *428,* 103–112.
13. Nicoll-Griffith, D.; Scartozzi, M.; Chiem, N. *J. Chromatogr.* **1993,** *653,* 253–259.
14. Oi, N.; Kitahara, T.; Doi, T.; Yamamoto, S. *Bunseki Kagaku* **1983,** *32,* 345–346.
15. Ahn, H.-Y.; Shiu, G. K.; Trafton, W. F.; Doyle, T. D. *J. Chromatogr.* **1994,** *653,* 163–169.
16. Doyle, T. D.; Brunner, C. A.; Vick, J. A. *Biomed. Chromatogr.* **1991,** *5,* 43–46.
17. Oi, N.; Kitahara, H.; Matsumoto, Y.; Nakajima, H.; Horikawa, Y. *J. Chromatogr.* **1989,** *462,* 382–386.
18. Pirkle, W. H.; Welch, C. J. *J. Liq. Chromatogr* **1992,** *15,* 1947–1955.
19. Pirkle, W. H.; House, D. W.; Finn, J. M. *J. Chromatogr.* **1980,** *192,* 143–158.
20. Doyle, T. D.; Brunner, C. A.; Smith, E. U.S. Patent 4 919 803, 1990.
21. *The United States Pharmacopeia XXIII and National Formulary XVIII;* The United States Pharmacopeial Convention: Rockville, MD, 1995.
22. Riegl, J.; Maddox, M. L.; Harrison, I. T. *J. Med. Chem.* **1974,** *17,* 377–378.
23. Doyle, T. D. In *Chiral Separations by Liquid Chromatography;* ACS Symposium Series 471; Ahuja, S., Ed.; American Chemical Society: Washington, DC, 1991; Chapter 2.

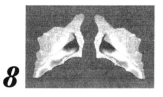

8

Stereoselective Renal Elimination of Drugs

Ronda J. Ott and Kathleen M. Giacomini

There have been many reports of stereoselective drug metabolism, but only recently have examples of stereoselective renal drug elimination been reported. This chapter reviews the literature on stereoselective renal drug elimination involving stereoselective plasma protein binding, filtration, secretion, or reabsorption mechanisms. We begin with a brief discussion of the basic mechanisms responsible for renal drug elimination, including tubular transport of drugs in either the secretory or the reabsorptive direction. Stereoselective filtration due to stereoselective plasma protein binding of disopyramide and carbenicillin is discussed. The stereoselective renal secretion of various organic cations, including pindolol, the diastereomers quinine and quinidine, and the organic anion DBCA is examined in detail. Finally, the stereoselective renal reabsorption of endogenous sugars and amino acids is discussed.

Considerable progress has been made in understanding stereoselective pharmacokinetics in terms of stereoselective metabolism, and there is a large body of information on stereoselective drug metabolism in various species, including humans. In contrast, our knowledge of stereoselective renal elimination of compounds is limited. The first studies demonstrating stereoselective renal elimination of drugs in humans appeared less than a decade ago, and since then it has become increasingly clear that, like drug metabolism, renal elimination may be stereoselective. In this chapter, we first review renal physiology relevant to drug elimination. We then discuss the processes involved in the tubular transport of drugs in either the secretory or reabsorptive direction. Finally, we discuss how stereoselective renal elimination of drugs is affected by stereoselective binding to

3407–8/96/0173$17.00/0

plasma proteins as well as stereoselective tubular reabsorption and secretion. Where possible, we use clinical examples to illustrate the principles of stereoselective renal handling.

Renal Physiology

Blood flow to the kidneys of an average adult is about 1200 mL/min, or 25% of the cardiac output. The kidney clears the blood of metabolic waste products and foreign substances and preserves water and electrolyte balance within the body. These functions are carried out by the filtration, reabsorption, and secretion of water and other plasma constituents from the blood. Each kidney contains approximately 1 million nephrons (Figure 1) (1). The nephron consists of a glomerular capillary network, Bowman's capsule, and an epithelial tubule that extends from the capsule to a collecting duct within the kidney. The glomerulus is composed of capillary loops derived from the afferent arteriole of the kidney. These capillaries are butted against the side of the Bowman's capsule, leaving a small space. The plasma ultrafiltrate (plasma water and all non-protein-bound constituents) flows into this space after being forced through the endothelial capillaries, the basement membrane, and the single layer of epithelial cells that rests on the basement membrane (1).

The ultrafiltrate is similar in composition and pH to the blood, except that the red blood cells and plasma proteins are retained in the capillary endothelium.

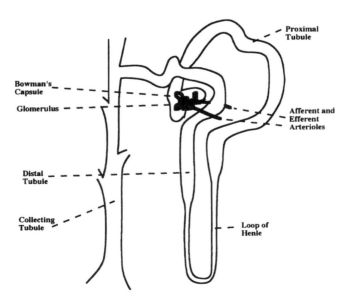

Figure 1. Diagram of a nephron.

Approximately 20% of the plasma entering the glomeruli is filtered through the Bowman's capsule, while the rest leaves the kidney circulation via the efferent arteriole. Glomerular filtration occurs by bulk flow due to the pressure differences within the glomerular capillary–Bowman's capsule network and large pores between the cells that allow for the free flow of solutes. The glomerular filtration rate, or the volume of plasma filtered per unit time by the glomeruli, is 120 mL/min for an adult (*1*), and the urine flow rate normally averages 1 mL/min. While passing through the nephron tubule, the ultrafiltrate is acidified (average pH = 6). More than 99% of the fluid, along with most nutrients, is reabsorbed into the extracellular fluid of the kidneys. In addition, metabolic waste products (creatinine and urate) and foreign substances are eliminated by active secretion into the tubular fluid as it flows through the tubule system.

The renal tubule consists of a single layer of epithelial cells that rests on a basement membrane of glycoproteins and mucopolysaccharides. The tubule can be divided into segments on the basis of structure and function. The segment beginning at the Bowman's capsule is the proximal tubule (which can be further subdivided into three limbs, histologically, as S1, S2, and S3). Next are the descending and ascending sections of the loop of Henle, and finally the convoluted and straight segments of the distal tubule. The tubular fluid leaving the distal tubule is pooled from many nephrons into collecting tubules and finally arrives as urine in the bladder by way of the ureter. Each segment of the nephron tubule functions to specifically change the overall composition of the tubular fluid as it flows past. For example, the proximal tube reabsorbs more than 65% of the sodium ions and water of the glomerular ultrafiltrate (*1*). Most, if not all, nutrients such as glucose and amino acids are reabsorbed (*1*), while organic cations and organic anions can be actively secreted at the proximal tubule. In addition, the ultrastructural features of each segment are different. For example, the epithelial cells forming the proximal tubule have many more mitochondria than the cells forming the loop of Henle, suggesting the high energy requirements for the cells that perform active transport.

The epithelial cells of the nephron tubules transport substances into and out of the tubular fluid, causing the net movement of essential ions, nutrients, and potential toxins. For this net vectorial transport to occur, the cells must have polarized membranes. The transport mechanism is different on each membrane surface. The luminal membrane (also referred to as the mucosal, brush border, or apical membrane) faces the lumen of the tubule and is in direct contact with the ultrafiltrate. The luminal membrane has many fingerlike projections, or microvilli, which increase its surface area dramatically and thus facilitate the transfer of large quantities of substances (Figure 2). The antiluminal membrane (also referred to as the basolateral, peritubular, or serosal membrane) faces the extracellular fluid (or blood). The antiluminal membrane lies on the basement membrane (a structural support of glycoproteins and collagen), which is not a significant barrier to the movement of substances. The luminal and antiluminal membranes are separated physically by the zona occludens (tight junction pro-

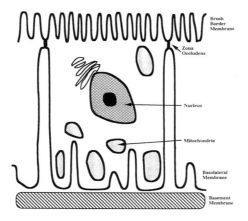

Figure 2. Epithelial cell, showing the polarized membranes and cellular components.

teins), which prevents the movement and mixing of membrane proteins between them (*1*).

Vectorial reabsorption or secretion of a substance requires three events: transport of the substance across the entry membrane, flow across the interior of the epithelial cell, and transport across the exit membrane. In the kidney, net movement of a substance from the antiluminal membrane (blood side) through the cell and out the luminal membrane (ultrafiltrate side) is secretion. The reverse movement, from the luminal to the antiluminal, is reabsorption. For vectorial reabsorption or secretion, the transporters on the membranes must be different yet function in concert to force the solute in the proper direction. For example, it would not be advantageous for both the luminal and the antiluminal membranes to have transporters to pump glucose into the cell; only cellular accumulation would result. Finally, cellular metabolism may complicate transport studies, since cellular metabolism of the substrate within the epithelial cells can appear to be stereoselective transport.

Renal Filtration, Secretion, and Reabsorption

Renal excretion can constitute a considerable fraction of the overall drug elimination from the body. For certain drugs, such as penicillin or ampicillin, elimination is conducted almost entirely by the kidneys (*2*). The renal excretion rate is determined by the net combination of three complex processes; filtration at the glomeruli, tubular reabsorption, and tubular secretion. Both secretion and reabsorption can occur throughout the nephron tubule. The renal excretion rate is defined as follows:

renal excretion rate = filtration rate − reabsorption rate + secretion rate

The filtration rate is equal to GFR · fu · C, where GFR is the glomerular filtration rate, C is the plasma concentration of the drug, and fu is the unbound fraction (the portion of the drug in plasma that is not bound to plasma proteins) (*3*). Because the nephron does not allow plasma proteins to pass into the ultrafiltrate, bound drug is not available for filtration and therefore cannot be excreted by the kidneys. Some drugs, such as chlorpromazine and diazepam (*2*), are extensively bound to plasma proteins (i.e., fraction bound is >99%, or fu < 0.01) and so are mostly unavailable for filtration by the kidneys. GFR can be experimentally determined by measuring the renal clearance of inulin, an exogenous polysaccharide that is not bound to plasma proteins (fu = 1) and is eliminated only by filtration in the glomerulus (*3*). Drugs that bind stereoselectively to plasma proteins will be eliminated stereoselectively by filtration.

The reabsorption rate is the rate at which the drug is reabsorbed into the bloodstream through the nephron tubule from the ultrafiltrate in the tubule lumen. Therefore, reabsorption of a compound decreases its overall excretion rate.

The secretion rate is the rate at which a drug is secreted from the blood (or extracellular fluid) through the renal tubule into the ultrafiltrate. Secretion increases the overall excretion rate of the drug by the kidney.

The renal clearance of a drug (the volume of plasma cleared of drug per unit time) is defined as

renal clearance = renal excretion rate ÷ plasma concentration of drug
renal clearance = (filtration rate − reabsorption rate + secretion rate) ÷ plasma concentration of drug

By comparing the renal clearance of a drug with its clearance by filtration (GFR · fu), we can determine whether it undergoes net reabsorption or secretion by the kidney. If the renal clearance is lower than GFR · fu, net reabsorption has occurred (simultaneous secretion is possible, but reabsorption must predominate). Conversely, if the renal clearance is higher than GFR · fu, the drug must be actively secreted by the kidney (simultaneous reabsorption is possible, but secretion must predominate). The absence of many organic nutrients, such as glucose and amino acids, from the urine suggests active reabsorption by the kidney. In addition, a drug may be handled differently at various segments of the nephron. It could be actively secreted in the proximal tubule (increased renal clearance) and passively reabsorbed in the distal tubule (decreased renal clearance).

The renal clearances of two enantiomers will be different if any of these processes are stereoselective (*4–6*). The filtration clearance (GFR · fu) will differ between enantiomers if a compound is bound stereoselectively to plasma proteins (4–6) and therefore filtered stereoselectively. The rates of reabsorption or secretion of enantiomers can be different if transport proteins at either the basolateral or the brush border membrane show stereoselective transport kinetics. Stereoselective binding to plasma proteins will also affect reabsorptive or secretory transport in the renal tubule. In addition, stereoselective cellular metabolism (including chiral inversion of enantiomers) within the real cells cannot be ignored as a potential mechanism responsible for differing enantiomeric renal clearances. Examples of stereoselective filtration, secretion, and reabsorption are discussed later.

Tubular secretion and reabsorption involve transport of molecules across the brush border and basolateral membranes of the renal tubule. The transport of molecules across a membrane can occur by four mechanisms: simple diffusion, facilitated diffusion, primary active transport, and secondary active transport (7, 8). Because enantiomers have identical physicochemical properties, their permeability characteristics in biomembranes are identical. Simple diffusion is not mediated by any cellular event (e.g., an interaction with membrane proteins), and the diffusion of one species is not influenced by the presence of another. Therefore, if a chiral substance is transported only by simple diffusion, the transport rate of the enantiomers will be equal and no stereoselective transport will occur.

In facilitated diffusion, a substance is transported faster than predicted by its permeability coefficient. The net flux is in the direction of the electrochemical gradient, and no energy is expended directly by the cell for transport. The substance binds to integral membrane proteins (transporters) that "facilitate" or enhance the transport across the membrane. This allows the substance to flux across the membrane at an accelerated rate. Because the number of transport proteins within a cell is limited, transport by facilitated diffusion is saturable (9).

Active transport (either primary or secondary) is similar to facilitated diffusion except that the substance is transported against its concentration gradient. This requires energy expenditure by the cell. In *primary* active transport, the movement of the substance against its concentration gradient is directly coupled to the breakdown of ATP or to the absorption of light energy (7, 8). Primary active transport systems are normally called pumps—for example, the $Na^+–K^+$ ATPase pump and the H^+ ATPase pump.

In *secondary* active transport, the flow of one substance against its concentration gradient is directly coupled to the flow of another down its concentration gradient. The substance moving down its concentration gradient supplies the energy for the other substance to be transported against its concentration gradient. When both substances are transported across the membrane in the same direction, the transporter is termed a *cotransporter*. When the substances are transported from opposite membrane faces, the transporter is termed an *antiporter*.

For a substrate that is transported by facilitated diffusion or active transport, the relationship between the transport rate and concentration can be described by a rectangular hyperbola and can be fit to the following Michaelis–Menten equation:

$$V = V_{max}S/(K_t + S)$$

where V is the flux, V_{max} is the maximal transport rate observed for the substrate at a given temperature, S is the substrate concentration, and K_t is the concentration of the substrate that produces half the maximal rate. V_{max} depends on the number of transporters and the transporter's turnover number (its efficiency in making a complete cycle—substrate binding, transport through the membrane, substrate release, and return to the original protein conformation). The substrate affinity ($1/K_t$) is a measure of how well the substrate binds to and interacts with the transporter (7, 8). Therefore, if the affinity for the transporter is high, K_t will be low. Because the substrate must bind to the transporter, structurally similar substances can inhibit binding and therefore transport (Figure 3) (10). The binding pocket (active site) of the transporter is a three-dimensional space, so the stereochemistry of the substrate is important. The transport rates of various stereoisomers have been shown to be significantly different (Figure 4) (11).

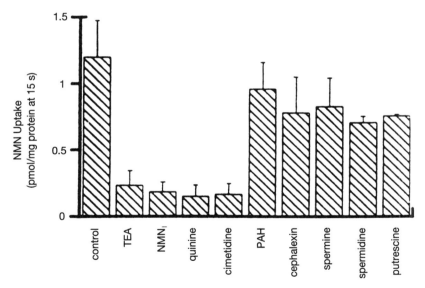

Figure 3. Effect of various compounds on the uptake of N-*methylnicotinamide (NMN) in human brush border membrane vesicles in the absence (control) and presence of 10 mM tetraethylammonium (TEA), 10 mM NMN, 1 mM quinine, 1 mM cimetidine, 10 mM* p-*aminohippuric acid (PAH), 10 mM cephalexin, 10 mM spermine, 10 mM spermidine, and 10 mM putrescine. Only TEA, NMN, quinine, and cimetidine produced significant inhibition. (Reproduced with permission from reference 10. Copyright 1991.)*

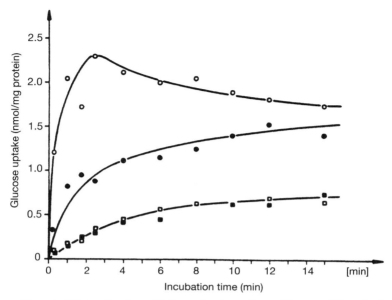

Figure 4. Uptake of D-glucose (circles) and L-glucose (squares) in rat renal brush border membrane vesicles in the presence (open) or absence (closed) of a sodium gradient. (Reproduced with permission from reference 11. Copyright 1975.)

For enantiomeric recognition to occur, the substrate must bind to the protein at three or more sites (*12*). The proximity of the chiral center to the critical structural feature of the substrate that promotes binding will affect the likelihood of enantiomeric recognition. Although substrate enantiomers would not affect the number of transporters, V_{max} could be affected stereoselectively if the enantiomers cause different translocation rates of the transporter. The rate constant K_t, which is a function of the rates of formation and breakdown of the substrate–transporter complex and the rate of membrane translocation of the complex itself, can be severely affected by stereoisomeric structural changes. The formation rate and breakdown rate of the substrate–protein complex may differ between enantiomers. In addition, the protein's ability to change its conformation may be enantiomerically different after substrate binding, thereby affecting the translocation rate of the transporter.

Stereoselective Plasma Protein Binding

In Situ Studies

Many drugs bind stereoselectively to plasma proteins, and this binding may result in stereoselective renal clearance of the drugs. For a drug that is filtered in the glomerulus only, the drug in plasma not bound to proteins is filtered. The major binding proteins for drugs—albumin and α_1-acid glycoprotein—are too large to undergo filtration. Thus, the renal clearance (CL_R) can be described by the following equation:

$$CL_R = fu \cdot GFR$$

The relative renal clearance of enantiomers that are eliminated only by filtration will depend on the plasma protein binding and can be predicted from the ratio of fu values for the two enantiomers.

For drugs that are actively secreted in the kidney, the relationship between plasma protein binding and renal clearance is more complex. If the drug is avidly secreted, its renal clearance will approach, and be limited by, renal blood flow (*4*). In fact, the renal clearance of highly secreted drugs such as *p*-aminohippurate (PAH) is used as a measure of renal blood flow. Plasma protein binding will not influence the renal clearance of a drug with a renal clearance approaching renal blood flow. In contrast, if a drug is secreted less avidly in the renal tubule, plasma protein binding does affect renal clearance. In an elegant study using isolated, perfused rat kidneys in situ, Hall and Rowland described the effect of plasma protein binding on the renal clearance of furosemide, a drug that is actively secreted in the proximal tubule and is not reabsorbed (*13*). Thus, the renal clearance of furosemide could be defined as follows:

$$CL_R = fu \cdot GFR + CL_{sec}$$

where CL_{sec} represents the secretory clearance. By varying the concentration of albumin while maintaining a constant osmolarity, Hall and Rowland were able to vary the binding of furosemide 28-fold. They observed a linear relationship between CL_R and fu. The data suggest that $CL_{sec} = fu \cdot CL_{intsec}$, in which CL_{intsec} represents an intrinsic clearance of secretion given by V_{max}/K_t (for a drug that is secreted by a single transporter). Thus, for a drug that is filtered and secreted,

$$CL_R = fu \cdot GFR + fu \cdot CL_{intsec} = fu \cdot (GFR + CL_{intsec})$$

When binding to plasma proteins is stereoselective, renal clearance will be stereoselective, and a plot of renal clearance versus fu for each enantiomer will be linear, with a zero intercept and a slope of $GFR + CL_{intsec}$.

Using a similar experimental approach, Hall and Rowland studied the influence of plasma protein binding on the reabsorption of digitoxin, a drug that is not secreted but is subject to extensive tubular reabsorption (*14*). They observed that the relationship between CL_R and fu could be described as follows:

$$CL_R = fu \cdot GFR - F(fu \cdot GFR)$$

where *F* is the fraction of drug entering the tubule lumen that is subject to passive reabsorption. As for a drug that is actively secreted, the renal clearance of a drug that is reabsorbed will depend on its plasma protein binding, and stereoselective binding to plasma proteins will result in a stereoselective renal clearance.

Clinical Studies

The finding from in situ organ perfusion studies that the renal clearance of a drug is dependent on plasma protein binding has been demonstrated clinically for disopyramide, an antiarrhythmic drug marketed as a racemic mixture. Both enantiomers of disopyramide exhibit concentration-dependent binding to plasma proteins in the therapeutic plasma concentration range (15–17). This stereoselective binding is due to stereoselective interactions with α_1-acid glycoprotein (15). The (S)-(+)-enantiomer has a lower fu at a given plasma concentration and a higher affinity for α_1-acid glycoprotein than the (R)-(−)-enantiomer. In humans, disopyramide is cleared in the kidneys by filtration, secretion, and possibly reabsorption. Giacomini et al. studied the effect of plasma protein binding on the renal clearance of the enantiomers of disopyramide in normal volunteers who received a deuterated pseudoracemic mixture of the drug intravenously (16). A plot of CL_R versus fu for each enantiomer in each subject resulted in a straight line with a zero intercept, suggesting that both secretion and filtration depend on plasma protein binding. As predicted, the renal clearance of (R)-(−)-disopyramide (65 mL/min) was significantly greater than that of (S)-(+)-disopyramide (37 mL/min), consistent with the higher fu of the (R)-(−)-enantiomer. These data in humans are consistent with the predictions of Hall and Rowland in the isolated perfused organ system and highlight the importance of plasma protein binding as a determinant of the stereoselective renal clearance of drugs.

Recently, the stereoselective disposition of carbenicillin, an organic anion marketed clinically as a racemic mixture, was studied in healthy volunteers (18). The data demonstrated that the fu of (R)-carbenicillin was significantly higher than that of (S)-carbenicillin; the ratio averaged 1.4. The ratio of renal clearances was 1.2, which was reasonably close to the fu ratio. The investigators interpreted the differences between the two ratios as a true stereoselective renal secretion, with the (S)-enantiomer being secreted preferentially in the renal tubule; that is, CL_{intsec} differed between the two enantiomers. Interestingly, following coadministration of probenecid, which inhibits the tubular secretion of organic anions, the ratio of renal clearances was 1.4, reflecting the stereoselective plasma protein binding of the two enantiomers.

Stereoselective Renal Secretion

Most drugs secreted by the kidney are handled by two transport systems: one for organic cations and one for organic anions. The stereoselectivity of each system is detailed in this section.

Organic Cations

Active tubular secretion of organic cations was first reported in the 1940s (*19*). Most active organic cation secretion occurs in the first segment (S1) of the proximal tubule (*20*). Initially, organic cations are transported into the proximal tubule cell across the basolateral membrane by a potential-sensitive facilitated diffusional system. Organic cations accumulate within the cell because of the favorable electrochemical gradient (*21*). The cations are transported into the lumen of the tubule across the brush border membrane through a secondary active transport mechanism driven by the counterflow of protons. This system is referred to as the *organic cation–proton antiporter* (*10, 21–27*). The $Na^+–H^+$ antiporter, also in the brush border membrane, may be responsible for the proton gradient between the lumen and the cell. The Na^+ gradient across the apical membrane is produced by the $Na^+–K^+$ ATPase system in the basolateral membrane (*21*). The organic cation–proton antiporter has been identified and examined in numerous species (rat, rabbit, and dog) (*21–27*), including humans, using brush border membrane vesicles (*10*). When an outward-directed proton gradient is present, the accumulation of organic cations within vesicles is enhanced transiently, producing an "overshoot" above the equilibrium concentration. This overshoot phenomenon reflects the transient accumulation, against a concentration gradient, of organic cations in response to the proton gradient and occurs because the downhill transport of protons is coupled directly to the uphill transport of the organic cations (Figure 5) (*10*). A schematic of the overall organic cation secretion is detailed in Figure 6. This system transports a wide range of structurally diverse substrates, but the basic nitrogen atom is a critical determinant of transport (*19*). There is a loose correlation between the transport affinity of the substrate and the lipophilicity of the basic nitrogen atom (*28*).

Many clinically important organic cations, such as cimetidine, procainamide, quinine, quinidine, triamterene, and pindolol, are actively secreted (renal clearance > GFR · fu) (*29–36*). Potential stereoselective renal secretion has been reported recently for several drugs, including pindolol (*37*) and the diastereomers quinine and quinidine (*38*). In addition, various other organic cations (such as verapamil, metoprolol, terbutaline, atenolol, and hydroxychloroquine), which are not extensively eliminated by the kidney, appear to show either stereoselective renal clearance or stereoselective metabolism (*39–43*). For stereoselective active secretion of organic cations, at least one step (facilitated diffusion across the basolateral membrane, drug metabolism within the renal cell, or secondary active transport across the brush border membrane) must be stereoselective. We describe the two most studied examples: the enantiomers of pindolol and the diastereomers quinine and quinidine (Figure 7).

Pindolol

Pindolol, a β-adrenergic blocking agent, is largely (>50%) eliminated unchanged in the urine and is actively secreted by the organic cation transport system within

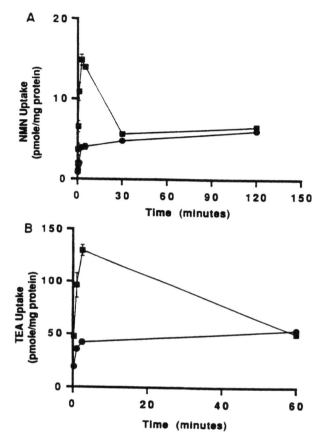

Figure 5. NMN (A) or TEA (B) uptake into human brush border membrane vesicles in the presence of an outward-directed proton gradient (squares), where $pH_{in} = 6.0$ and $pH_{out} = 7.4$, and in the absence of a proton gradient (circles), where $pH_{in} = 7.4$ and $pH_{out} = 7.4$. An overshoot phenomenon is evident under proton-gradient conditions. (Reproduced with permission from reference 10. Copyright 1991.)

the kidney (unbound renal clearance is approximately 5 times GFR) (*44*). In addition, several recent clinical studies have suggested that pindolol is stereoselectively eliminated (*37, 45*). After oral administration of 20 mg racemic pindolol to humans, the renal clearance of (*S*)-(−)-pindolol was 240 ± 55 mL/min, significantly higher than that of (*R*)-(+)-pindolol (200 ± 51 mL/min). Likewise, the net clearance by secretion of (*S*)-(−)-pindolol (196 ± 47 mL/min) was 30% higher than that for (*R*)-(+)-pindolol (157 ± 48 mL/min) (Figure 8) (*37*). The unbound fraction was not significantly different between the enantiomers, suggesting that stereoselective renal metabolism or secretion must account for the enan-

Lumen

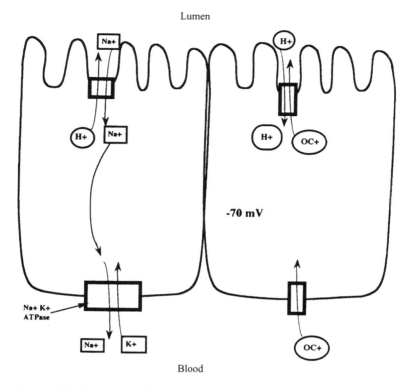

Figure 6. Model of the transepithelial flux of organic cations (OC⁺) in the proximal tubule.

tiomeric differences in clearance, with the (S)-(–)-enantiomer being preferentially cleared.

Somogyi et al. investigated the effect of coadministration of cimetidine on the renal clearance of the enantiomers of pindolol in humans (45). The study confirmed the stereoselective renal clearance of pindolol (renal clearance was 222 ± 66 mL/min for the (S)-(–)-enantiomer, versus 170 ± 55 mL/min for the (R)-(+)-enantiomer). Once again, these data are consistent with a stereoselective renal secretion mechanism, with the (S)-enantiomer being preferentially cleared. Cimetidine coadministration significantly reduced the renal clearance of both enantiomers (renal clearance was 155 ± 38 mL/min for (S)-(–)-pindolol and 104 ± 18 mL/min for (R)-(+)-pindolol), suggesting renal inhibition at the organic cation secretory system. Cimetidine coadministration reduced the renal clearance more for (R)-(+)-pindolol than for (S)-(–)-pindolol (34% versus 26%) (Figure 9) (45). These studies suggest that pindolol is stereoselectively secreted in the kidney and that cimetidine can have a stereoselective inhibitory effect on the renal tubular secretion of pindolol.

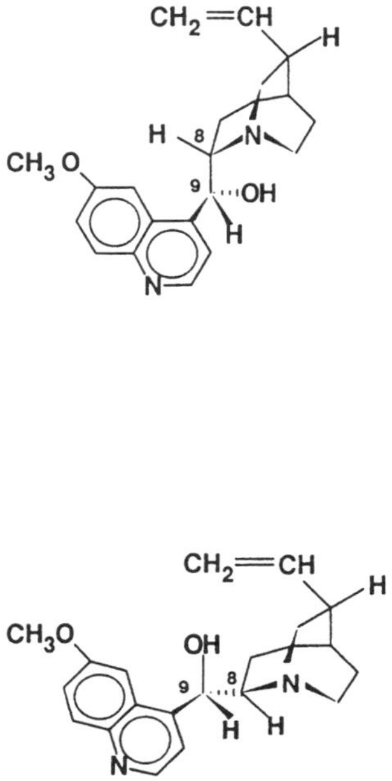

Figure 7. Chemical structures of quinine and quinidine.

We hypothesized that the stereoselective interaction might be occurring at the brush border membrane, the site of active organic cation secretion. The use of isolated plasma membranes eliminates the metabolic machinery of the renal cell (46) and allows us to directly examine the interactions of enantiomers with the organic cation–proton antiporter at the brush border membrane. Our reasoning was that if the enantiomers of pindolol interacted stereoselectively with the organic cation–proton antiporter, their IC_{50} values should reflect this interaction. Therefore, we examined the individual inhibitory effects of (R)-(+)- and (S)-(–)-pindolol on the uptake of N-methylnicotinamide (NMN, an unmetabolizable model organic cation) in rabbit renal brush border membrane vesicles (47).

Both enantiomers of pindolol significantly inhibited NMN transport under proton-driven conditions (Figure 10). The IC_{50} values were 140 ± 20 µM for (R)-(+)-pindolol and 120 ± 20 µM for (S)-(–)-pindolol. Although the trend matches the previously reported clinical data (with the (S)-(–)-enantiomer having a slightly higher affinity), there was no significant difference in the IC_{50} values for the enantiomers of pindolol, suggesting no stereoselective binding with the organic cation–proton antiporter in the rabbit. In a similar study, Gross and Somogyi examined the effect of 90 µM of either enantiomer of pindolol on the proton-driven transport of tetraethylammonium (TEA, an unmetabolized organic cation) across rat brush border membrane vesicles. They also detected an inhibitory effect that was not significantly different between (R)-(+)-pindolol (% of control

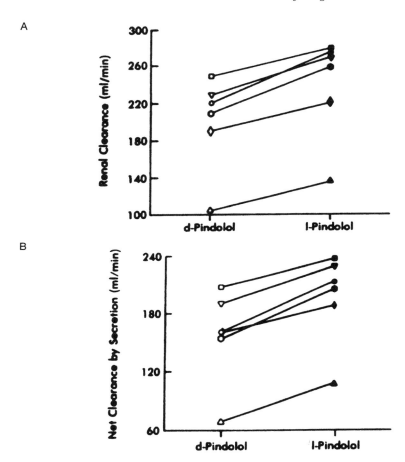

Figure 8.   Renal clearance (A) and net clearance by secretion (B) of (R)-(+)- and (S)-(−)-pindolol in six subjects. (Reproduced with permission from reference 37. Copyright 1985.)

TEA uptake = 52.3 ± 7.3) and (*S*)-(−)-pindolol (% of control TEA uptake = 60.5 ± 14.9) (*28*). Both of these studies suggest that the organic cation–proton antiporter in the brush border membrane may not be responsible for the stereoselective renal clearance of pindolol noted clinically.

The results of one other study, which employed cultured renal epithelial cells, conflict with the brush border membrane studies (*48*). Opossum kidney (OK) cells are a continuous renal epithelial cell line that expresses an organic cation transporter within the brush border membrane with characteristics closely matching those in other mammalian species (*49*). In addition, this cell line expresses sodium–glucose and sodium–phosphate cotransporters (*50, 51*). Using this cell line, we examined the individual inhibitory effects of (*R*)-(+)- and (*S*)-(−)-pindolol on the uptake of TEA across the brush border membrane. We deter-

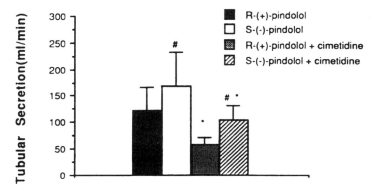

Figure 9. Mean net renal clearance by tubular secretion for (R)-(+)- *and* (S)-(−)-*pindolol when racemic pindolol was given alone and when coadministered with cimetidine in eight healthy subjects.* *p < 0.001, *enantiomer alone versus enantiomer plus cimetidine.* #p < 0.001, (R)-(+)- *versus* (S)-(−)-*pindolol. (Reproduced with permission from reference 45. Copyright 1992.)*

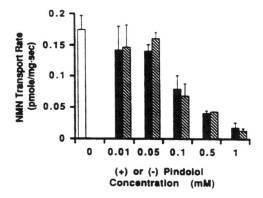

Figure 10. Initial uptake of NMN in rabbit brush border membrane vesicles in the presence of various concentrations of pindolol. The open bar represents the control, the black bars represent (R)-(+)-*pindolol, and the striped bars represent* (S)-(−)-*pindolol. (Reproduced with permission from reference 47. Copyright 1991.)*

mined a statistically significant difference between the IC_{50} values: 30 ± 4 μM for (R)-(+)-pindolol and 23 ± 4 μM for (S)-(−)-pindolol ($p < 0.02$) (Figure 11) (48). These data suggest a stereoselective interaction of the enantiomers of pindolol with the organic cation–proton antiporter.

All three in vitro studies correlate with the clinical data (that of the (S)-(−)-pindolol being more avidly cleared), but the enantiomeric differences were significant only in the whole-cell study. This may be explained somewhat by the increased affinity of the OK cell transporter for both pindolol (5-fold) (48) and

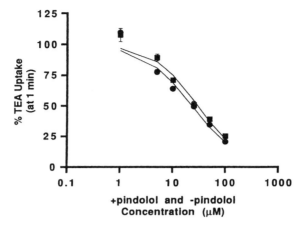

Figure 11. Inhibition of TEA uptake in OK cells by (R)-(+)-pindolol (squares) and (S)-(–)-pindolol (circles), expressed as the percentage of maximum TEA uptake in the absence of inhibitor versus the logarithm of the inhibitor concentration. The curves were generated by computer fit of the data. (Reproduced with permission from reference 48. Copyright 1993.)

TEA (*49*) as compared with isolated brush border membrane vesicles. In addition, different species and techniques were employed. It has been suggested that the leakiness of the vesicles and their variability between preparations (*52, 53*) lead to large standard errors and thereby lessen the probability of finding a significant difference between enantiomers. With OK cells, the cell membranes are probably not damaged during the experiments, and a continuous cell line is more likely to give a homogeneous population. However, metabolism of pindolol within the OK cells was not ruled out, and metabolism could interfere with the interpretation of this study. To fully understand the mechanisms responsible for the stereoselective renal secretion of pindolol, further work is required. Transport at the basolateral membrane and drug metabolism within the renal cells need to be addressed. In addition, the stereoselective transport of pindolol itself (determination of K_t) must be examined.

Quinine and Quinidine

The unbound renal clearances of the diastereomers quinine and quinidine are both greater than inulin clearance, indicating net renal tubular secretion of these compounds. In 1986, Notterman et al. determined that the human renal clearance of quinidine was 4 times that of quinine (Figure 12) (*38*). A more recent study confirmed these results (*54*). Both quinine and quinidine are highly protein-bound (90%), but no stereoselective differences in plasma protein binding were observed. The diastereomers have similar physical properties, such as octanol–water partition coefficients and p*K*a values, so their renal filtration and

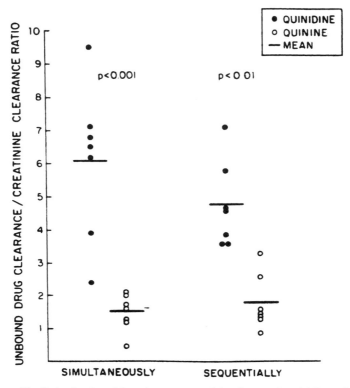

Figure 12. Ratio of unbound drug clearance to creatinine clearance for quinidine (solid circles) and quinine (open circles) in healthy adult subjects receiving these drugs simultaneously and sequentially on different days. (Reproduced with permission from reference 38. Copyright 1986.)

passive reabsorption should be similar (*38*). Stereoselective active renal secretion may be responsible for the observed differences in renal clearances. Once again, however, the potential concomitant stereoselective renal metabolism of these drugs was not addressed in these studies.

Eight in vitro studies examining this potential stereoselective transport mechanism have been reported. Three studies have used brush border membrane vesicles, while the rest have relied on cortical slices, tubule suspensions, and cultured renal cells. Combined, the results are fairly confusing and are detailed in Table I. Using rabbit brush border membrane vesicles, we examined the effect of quinine and quinidine on the uptake of NMN (*47*). Both compounds significantly inhibited the NMN transport, suggesting an interaction at the organic cation–proton antiporter (Figure 13). However, the IC_{50} values were not significantly different (2.5 ± 0.8 μM for quinine and 2.4 ± 0.5 μM for quinidine), suggesting no stereoselective interaction with the organic cation–proton antiporter.

Table I. Stereoselective Differences in the K_I or IC_{50} of Quinine and Quinidine with In Vitro Studies (Mean ± SE)

Authors	Ref.	Technique	Quinine (μM)	Quinidine (μM)
Ott et al.	47	Rabbit BBMV	2.5 ± 0.8	2.4 ± 0.5
Gross and Somogyi	28	Rat BBMV	1.3 ± 0.2[a]	1.4 ± 0.4[a]
Bendayan et al.	55	Dog BBMV	7.0	0.7
Wong et al.	57	Human slices	261 ± 44	586 ± 68
Wong et al.	56	Rat slices	288 ± 21	861 ± 79
Wong et al.	56	Rat tubules	32 ± 3	84 ± 11
Wong et al.	58	Rat tubules	42 ± 7	112 ± 8
Ott and Giacomini	48	OK cells	17 ± 2[a]	51 ± 13[a]

[a]Mean ± SD
BBMV is brush border membrane vesicles

Two additional studies have used this same approach with rat and dog renal brush border membrane vesicles (*28, 55*). In the rat, Gross and Somogyi found that quinine inhibited TEA uptake with an IC_{50} value of 1.3 ± 0.2 μM, while quinidine had an IC_{50} value of 1.4 ± 0.4 μM, once again suggesting no stereoselective interaction at the brush border membrane (*28*). In the dog, however, Bendayan et al. found that quinine inhibited the NMN uptake with a K_I of 7.0 μM, while the K_I for quinidine was 0.7 μM (*55*). These data suggest that quinidine is the more potent inhibitor and that a potential stereoselective inhibition at the organic cation–proton antiporter occurs in the dog.

The results from the other five reports conflict with the previously detailed

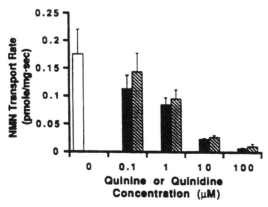

Figure 13. Initial uptake of NMN in rabbit brush border membrane vesicles in the presence of various concentrations of quinine or quinidine. The open bar represents the control, the black bars represent quinine, and the shaded bars represent quinidine. (Reproduced with permission from reference 47. Copyright 1991.)

studies. Wong et al. examined the effect of quinine and quinidine on the uptake of the organic cation amantadine (56, 57) using both human and rat renal cortical slices. In both instances, quinine (human $K_I = 261 \pm 44$ μM; rat $K_I = 288 \pm 21$ μM) was a significantly more potent inhibitor of amantadine than quinidine was (human $K_I = 586 \pm 68$ μM; rat $K_I = 861 \pm 79$ μM). Likewise, in two studies of rat proximal tubule suspensions, quinine ($K_I = 32 \pm 3$ μM; $IC_{50} = 42 \pm 7$ μM) was significantly more potent than quinidine ($K_I = 84 \pm 11$ μM; $IC_{50} = 112 \pm 8$ μM) (56, 58). With the OK cell line, we also determined that quinine ($IC_{50} = 17 \pm 2$ μM) was significantly more potent than quinidine ($IC_{50} = 51 \pm 13$ μM) at inhibiting TEA uptake across the brush border membrane (Figure 14). Combined, these results are in direct conflict with the clinical and membrane vesicle studies.

Clearly the results to date using in vitro methods are conflicting. Of the eight studies detailed, those using isolated plasma membrane vesicles showed either that there was no stereoselective difference between quinine and quinidine or that quinidine was the more potent inhibitor. This last result, in the dog, closely matches the clinical data. The five studies using cortical slices, tubule suspensions, or cultured cells all produced similar findings on stereoselectivity of the organic cation transport system for quinine and quinidine. However, these studies all suggest that quinine is the more potent inhibitor. This confusing data may be a result of species differences in transport mechanisms or differences in experimental methods. In addition, the stereoselective interactions of quinine and quinidine with the organic cation transport system may be quite complex and in-

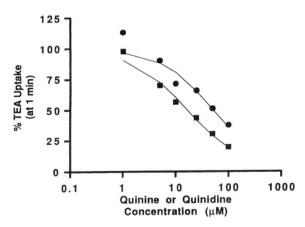

Figure 14. Inhibition of TEA uptake in OK cells by quinidine (circles) and quinine (squares), expressed as the percentage of maximum TEA uptake in the absence of inhibitor versus the logarithm of the inhibitor concentration. The curves were generated by computer fit of the data. (Reproduced with permission from reference 48. Copyright 1993.)

volve stereoselective interactions at both the basolateral and brush border membranes. Renal cellular metabolism may help explain the conflicting experimental data. As with pindolol, further studies are required to determine potential stereoselective interactions for this system. The studies to date have relied on inhibition potencies (K_I or IC_{50}) of the diastereomers. Studies directly examining the transport of the enantiomers are needed, which may reveal differences in binding and translocation rates, and the basolateral membrane transport must be investigated to fully understand the active renal secretion of quinine and quinidine.

Finally, one recent report has detailed the stereoselective transport at the brush border membrane of other organic cations—namely, a series of chiral sympathomimetic amines (*28*). This study strongly suggests that the proximity of the chiral center to the basic nitrogen atom (a requirement for organic cation secretion) is indicative of stereoselective interactions at the organic cation–proton antiporter (Table II). Further work on substrate series, with varying distances between the chiral center and the positive charged group, must be conducted at both membrane faces.

Organic Anions

The active secretion of organic anions has been extensively examined in studies using PAH. The active step occurs in the basolateral membrane, but the exact mechanism is still controversial. Some studies have suggested that the active transport of organic anions involves a PAH–dicarboxylate exchange mechanism, which is driven by a sodium gradient through a Na^+–dicarboxylate cotransporter (*59–62*). Transport at the brush border membrane is thought to occur by a

Table II. Influence of Stereoisomeric Basic Drugs (90 μM) on the Uptake of TEA in Rat Renal Brush Border Membrane Vesicles (Mean ± SD)

Substrate	% of Control TEA Uptake
(*R*)-(+)-Bupivacaine	66.1 ± 9.7
(*S*)-(–)-Bupivacaine	59.1 ± 13.9
(*R*)-(+)-Disopyramide	43.7 ± 4.3
(*S*)-(–)-Disopyramide	43.5 ± 16.8
(1*S*,2*R*)-(+)-Phenylpropanolamine	46.4 ± 13.2[a]
(1*R*,2*S*)-(–)-Phenylpropanolamine	64.7 ± 20.1[a]
(1*S*,2*S*)-(+)-Norpseudoephedrine	68.0 ± 12.9[a]
(1*R*,2*R*)-(–)-Norpseudoephedrine	49.5 ± 9.8[a]
(1*S*,2*R*)-(+)-Ephedrine	47.1 ± 16.0[a]
(1*R*,2*S*)-(–)-Ephedrine	67.7 ± 16.0[a]
(1*S*,2*S*)-(–)-Pseudoephedrine	70.3 ± 15.3
(1*R*,2*R*)-(+)-Pseudoephedrine	77.2 ± 14.4

[a]Significantly different ($p < 0.05$) between pairs of enantiomers.
SOURCE: Data derived from reference 28.

downhill facilitated step involving either an anion exchanger or a potential-sensitive transport system (Figure 15) (*59–62*). Probenecid has become the standard inhibitor of the organic anion transporter (*63*). Organic anions transported by this system include salicylates, penicillins, and cephalosporins (*63–65*). Much work has been conducted on the system's structural specificity but little on its stereoselectivity, probably because PAH and probenecid are achiral. One study has demonstrated a higher clearance for (*R*)-(–)-ibuprofen than for (*S*)-(+)-ibuprofen in the isolated perfused rat kidney model, suggesting possible stereoselective tubular transport (*66*). According to another study, however, ibuprofen may be cleared stereoselectively in the kidney through stereoselective formation of the glucuronide rather than stereoselective transport (*66*).

Several recent reports suggest stereoselective transport of the organic anion 5-dimethylsulfamoyl-6,7-dichloro-2,3-dihydrobenzofuran-2-carboxylic acid (DBCA) by the organic anion transport system in the proximal tubule. DBCA, or S-8666 (Figure 16), is a novel uricosuric antihypertensive diuretic whose (*S*)-(–)-enantiomer is responsible for the diuretic and antihypertensive activity by promoting reduced Cl^- and Na^+ transport in the thick ascending segment of the loop of Henle (*67*). As with various diuretics, DBCA is actively secreted by the proximal tubule organic anion transport system in the rat (*68*), rabbit, dog (*67*), and monkey (*69*). In the monkey, the unbound drug clearance was 14 to 29 times the creatinine clearance (*69, 70*). In addition, the active secretion of DBCA is inhibited by probenecid (*67*).

DBCA appears to be handled stereoselectively in protein binding, hepatic extraction, metabolism (*68, 71*), and active renal secretion at the organic anion transport system (*69, 70, 72*). In the monkey (where 78% of the drug is excreted unchanged in the urine), probenecid preferentially inhibited the (*S*)-(–)-enan-

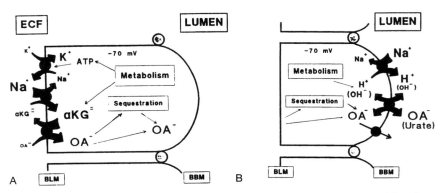

Figure 15. Schematic diagram of antiluminal (A) and luminal (B) events that participate in organic anion (OA⁻) secretion and reabsorption. ECF, extracellular fluid; BLM, basolateral membrane; BBM, brush border membrane; α-KG, α-ketoglutarate. (Reproduced with permission from reference 62. Copyright 1993.)

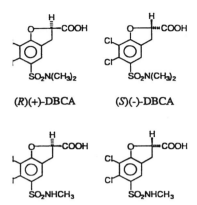

Figure 16. Chemical structures of DBCA and its demethylated metabolite. (Reproduced with permission from reference 70. Copyright 1993.)

tiomer by 53% and the (R)-$(+)$-enantiomer by 14% (*69, 70*), and the unbound renal clearance of the (S)-$(-)$-enantiomer was 1.3 to 1.6 times that of the (R)-$(+)$-enantiomer (*69*). Excluding possible effects of probenecid on protein binding and metabolism of DBCA, these data suggest stereoselective renal transport at the organic anion transport system.

When racemic DBCA was perfused in a rat kidney, however, Higaki et al. determined that the unbound renal clearance of the (R)-$(+)$-enantiomer was 3.3 times that of the (S)-$(-)$-enantiomer (*72*). When the enantiomers of DBCA were perfused individually, the unbound renal clearances were not significantly different; in similar experiments with the N-demethylated metabolite, however, the (R)-$(+)$-enantiomer had 6 times the unbound clearance of the (S)-$(-)$-enantiomer. The results from the perfused rat kidney are in direct conflict with those seen in the monkey but may be partially explained by a clear species difference in the metabolism of DBCA. Only traces of the N-demethylated metabolite are seen in the monkey, while large amounts are excreted in the urine of the rat (where only 20% to 50% of the dose is excreted unchanged in the urine) (*72*). In addition, the rat preferentially and selectively N-demethylates (R)-$(+)$-DBCA within the kidney itself; when (S)-$(-)$-DBCA alone was infused in the isolated kidney, no N-demethylated metabolite was detected (*68, 71, 72*). However, the (R)-$(+)$-enantiomer of the N-demethylated metabolite appears to preferentially inhibit the active secretion of (S)-$(-)$-DBCA over (R)-$(+)$-DBCA (*72*). Upon administration of the racemate in the rat, this preferential inhibition would lead to a reversal of the relative unbound renal clearances seen in the monkey. Clearly, the data to date suggest that at least this organic anion is stereoselectively transported within the kidney, most likely by the organic anion transport system. Further mechanistic work is needed to determine the site of this stereoselective interaction.

Stereoselective Renal Reabsorption

Secretion and reabsorption can occur simultaneously and in different segments of the nephron, complicating the net handling of drugs by the kidney. In general, for many exogenous compounds, reabsorption appears to occur by simple diffusion. The rate of reabsorption depends on the degree of ionization and lipophilicity of the drug and on the urine pH and flow. The driving force for the passive reabsorption of drugs is water movement. As water is reabsorbed, the drug concentration within the tubule lumen increases, and drug reabsorption occurs passively by simple diffusion due to this imposed gradient. However, the kidney actively reabsorbs most endogenous nutrients, including inorganic ions, sugars, amino acids, and vitamins. Systems dealing with chiral substrates, such as sugars and amino acids, are highly stereoselective.

Sugars

Glucose is actively and completely reabsorbed by the first segment of the proximal tubule (*1*). The transport of glucose across the brush border membrane is active and coupled to the downhill movement of Na^+ across the membrane by the Na^+–glucose cotransporter, with the Na^+ gradient supplied by the basolateral Na^+–K^+ ATPase pump. Glucose is transported down its concentration gradient across the basolateral membrane by a passive uniporter, a transporter that carries one substrate molecule per cycle (*73*). This classic pump-leak arrangement for glucose is so efficient that only in cases such as diabetes does any glucose escape reabsorption.

The Na^+–glucose cotransporter is selectively inhibited by phloridzin, and the basolateral uniporter is selectively inhibited by phloretin (*11, 74–81*). The stereoselective requirements for the transport of sugars by these transporters are quite restrictive. The D-isomers of glucose, mannose, fructose, and galactose appear to interact with both systems, but only D-glucose, D-galactose, α-methyl-D-glucoside, and 6-deoxy-D-glucose appear to be transported by the Na^+ system. The L-isomers of glucose and other sugars are not transported across either membrane by mediated mechanisms and appear to cross the membrane by simple diffusion only (*11, 74–81*). Therefore, the structural and stereoselective requirements for sugar transport in the kidney are very restrictive.

Amino Acids

All amino acids are efficiently reabsorbed by the early proximal tubule via a pump-leak arrangement, with Na^+ cotransporters located in the brush border membrane (*82*). The membrane contains one transporter for acidic amino acids, one for basic amino acids, and at least two for neutral amino acids (*83–91*). In general, amino acid transporters are less stereospecific than glucose transporters.

The naturally occurring L-isomers are transported through the brush border membrane at a much greater rate than the D-isomers (*92*). At high concentrations, however, the D-isomers can inhibit the transport of L-isomers, and vice versa. Therefore, the D-isomers do appear to be actively transported by these systems, albeit with a much lower affinity (*92*). The transporter for acidic amino acids appears to exhibit a more pronounced stereoselectivity. D-Glutamate has almost no affinity for this transporter; however, D-aspartate inhibits the transport of both L-glutamate and L-aspartate (*83*).

Conclusions and Future Directions

It is clear that the processes involved in the renal elimination of drugs may be stereoselective. First, stereoselective binding to plasma proteins is a major determinant of the renal clearance of compounds that undergo filtration as well as either tubular secretion or reabsorption. The effects of stereoselective binding to plasma proteins on renal handling of drugs have been studied in isolated perfused kidneys and verified in clinical studies with racemic drugs. Active tubular secretion has been shown to be stereoselective for both acidic and basic drugs, indicating that the two major systems involved in the secretion of drugs—the organic anion and organic cation systems—are stereoselective. Finally, stereoselective reabsorption of compounds has been demonstrated; however, this appears to apply mainly to endogenous compounds.

   Although in the past few years, new information about the mechanisms of stereoselective renal excretion of drugs at a cellular level has been obtained, there is much that we do not know. Information is available on stereoselective renal elimination of organic cations at the brush border membrane of the proximal tubule of various species, but we have little information on brush border membrane vesicles derived from human kidney. Thus, it is unclear whether the organic cation–proton antiporter in the human renal brush border membrane is stereoselective. Second, we know very little about stereoselective organic cation transport at the basolateral membrane of the human kidney. At a molecular level, it is unclear whether stereoselective transport of organic cations implies stereoselective binding to the transporter, translocation of the organic cation, or stereoselective release. Furthermore, a number of studies suggest that organic cations are transported by multiple transporters in the proximal tubule, including a second subtype of the organic cation–proton antiporter and multidrug-resistance proteins. It is not known whether these transporters exhibit stereoselectivity. Much less is known about stereoselective transport of organic anions, and we have virtually no information about the stereoselective reabsorption of drugs. Future studies will no doubt focus on the mechanisms of stereoselective transport and the role of other transporters and subtypes of transporters in the stereoselective renal handling of drugs.

References

1. Vander, A. J. *Renal Physiology*, 3rd ed.; McGraw-Hill: New York, 1985.
2. Gilman, A. G.; Goodman, L. S.; Gilman, A. *The Pharmacological Basis of Therapeutics*, 6th ed.; MacMillan Publishing: New York, 1980.
3. Rowland, M.; Tozer, T. N. *Clinical Pharmacokinetics*; Lea and Febiger: Philadelphia, PA, 1980.
4. Blaschke, T. F.; Giacomini, K. M. *Pharmacodynamic Consequences of Enantioselective Drug Disposition*; Elsevier Science Publishers: Amsterdam, Netherlands, 1987; pp 803–806.
5. Drayer, D. E. *Clin. Pharmacol. Ther.* **1986,** *40,* 125–133.
6. Levy, R. H.; Boddy, A. V. *Pharm. Res.* **1991,** *8,* 551–556.
7. Hofer, M. *Transport Across Biological Membranes*; Pitman: Boston, MA, 1981.
8. Stein, W. D. *Channels, Carriers and Pumps*; Academic Press: Orlando, FL, 1990.
9. Silverman, M. *Hosp. Pract.* **1989,** *24,* 180–204.
10. Ott, R. J.; Hui, A. C.; Yuan, G.; Giacomini, K. M. *Am. J. Physiol.* **1991,** *261,* F443–F451.
11. Kinne, R.; Murer, H.; Kinne-Saffran, E.; Thees, M.; Sachs, G. *J. Mem. Biol.* **1975,** *21,* 375–395.
12. Christensen, H. N. *Biological Transport*, 2nd ed.; W. A. Benjamin: London, 1975.
13. Hall, S.; Rowland, M. *J. Pharmacol. Exp. Ther.* **1984,** *232,* 263–268.
14. Hall, S.; Rowland, M. *J. Pharmacol. Exp. Ther.* **1983,** *227,* 174–179.
15. Valdivieso, L.; Blaschke, T. F.; Giacomini, K. M. Paper presented at the 2nd World Conference on Clinical Pharmacology and Therapeutics, Australia, 1983; ABS No. 674.
16. Giacomini, K. M.; Nelson, W. L.; Pershe, R. A.; Valdivieso, L.; Turner-Tamiyasu, K.; Blaschke, T. F. *J. Pharmacokin. Biopharm.* **1986,** *14,* 335–356.
17. Lima, J. J.; Jungblath, G.; Devina, T.; Robertson, L. *Life Sci.* **1984,** *85,* 834–838.
18. Itoh, T.; Ishida, M.; Onuki, Y.; Tsuda, Y.; Shimada, H.; Yamada, H. *Antimicrob. Agents Chemother.* **1993,** *37,* 2327–2332.
19. Rennick, B. R. *Am. J. Physiol.* **1981,** *240,* F83–F89.
20. McKinney, T. D. *Am. J. Physiol.* **1982,** *243,* F404–F407.
21. Holohan, P. D.; Ross, C. R. *J. Pharmacol. Exp. Ther.* **1981,** *216,* 294–298.
22. Dantzler, W. H.; Brokl, O. H.; Wright, S. H. *Am. J. Physiol.* **1989,** *256,* F290–F297.
23. Hsyu, P.-H.; Giacomini, K. M. *J. Biol. Chem.* **1987,** *262,* 3964–3968.
24. Maegawa, H.; Kato, M.; Inui, K.-I.; Hori, R. *J. Biol. Chem.* **1988,** *263,* 11150–11154.
25. Rafizadeh, C.; Roch-Ramel, F.; Schali, C. *J. Pharmacol. Exp. Ther.* **1987,** *240,* 308–313.
26. Wright, S. H. *Am. J. Physiol.* **1985,** *249,* F903–F911.
27. Wright, S. H.; Wunz, T. M. *Am. J. Physiol.* **1987,** *253,* F1040–F1050.
28. Gross, A. S.; Somogyi, A. A. *J. Pharmacol. Exp. Ther.* **1994,** *268,* 1073–1080.
29. Takano, M.; Inui, K.-I.; Okano, T.; Hori, R. *Life Sci.* **1985,** *37,* 1579–1585.
30. Wright, S. H.; Wunz, T. M. *Am. J. Physiol.* **1989,** *256,* F462–F468.
31. Gisclon, L. G.; Wong, F.-M.; Giacomini, K. M. *Am. J. Physiol.* **1987,** *253,* F141–F150.
32. McKinney, T. D.; Kunnemann, M. E. *Am. J. Physiol.* **1985,** *249,* F532–F541.
33. McKinney, T. D.; Kunnemann, M. E. *Am. J. Physiol.* **1987,** *252,* F525–F535.
34. Somogyi, A. *Trends Pharmacol. Sci.* **1987,** *8,* 354–357.
35. van Crugten, J.; Bochner, F.; Keal, J.; Somogyi, A. *J. Pharmacol. Exp. Ther.* **1986,** *236,* 481–487.
36. Muirhead, M.; Bochner, F.; Somogyi, A. *J. Pharmacol. Exp. Ther.* **1988,** *244,* 734–739.
37. Hsyu, P.-H.; Giacomini, K. M. *J. Clin. Invest.* **1985,** *76,* 1720–1726.
38. Notterman, D. A.; Drayer, D. E.; Metakis, L.; Reidenberg, M. M. *Clin. Pharmacol. Ther.* **1986,** *40,* 511–517.

39. Mikus, G.; Eichelbaum, M.; Fischer, C.; Gumulka, S.; Klotz, U.; Kroemer, H. K. *J. Pharmacol. Exp. Ther.* **1990,** *253,* 1042–1048.
40. Brogstrom, L.; Nyberg, L.; Jonsson, S.; Lindberg, C.; Paulson, J. *Br. J. Clin. Pharmacol.* **1989,** *27,* 49–56.
41. Lennard, M. S.; Tucker, G. T.; Freestone, S.; Ramsay, L. E.; Woods, H. F. *Clin. Pharmacol. Ther.* **1983,** *34,* 732–737.
42. Mehvar, R.; Gross, M. E.; Kreamer, R. N. *J. Pharm. Sci.* **1990,** *79,* 881–885.
43. McLachlin, A. J.; Tett, S. E.; Cutler, D. J.; Day, R. O. *Br. J. Clin. Pharmacol.* **1993,** *36,* 78–81.
44. Meier, J. *Am. Heart J.* **1982,** *104,* 364–372.
45. Somogyi, A. A.; Bochner, F.; Sallustio, B. C. *Clin. Pharmacol. Ther.* **1992,** *51,* 379–387.
46. Ross, C. R.; Holohan, P. D. *Ann. Rev. Pharmacol. Toxicol.* **1983,** *23,* 65–85.
47. Ott, R. J.; Hui, A. C.; Wong, F.-M.; Hsyu, P.-H.; Giacomini, K. M. *Biochem. Pharmacol.* **1991,** *41,* 142–145.
48. Ott, R. J.; Giacomini, K. M. *Pharm. Res.* **1993,** *10,* 1169–1173.
49. Yuan, G.; Ott, R. J.; Salgado, C.; Giacomini, K. M. *J. Biol. Chem.* **1991,** *266,* 8978–8986.
50. Malstrom, K.; Murer, H. *Am. J. Physiol.* **1986,** *251,* C23–C31.
51. Malstrom, K.; Stange, G.; Murer, H. *Biochim. Biophys. Acta* **1987,** *902,* 269–277.
52. Murer, H.; Kinne, R. *J. Membr. Biol.* **1980,** *55,* 81–95.
53. Murer, H.; Gmaj, P. *Kidney Int.* **1986,** *30,* 171–186.
54. Gaudry, S. E.; Sitar, D. S.; Smyth, D. D.; McKenzie, J. K.; Aoki, F. Y. *Clin. Pharm. Ther.* **1993,** *54,* 23–27.
55. Bendayan, R.; Sellers, E. M.; Silverman, M. *Can. J. Physiol. Pharmacol.* **1990,** *68,* 467–475.
56. Wong, L. T. Y.; Smyth, D. D.; Sitar, D. S. *J. Pharmacol. Exp. Ther.* **1990,** *255,* 271–275.
57. Wong, L. T. Y.; Smyth, D. D.; Sitar, D. S. *Br. J. Clin. Pharmacol.* **1992,** *34,* 438–440.
58. Wong, L. T. Y.; Escobar, M. R.; Smyth, D. D.; Sitar, D. S. *J. Pharm. Exp. Ther.* **1993,** *267,* 1440–1444.
59. Takano, M.; Hirozane, K.; Okamura, M.; Takayama, A.; Nagai, J.; Hori, R. *J. Pharm. Exp. Ther.* **1994,** *269,* 970–975.
60. Pritchard, J. B. *Biochim. Biophys. Acta* **1987,** *906,* 295–309.
61. Burckhardt, G.; Ullrich, K. J. *Kidney Int.* **1989,** *36,* 370–377.
62. Pritchard, J. B.; Miller, D. S. *Physiol. Rev.* **1993,** *73,* 765–796.
63. Moller, J. V.; Sheikh, M. I. *Pharmacol. Rev.* **1983,** *34,* 315–358.
64. Ullrich, K. J.; Rumrich, G.; Kloss, S. *Kidney Int.* **1989,** *36,* 78–88.
65. Pancoast, S. J.; Neu, H. C. *Clin. Pharmacol. Ther.* **1978,** *24,* 108–116.
66. Ahn, H.-Y.; Jamali, F.; Cox, S. R.; Kittayanond, D.; Smith, D. E. *Pharm. Res.* **1991,** *8,* 1520–1524.
67. Nakumura, M.; Kawabata, T.; Itoh, T.; Miyata, K.; Harada, H. *Drug Dev. Res.* **1990,** *19,* 23–36.
68. Higaki, K.; Nakano, M. *Drug Metab. Dispos.* **1992,** *20,* 350–355.
69. Nakano, M.; Kawahara, S. *Drug Metab. Dispos.* **1992,** *20,* 179–185.
70. Nakano, M.; Higaki, K.; Kawahara, S. *Xenobiotica* **1993,** *23,* 525–536.
71. Higaki, K.; Kadono, K.; Nakano, M. *J. Pharm. Sci.* **1992,** *81,* 935–939.
72. Higaki, K.; Kadono, K.; Goto, S.; Nakano, M. *J. Pharm. Exp. Ther.* **1994,** *270,* 329–335.
73. Silverman, M. *Can. J. Physiol. Pharmacol.* **1981,** *59,* 209–224.
74. Kinne, R. *Curr. Top. Membr. Transp.* **1976,** *8,* 209–267.
75. Aronson, P. S.; Sacktor, B. *J. Biol. Chem.* **1975,** *250,* 6032–6039.
76. Barnett, J. E. G.; Jarvis, W. T. S.; Munday, K. A. *Biochem. J.* **1968,** *109,* 61–67.
77. Kleinzeller, A. *Biochim. Biophys. Acta* **1970,** *211,* 264–276.

78. Silverman, M. *Biochim. Biophys. Acta* **1974,** *332*, 248–262.
79. Ullrich, K. J.; Rumrich, G.; Kloss, S. *Pfluegers Arch.* **1974,** *351*, 35–48.
80. Glossman, H.; Neville, D. M. *J. Biol. Chem.* **1972,** *247*, 7779–7789.
81. Aronson, P. S.; Sacktor, B. *Biochim. Biophys. Acta* **1974,** *356*, 231–243.
82. Mircheff, A. K.; Kippen, I.; Hirayama, B.; Wright, E. M. *J. Membr. Biol.* **1982,** *64*, 113–122.
83. Silbernagl, S. *Amino Acids and Oligopeptides;* Raven Press: New York, 1985; Vol. 71, pp 1677–1701.
84. Lerner, J. *Comp. Biochem. Physiol.* **1987,** *87B*, 443–457.
85. Sigrist-Nelson, K.; Murer, H.; Hopfer, U. *J. Biol. Chem.* **1975,** *250*, 5674–5680.
86. Lynch, A. M.; McGivan, J. D. *Biochem. Biophys. Acta* **1987,** *899*, 176–184.
87. Evers, J.; Murer, H.; Kinne, R. *Biochim. Biophys. Acta* **1976,** *426*, 598–615.
88. McNamara, P. D.; Rea, C. T.; Segal, S. *Am. J. Physiol.* **1986,** *251*, F734–F742.
89. Hammerman, M. R.; Sacktor, B. *J. Biol. Chem.* **1977,** *252*, 591–595.
90. Hammerman, M. R.; Sacktor, B. *Biochim. Biophys. Acta* **1982,** *686*, 189–196.
91. Stieger, B.; Stange, G.; Biber, J.; Murer, H. *J. Membr. Biol.* **1983,** *73*, 25–37.
92. Fass, S. J.; Hammerman, M. R.; Sacktor, B. *J. Biol. Chem.* **1977,** *252*, 583–590.

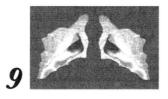

9

Stereochemical Analyses of Food Components

K. Helen Ekborg-Ott and Daniel W. Armstrong*

Most of the major nutritional, organic components of foods and beverages are chiral, including proteins, amino acids, carbohydrates, fats, and some vitamins. Many flavor and fragrance components of foods and beverages are chiral as well. Until recently, however, the stereochemical composition of food and beverage components has not been considered. One reason is that adequate analytical methods for the efficient resolution of enantiomers were just developed in the past decade. The enantiomeric composition of food components can affect aroma, taste, and nutritional value. Enantioselective analyses can be used to (1) evaluate age, treatment, and storage effects; (2) control or monitor fermentation processes; (3) identify adulterated products; (4) more exactly understand and control flavors and fragrances; and (5) fingerprint complex mixtures. Thus far, more enantioselective analyses have been done on amino acids than on any other class of compounds. However, sugars, lactones, alcohols, esters, acids, aldehydes, ketones, and monoterpenoids found in foods and beverages have also been subjected to enantioselective analyses. This chapter reviews the publications in this field as well as the relevant methods through 1995.

One of the most important substances in our chemical environment is food. Whether we accept or reject commercially available foods and beverages depends on their taste, smell, and texture. Food chemistry has become a large and sophisticated field of research. Recent advances in analytical instrumentation

*Corresponding author.

3407–8/96/0201$27.50/0

have given impetus to the study of flavor and fragrance chemistry. The correlation between chemical structures and sensory perception has been investigated, and increasingly complex studies are being done today. Another area of great interest is the chemical and biological origin and mechanism of chiral flavor and fragrance formation. These studies provide better insight into why foods and beverages taste and smell the way they do. Stereochemical analyses, in particular, can provide information on the processing and storage of food and beverage products as well as the loss of desirable flavors or the production of undesirable ones.

Flavors and fragrances can result from complex mixtures of hundreds of compounds, many occurring at trace levels. Many of the organic components of foods and beverages are chiral molecules, some of which are key to characteristic tastes and textures. In the past, most publications on aroma and fragrance research neglected enantiomeric compositions, even though many of the major flavor or fragrance components were known to be chiral. Even some recent investigations of food and beverage components, especially amino acids, lactones, and alcohols, have not considered chirality (1–9). Chiral evaluation and discrimination are important parts of odor perception as well as stereochemical structure–function relationships relating to biological activity (10–18). Smell and taste are individual experiences that are sometimes hard to express in words. Odor perception starts when odor molecules interact with receptors, which studies have shown to be proteins. Since proteins are chiral molecules, the receptors can interact differently with two enantiomers (19). Stereoisomeric flavor components are known to differ in both odor quality and odor intensity (20). The odor quality of optical isomers has been analyzed by many researchers, including Russell (11), Friedman (12), Ohloff (13), and Mosandl (21). 1-p-Menthene-8-thiol (22), one flavor component of grapefruit juice, is thought to be the most powerful aroma constituent cited in the literature (23). The biological activity of low molecular weight compounds is known to be related to specific configurations. For example, D-asparagine tastes sweet, while L-asparagine tastes bitter; (S)-limonene smells like lemons, while (R)-limonene smells like oranges; and (S)-(+)-carvone smells like caraway, while (R)-(−)-carvone smells like spearmint (11, 12, 23–26). The eight stereoisomers of menthol have been characterized and separated by many methods, but only the (1R,3R,4S)-isomer has the well-known minty odor (23, 24, 27–29).

Often, stereochemistry can be used to differentiate between compounds of natural and synthetic origin. Since this requires information about plant-specific distributions and optical purities of compounds in various foods, stereochemical analyses will allow more accurate evaluations of flavors and fragrances (30, 31). The increased demand for optically pure reference compounds has not been adequately addressed by specialty chemical companies or government agencies. Enantiomeric analyses are also useful in bioflavor research for (1) determining the configuration of the naturally occurring enantiomers in foods, beverages, and plants; (2) following and optimizing the biosynthesis and metabolization of enantiomeric compounds using microorganisms as model systems; and (3) screening enzymes for their stereospecificity in the biosynthesis of enantiomeric compounds (*32*). Biochemical pathways—as in microbial reactions, enzyme reactions, and plant metabolism processes—are important in the origination of food aromas. Selected microorganisms are known to improve the texture, nutritional value, and shelf life of some consumer products (*33*); they also produce flavor components such as amino acids, acidulants, and 5′-nucleotides in industrial-scale fermentations. In spite of all the recent experimental advances in bioengineering and cell biology, few industrial-scale microbial processes relating to flavors and fragrances are in operation. Microorganisms (of known strains) are used in large-scale processes, to make dairy products, beer, wine, baked goods, and other products.

Another important area of enantiomeric analysis is the identification of adulterated consumer products. The flavor of commercial foods and beverages can change during handling of the raw materials, processing, or storage. The loss of primary flavor compounds and the formation of secondary ones can either improve or spoil the final products. Both unintentional and intentional flavor changes can take place during production. Deliberate changes include de novo synthesis of flavor compounds, masking of flavors, conversion of one flavor compound into another, removal of a particular flavor compound, and inhibition of the formation of a particular flavor compound (*34*). For example, the roasting of coffee is known to reduce the levels of most pesticides (*35, 36*). Spray- and freeze-drying can cause coffee and other foods to lose volatile flavor and aroma components. Many techniques are then applied to add the flavor and aroma back to the soluble coffee before it is packaged. These compounds are usually added as late as possible before packaging, since they are very sensitive to moisture and oxygen. Thus, enantiomeric purity can be used to evaluate the storage, shipping, handling, age, and adulteration of various foods, beverages, and other consumer products. Although amino acids are important in the analysis of foods and beverages for adulteration, other chiral molecules are even more important, including lactones, alcohols, diols, organic acids, esters, aldehydes, and ketones. Werkhoff et al. have reviewed research on flavor chemistry (*37–39*).

Two enantiomers may differ in taste, fragrance, or pheromonal or insecticidal activities. Often, one enantiomer displays the desired effect while the other is inactive or displays undesirable effects (*40*). Also, many plants are known to contain natural insecticides (*41–43*).

Enantiomeric separations are useful for evaluating not only flavor and fragrance components but also a variety of other compounds important to the food and beverage industry, including preservatives, growth regulators, pesticides, herbicides, fumigants, and pheromones (*44*). The synthesis of enantiomerically pure chemicals and pharmaceuticals is highly desirable. Since enzymes are diastereoselective and enantioselective, they are important catalysts for chiral reactions (*40*).

To obtain insights into the physiologic, sensory, and toxicologic properties of chiral flavor compounds, many of which are present at trace levels, requires analytical and instrumental methods with suitable selectivity and sensitivity (*30*). These methods include gas chromatography (GC), high-performance liquid chromatography (HPLC), capillary electrophoresis (CE), supercritical fluid chromatography (SFC), absorbance spectroscopy using infrared (IR), nuclear magnetic resonance (NMR), gas chromatography–mass spectrometry (GC–MS), high-performance liquid chromatography–mass spectrometry (HPLC–MS), and circular dichroism. The methods involving GC or HPLC can use chiral or achiral stationary phases. Many of the chiral selectors used for HPLC separations are also useful as chiral shift reagents in NMR (*45*).

The most prevalent chiral flavor components in foods and beverages are L-amino acids, D-sugars, carbohydrates composed of D-sugars, and lactones. The D-amino acids, which are less common, are found in processed foods and beverages subjected to (1) severe thermal treatment (e.g., roasted coffee), (2) fermentation processes (e.g., beer, mead, wine, sake, vinegar, cocoa powder, coffee, dried tea leaves, dairy products, lactic acid fermented vegetables, fermented sausage, and soy sauce) (*46–52*), (3) high solution temperatures (e.g., milk products), and (4) extremes in pH (*53*). Although all amino acids in the proteins and peptides of processed food racemize at the same time, they do so at very different rates depending on their nature, the temperature, pH, physical state, and many other factors. Research in this area will continue to expand as new compounds are identified and related to characteristic flavors.

Following the invention of stable, widely applicable chiral stationary phases (CSPs) for liquid chromatography, their use in the analysis of pharmaceutical compounds has become routine (*54, 55*). Approximately 75% of all known pharmaceutical compounds are chiral, containing one or more enantiomeric centers. Only 20% of these are distributed as pure enantiomers (*56*). As will be seen in this review, many CSPs used for pharmaceutical analyses are also useful in resolving the nonvolatile stereoisomers in foods and beverages. The advent of a variety of commercial cyclodextrin-based GC stationary phases around 1990 proved tremendously useful for the analysis of more volatile chiral fragrance compounds. This chapter reviews the work reported in this area through 1994.

Amino Acids

Stereochemistry of Amino Acids in Foods and Beverages

Amino acids produce complex sensations in humans, although those of sweet, sour, salty, bitter, and umami (a basic "meaty" taste or brothy mouth feel) (*57, 58*) predominate (*59, 60*). Many thousands of publications address the subject of amino acids, but only a small fraction deal with the enantiomeric composition of amino acids in foods and beverages.

Most α-amino acids contain a chiral center and exist as both D- and L-enantiomers. The free amino acids and proteins found in higher living organisms are generally composed of L-enantiomers. The mirror-image D-enantiomers are less common, but D-alanine, D-glutamine, D-glutamic acid, D-asparagine, and D-aspartic acid are constituents of the peptidoglycan in bacterial cell walls (*61–66*). D-Amino acids also occur in several peptide antibiotics. Some insects and marine invertebrates have D-amino acids as major components of their cellular fluids (*67–71*), and a few higher plants also contain D-amino acids (*72*).

Today, there is an expanding interest in the occurrence of D-amino acids in commercial foods and beverages. These D-amino acids may be of interest because many have sweet tastes, in contrast to many L-amino acids, which have bitter tastes (*73*). In particular, most hydrophobic L-amino acids are bitter, while the corresponding D-amino acids are sweet (*74*). For example, D-tryptophan, D-phenylalanine, D-histidine, D-tyrosine, and D-leucine are 35, 7, 7, 6, and 4 times as sweet as sucrose, respectively (*60*). Glycine and L-alanine also have sweet tastes. These amino acids seem to bind to certain receptors on the tongue (*74*). L-Glutamic acid and L-aspartic acid taste sour in their undissociated state, but their sodium salts have umami tastes. L-Glutamate especially elicits a savory taste in foods. In addition, the sodium salts of ibotenic and tricholomic acids (derivatives of oxyglutamic acid) are also umami substances, with taste intensities 4 to 25 times that of monosodium glutamate (MSG). Since these amino acid sodium salts are not commonly found in animals, they have not been used as flavor enhancers (*74*). Furthermore, free L-theanine and glutamic acid are the most important umami substances in green tea (discussed later) (*75–77*).

Diet seems to be the main source of D-amino acids (*78*). Another source may be the action of certain intestinal (gut) bacteria (*79, 80*). D-Amino acids seem to be more prevalent in certain types of foods and beverages: (1) those that have been exposed to microbial activity (fermentation, aging, etc.) (*52, 53, 81–86*), (2) highly processed foods and beverages (those that have been exposed to extremes of pH, heat, etc.) (*81–84, 87*), and (3) foods that naturally contain D-amino acids (e.g., certain seafoods) (*67–71*). Modern food technology uses a variety of processes to modify natural proteins to improve taste, texture, and shelf life of commercial foods and beverages. In food treated with acid, alkali, or heat during processing and preparation, L-amino acids may undergo acid- or base-catalyzed racemization (*87–95*). Racemization may impair the biological value and safety of foods, since the D-amino acids formed may be nutritional antagonists or act as antimetabolites (*86, 93, 94, 96–98*). During preparation, milk, meat, and grain proteins (which originally contain only trace amounts of D-amino acids) are often exposed to conditions that could cause racemization (*95*). Frank has published an excellent review on the mechanistic aspects of amino acid racemization (*99*).

Some earlier studies on the racemization of essential amino acids have been extensively reviewed by Neuberger (*100*) and Berg (*101*), who demonstrated that the D-isomers in mammals were poorly utilized, sometimes inhibited growth, and were mainly excreted in the urine. More recent investigations have confirmed these findings (*78, 102–105*). Racemization half-lives for the essential amino acids in humans have been studied (*106*). Neighboring amino acids can affect racemization in peptides and proteins. Several research groups have used GC to study racemization in processed foods treated with heat or alkali (*49–52, 81–83, 107–109*).

The occurrence of D-amino acids in foods and beverages has been investigated by several research groups, including Armstrong and co-workers (*78, 110–113*), Brückner and co-workers (*46–52, 86, 98, 114–118*), and Mosandl and Kustermann (*119*). Liardon et al. converted amino acids into *N*-trifluoroacetyl-*O*-isopropyl esters after hydrolysis of the protein in deuterated hydrochloric acid and then analyzed the derivatives by capillary GC–MS on a CSP (*120*). This technique was also used to measure the racemization of 14 amino acids in stored and heated milk powders and in alkali-treated casein (*81*). Amino acid racemization had little effect on the nutritional value of severely heated proteins but was partly responsible for the nutritional losses induced by alkali treatment.

Palla et al. used both capillary GC and reversed-phase HPLC to investigate the presence of D-amino acids in processed foods that had undergone severe heat treatment (*53*). They analyzed roasted coffee, ultrahigh-temperature (UHT) milk, yogurt, cheese, and other products (*53, 85*). Table I shows the percent racemization of free amino acids found. A very stable tetraamidic selector, L-phenylalanine trioxaundecanoyl tetraamide was used for the capillary GC analysis. In the HPLC portion of the study, amino acids were derivatized with dansyl chloride (DNS-Cl) and separated on a reversed-phase C_{18} column using a mobile phase containing copper(II) acetate and L-phenylalaninamide as the chiral eluent.

Several papers have been published on the amount of D-aspartate in various foods. Masters and Friedman (*109*) detected high levels of D-aspartate in processed commercial foods such as soy protein (9%), simulated bacon (13%),

Table I. Percent Racemization of Free Amino Acids in Food Products, Detected by Capillary GC using the Chiral Phase L-Phenylalanine Trioxaundecanoyl Tetramide

Food	Percent Racemization of Free Amino Acids, $[D/(D + L)] \times 100$				
	Ala	*Leu*	*Asp*	*Glu*	*Phe*
New milk	3–4		2–3	2–3	
Pasteurized milk	3–4		1–2	3–5	
Ultrahigh-temp. milk	4–6		2–3	3–5	
Dried milk	8–12		4–5	3–4	
Yogurt	64–68		20–32	53–66	
Aged cheese	20–45	2–7	8–35	5–22	2–13
Raw ham	0–1	<0.2		0–0.5	0–0.3
Roasted coffee			23–38	32–41	9–12
Green coffee				<0.2	

SOURCE: Data taken from reference 53.

and nondairy creamer (17%). Man and Bada (*95*) detected high levels of D-aspartate (13%) in a soy-protein-based liquid nutritive formulation found in a health food store. This level is significantly higher than that found for soy-based infant formulas. Finley (*121*) found high amounts of D-aspartate in wheat crackers (9.5%), wheat cereal (11.0%), tortillas (11.6%), and hominy (15.4%). He also found high levels of D-serine (50%), D-aspartate (37%), and D-phenylalanine (26%) in a weight-loss product that had been mistakenly treated with excess alkali.

Fermented foods such as dairy products (yogurt, sour milk, and cheese), alcoholic beverages (beer, wine, and mead), sourdough bread, vinegar, soy sauce, and other liquid spices have been known from the beginning of civilization (*86*). All fermented foods have one thing in common: The texture and flavor are formed by the action of certain strains of microorganisms (such as yeasts, bacteria, or molds) on raw food products (*122–125*). Selective fermentation results in desired tastes, odors, and textures of the final products. In addition, many compounds appear to racemize during fermentation. Some bacterial amino acids may be involved in the formation of D-amino acids in fermented foods (*125*). It is believed that some D-amino acids originate as microbial metabolites and through microbial autolysis. These D-amino acids do not contribute to the nutritional value of food for mammals. Some reviews and other reports exist on the utilization of D-amino acids (*50, 53, 87, 94, 95, 109, 126–130*).

Significant amounts of D-amino acids have been found in fermented soy sauce by HPLC (*50, 51, 86, 98, 111, 114*). Before separation, the primary amino acids were derivatized with either *o*-phthalaldehyde (OPA)/*N*-isobutyryl-L-cysteine or OPA/*N*-isobutyryl-D-cysteine (*131*). This method was then used to determine the enantiomeric ratios of amino acids in the soy sauce (*114*), and the results were compared with previous results found by capillary GC with a CSP (*116*). The HPLC results were in good agreement with those found by GC on Chirasil-L-Val (the HPLC results are listed first): D-Asp (4.4% vs. 4.7%), D-Glu (1.9% vs. 2.3%), D-Ala (2.7% vs. 3.0%), D-Phe (1.2% vs. 2.5%), and D-Leu (1.2% vs. 1.3%). The enantiomeric composition of soy sauce amino acids depends on how long the sample has been exposed to air and whether microbial degradation has occurred before derivatization or analysis.

Brückner and co-workers analyzed both enantiomeric composition and total concentrations of amino acids in fermented dairy products (various cheeses, curdled milk, and kefir), fermented vegetables (beet juice, beetroot, cabbage, carrots, and celery), fermented alcoholic beverages (beer, wine, sake, and vinegar), fermented Asian food (black beans), fermented sausage, liquid spices (soy sauce and miso), cacao, instant coffee, and tea (*46–52, 86*). The amino acids were separated by capillary GC on Chirasil-L-Val or XE-60-L-Val-(*S*)-α-phenylethylamide CSPs. All the fermented foods contained significant amounts of D-amino acids. The largest amounts were found in the liquid spices, fermented black beans, and Emmentaler cheese. Additional fermented foods were previously investigated by Brückner and Hausch (*50, 51*). Free amino acids have also been reported in dairy products such as yogurt and kefir (*49, 50*). In the same study, tea was found to contain about 5 times the level of free amino acids found in cacao, while the amount of D-amino acids in cacao was twice that in tea (*52*). This finding is understandable, since teas are not produced through true fermentation involving microorganisms.

Free amino acids exist in many animal and plant tissues, and several are major aroma components in foods and beverages. Armstrong et al. recently reported the quantitative stereochemical analysis of various free amino acids in processed foods and beverages, including tea (*110*), beer (*112*), honey (*113*), and soy sauce (*111*).

Tea is the most widely consumed beverage in the world. Types include green (nonfermented) tea, half-green or oolong (semifermented) tea, and black (fermented) tea. The fermentation step in the processing of tea has been somewhat unfortunately named, since it is not a true fermentation requiring microorganisms, although they may be present in some green teas. Instead, the fermentation step involves the oxidation of two or more tea flavanols (polyphenolic compounds) belonging to the catechin group (*132–134*). During black tea manufacture, the amino acids may be oxidized by catechin *o*-quinones and subsequently undergo Strecker degradation, which leads to new aroma components (*134–136*). The most important nonvolatile components are complex products formed by oxidation and polymerization of polyphenols during fermentation (*137*).

No single identified aroma constituent is characteristic of green, oolong, or black tea. Instead, the aroma and taste of tea result from many different volatile components as well as chiral amino acids. Of all the amino acids in tea, one is unique and needs special attention. It is the amino acid theanine, also known as glutamic acid γ-ethylamide or 5-*N*-ethylglutamine (Figure 1), which exists only in its free, nonprotein form (*138, 139*). Theanine is the major free amino acid in tea and makes up more than 1% of the total dry weight of tea leaves (*133, 134, 137, 140*). It is known to enhance the quality and characteristic taste of green tea (*141–143*). Green tea aroma has four characteristic taste elements: bitterness; astringency, sweetness, and brothiness (or umami) (*57, 58, 143*). Theanine was first found in the leaves of tea (*Camellia sinensis* and other camellia species) in the early 1950s by Sakato (*141*) and later found in the mushroom *Xerocomus badius* in the 1960s by Casimir and co-workers (*144*). No other sources of theanine have been reported.

A recent investigation of total amount of theanine and percent D-theanine in 17 tea types shows that the amounts vary with the type of tea analyzed (*110*).

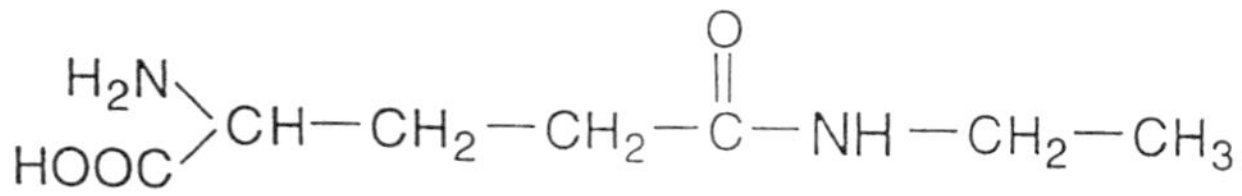

$$H_2N\diagdown \atop HOOC\diagup CH-CH_2-CH_2-\overset{\overset{\textstyle O}{\|}}{C}-NH-CH_2-CH_3$$

Figure 1. Molecular structure of DL*-theanine.*

Moisture and heat can cause stored tea to deteriorate. Astringency and flavor may be lost through lipid hydrolysis and autoxidative reactions that cause loss of amino acids (particularly theanine), sugars, and other flavor compounds (*145*). The tea samples in this study were derivatized with 9-fluorenylmethoxycarbonyl-glycine chloride (FMOC-Gly-Cl) and then separated on a coupled HPLC system. The coupled HPLC system consisted of both an achiral C_{18} column used in the reversed-phase mode and a chiral γ-cyclodextrin column used in the polar-organic mode (*110*). Figure 2 shows the enantiomeric separation of theanine in a tea sample on a γ-cyclodextrin column.

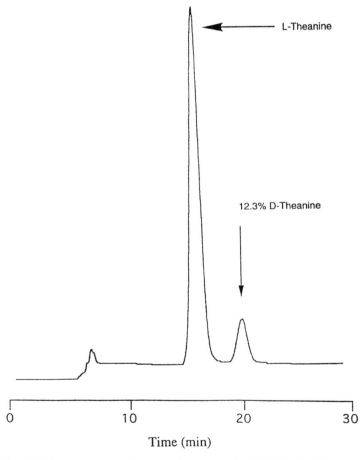

Figure 2. Chromatogram showing the enantioseparation of the FMOC-glycyl derivative of thea-nine in Formosa oolong tea on a γ-cyclodextrin column. Mobile phase: acetonitrile–methanol–tri-ethylamine–acetic acid (1050:35:4:0.5, v/v/v/v). Fluorescence detection: excitation wavelength at 266 nm, emission wavelength at 315 nm. Flow rate: 0.9 mL/min.

Each of the pure enantiomers of theanine as well as the racemic mixture has a sweet flavor without any bitter aftertaste. Theanine could therefore be used as a sweetening agent, but so far there is no economical method for producing it on a commercial scale. This has limited the number of studies on theanine's effects in humans and animals. In one of the few studies done, theanine inhibited the convulsive action of caffeine in mice but was ineffective against other convulsants such as strychnine, pentylenetetrazole, and picrotoxin (*146–149*).

Beer is an example of a fermented beverage in which the action of microorganisms produces the desired characteristic flavors. During fermentation, sugars in the wort are converted into alcohol and carbon dioxide by yeasts such as *Saccharomyces cerevisiae* and *S. carlbergenesis* (*150*). One of the most important problems for brewers is the quality and shelf life of packaged beer (*151*). Brewers use many biological, chemical, and physical methods of analysis in beer processing and quality control (*152*).

Oxidation of beer produces stale flavors as well as haze and browning. To deal with this problem, brewers often decrease the oxygen level in packaged beer and add ascorbic acid as an antioxidant. The last step of brewing involves filtering the beer and packaging it into cans, kegs, or bottles.

Low-alcohol beers, having less than 3.5% alcohol by volume, have had problems with loss of desirable flavor components during processing (*153*). Such beers are made either by stopping fermentation early or by removing some of the ethanol after fermentation is completed. Premature termination of the fermentation step may disturb the microbial degradation of sugars, which may result in a loss of desirable flavors and the production of off flavors. The removal of alcohol (after fermentation is completed) can be expensive, and desirable volatile flavor molecules may be lost in the process (*154*).

New yeast strains are continuously being developed, especially with the emergence of recombinant DNA technology. For example, "lite" (low-carbohydrate) beer can be produced with new strains of yeasts (*154*). Ekborg-Ott and Armstrong separated free enantiomers of proline, leucine, and phenylalanine in 25 different beers (*112*). The leucine and phenylalanine standards were derivatized with FMOC-Gly-Cl and then separated on a coupled HPLC system (an achiral C_{18} column in the reversed-phase mode and a chiral β-cyclodextrin column in the polar-organic mobile-phase mode). Proline was derivatized with 9-fluorenylmethyl chloroformate (FMOC) and separated on the same coupled HPLC system as mentioned earlier, except that the second column was a β-RN-cyclodextrin (*112*). Of all the amino acids enantiomerically separated in beer, proline showed the least racemization (less than 1% in all samples analyzed) but the highest total concentration. One exception is Budweiser beer, in which the total amounts of both leucine and phenylalanine far exceeded proline (1.0, 1.7, and 0.2 mg per 100 mL, respectively). Leucine and phenylalanine consistently had the highest levels of the D-enantiomers in all samples analyzed (*112*). Figure 3 shows the enantiomeric separation of some amino acids in different beers.

Honey is a complex natural product containing many classes of compounds

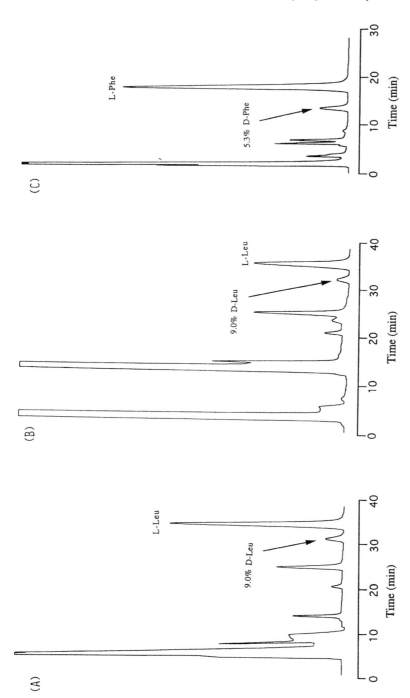

Figure 3. Chromatograms showing the enantiomeric composition of FMOC-glycyl derivatives of leucine and phenylalanine in different beer samples analyzed on a β-cyclodextrin column. (A) Miller Reserve, (B) Coors, and (C) Labatt's Canadian Ale. Mobile phases: for DL-Leu, acetonitrile–triethylamine–acetic acid (1000:8:1.5, v/v/v); for DL-Phe, acetonitrile–triethylamine–acetic acid (1000:8.5:1.5, v/v/v). Fluorescence detection: same as in Figure 2. Flow rates: 1 mL/min for (A) and (B), 1.5 mL/min for (C).

in addition to carbohydrates, including hydrocarbons, amino acids, alcohols, and vitamins. The main attractant for honeybees, as for many other insects, is the scent of a flower, which induces a selective visitation to certain flower species (155). The most popular way to identify honey is by amino acid analysis. This analysis gives information about both botanical and geographic origin (156–159), since the composition of honey varies with plant type, geographic location, and season (156–163). The most prominent amino acid in honey is proline, which can account for 50% to 85% of the total free amino acid content. While most amino acids in honey originate from both the bees and their food sources, proline is made predominantly by the bees themselves.

Pawlowska and Armstrong analyzed the enantiomeric composition of the free amino acids proline, leucine, and phenylalanine in honeys of various origins (113). Both the total concentrations and the enantiomeric ratios were determined for these three amino acids in 13 honey samples of different botanical and geographic origin. The separations were performed on a HPLC system with the same derivatizing agents (FMOC and FMOC-Gly-Cl) and column-switching method used for the beer study (112, 164).

Table II shows the total amount and the enantiomeric ratios of all three amino acids in different honey samples. As expected, proline was the major amino acid in most samples, followed by phenylalanine, while the total amount of leucine varied greatly from sample to sample. The level of D-proline was lower than 0.5% in all honey samples analyzed. In contrast, the amounts of D-leucine and D-phenylalanine ranged from 0.7% to 3.6%.

The differences among the various honey samples could not be explained solely by botanical and geographic origin. Processing (including heat treatment), storage, and handling must also be taken into consideration. Several honey samples were tested for heat treatment in a water bath and in a microwave oven. Heat treatment decreased the total level of amino acids, darkened the honey, and increased the relative amount of D-amino acids.

Ikeda was the first person to recognize the importance of amino acids in food taste (165). He discovered in 1908 that the essential taste component of Japanese sea tangle seasoner was MSG, which is known to produce the umami flavor typical of some amino acids. Glutamic acid is a nonessential, naturally occurring amino acid that is commonly used as a food additive in the form of its monosodium salt, MSG. MSG is added to many processed food products to increase palatability, but only the L-enantiomer has been shown to be involved in flavor enhancement (166). As much as 0.2% to 0.9% by weight of MSG is added to processed foods. In nature, free glutamate is present in fairly high levels in potatoes, tomatoes, peas, and mushrooms.

Rundlett and Armstrong studied the enantiomeric composition of free glutamic acid in several fermented food products, in nonfermented food with no added MSG, and in food containing added MSG (111). All food samples analyzed contained detectable levels of free D-glutamate. To avoid bacterial contamination, the food samples were derivatized immediately upon exposure to the air.

Table II. Total Amount and Enantiomeric Ratios of Proline, Phenylalanine, and Leucine in Honey Samples of Different Botanical and Geographic Origins

No.	Botanical	Geographic	Proline Total mg/100 g	Proline %D	Phenylalanine Total mg/100 g	Phenylalanine %D (SD)	Leucine Total mg/100 g	Leucine %D (SD)
1	*Phacelia tanacetifolia*	Poland	9.9	<0.02[g]	<0.5	—	<0.5	—
2	*Melilotus alba*	Poland	8.3	0.05	13.9	2.7	10.5	1.4
3	*Filia sp.*	Poland	33.1	0.07	6.5	0.9	<0.5	—
4	*Fagopyrum esculentum*	Poland	62.8	0.06	14.3	1.1	18.2	1.8
5	Mixed flowers	Poland	42.1	0.09	5.8	0.7	6.1	1.5
6	Honeydew	Poland	2.8	0.3	<0.5	—	<0.5	—
7	Mixed flowers[a]	Argentina + USA + Mexico + Canada	24.8	0.2	3.5	3.1	2.9	3.6
8	Mixed flowers[a]	Argentina + USA + Mexico + Canada	17.6	0.2	5.5	2.3	2.9	2.6
9	Mixed flowers[b]	Unspecified	38.2	0.2	4.1	2.0	<0.5	—
10	Mixed flowers[c]	USA + Canada + China	32.4	0.2	2.8	2.3	<0.5	—
11	Mixed sweet clover lowers[d]	USA + Canada	21.5	0.2	1.8	2.9	<0.5	—
12	Mixed wild flowers[e]	USA + Canada + Argentina	10.4	0.4	7.5	1.3	7.9	2.3
13	White blossoms of the orange trees[f]	USA	54.3	0.3	9.4	2.8	8.3	2.4

[a] Golden Nectar (Aldi Inc., Batavia, IL).
[b] Hyde Park (Malone Hyde Inc., Memphis, TN).
[c] Cucamonga (Western Commerce Corp., CA).
[d] Clover pure honey (International Bazaar Inc., Dayton, OH).
[e] Wild flowers (International Bazaar Inc., Dayton, OH).
[f] Orange blossom (International Bazaar Inc., Dayton, OH).
[g] Less than detection limit according to reference 164.
Source: Data taken from reference 113.

The enantiomers of glutamic acid and all prepared food samples were derivatized with FMOC-Gly-Cl and analyzed on the same type of coupled HPLC system used for the tea and beer studies described above. The chiral separations were performed on a γ-cyclodextrin column in the polar-organic mobile-phase mode. Figure 4 shows enantiomeric separations of glutamic acid in various food sources. D-Glutamate probably does not contribute to the nutritional value of food. Thus far, no adverse effects resulting from the frequent intake of D-glutamate by humans have been reported.

The food supply of laboratory rodents used in an amino acid study was analyzed for free D-amino acids as well (78). D-Amino acids were found in this highly processed food material. Their presence is probably due to microbial action on proteins, peptides, and amino acids during manufacture or to physicochemical effects of the manufacturing process (high temperature, pH, etc.). D-Amino acids in animal feed can also come from racemic amino acid supplements (sometimes added to cattle feed).

Sandra et al. (167) and Kuneman et al. (168) discussed the importance of separating D- and L-amino acids in fruit juices, since the presence of the synthetically produced D-enantiomer could be used to detect adulterated products. DL-Amino acids can be derivatized with OPA reagent or other reagents and then separated by HPLC (46, 114, 169–177). Amino acids in fruit juices and gelatin have been profiled and analyzed to detect possible adulteration. The GC results of a previous study on amino acids in orange juice (bulk produce of European origin) and gelatin were compared with the HPLC findings (116). The results from the HPLC fruit juice study agreed with those found by GC on Chirasil-L-Val (the HPLC results are listed first): D-Ala (28.5% vs. 27.5%) and D-Ser (17.0% vs. 14.3%) (114). The HPLC analysis also produced results for some additional amino acids, which were not resolved by GC: D-Asp (0.6%) and D-Asn (6.7%). The HPLC results for the gelatin samples also showed good agreement with the previous GC analysis (the HPLC results are listed first): D-Asp (22.9% vs. 24.6%), D-Glu (2.5% vs. 3.4%), D-Ser (4.5% vs. 8.4%), D-Ala (1.5% vs. 1.6%), D-Tyr (14.6% vs. 15.0%), D-Phe (2.6% vs. 3.6%), and D-Lys (1.7% vs. 2.3%). D-Arginine (3.0%) was also analyzed by HPLC but not previously resolved by GC. The amino acids in the totally hydrolyzed gelatin sample showed (after heating) extensive racemization. Even relatively mild treatment, such as heating gelatin in water, causes the formation of high amounts of D-aspartic acid in the food protein (49, 50). Both aspartic acid and phenylalanine racemize rapidly at 100 °C and at neutral pH (178).

Carisano described the use of the precolumn derivatizing agent 4-chloro-7-nitrobenzo-2-oxa-1,3-diazole (NBD-Cl or 4-chloro-7-nitrobenzofurazan) for the determination of proline in fruit juices and protein hydrolysates by reversed-phase HPLC (179). NBD-Cl can be used to derivatize both primary and secondary amino acids, but the reaction with secondary amino acids is favored because it is much faster. This method is thus complementary to the analysis of amino acids using precolumn derivatization with OPA, which works only for pri-

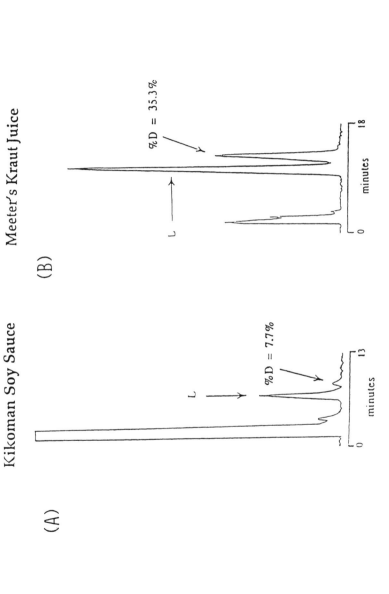

Figure 4. Chromatograms showing the enantioseparation of FMOC-glycyl derivatives of monosodium glutamate (MSG) in commercial food products and in commercial MSG products analyzed on a γ-cyclodextrin column. (A) Kikkoman soy sauce, (B) Meeter's kraut juice,

(continued on next page)

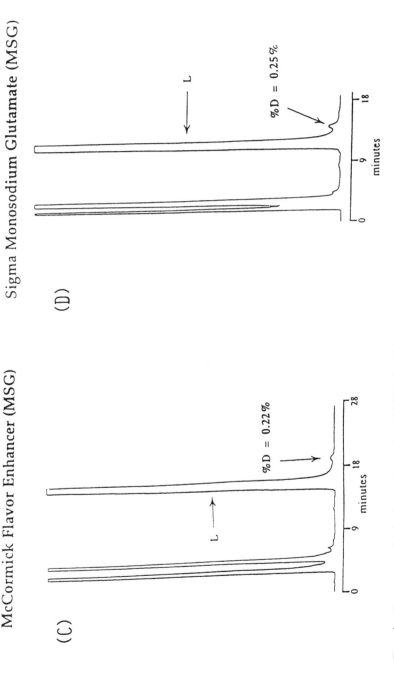

Figure 4. Chromatograms showing the enantioseparation of FMOC-glycyl derivatives of monosodium glutamate (MSG) in commercial food products and in commercial MSG products analyzed on a γ-cyclodextrin column. (C) McCormick flavor enhancer (MSG), and (D) Sigma MSG. Mobile phase: acetonitrile–methanol–triethylamine– acetic acid (850:150:12:3, v/v/v/v). Flow rates: 3.0 mL/min for the soy sauce, 2.0 mL/min for the kraut juice, 1.0 mL/min for McCormick MSG, and 1.5 mL/min for Sigma MSG. Fluorescence detection: same as in Figure 2. (Reproduced with permission from reference 111. Copyright 1994 Wiley–Liss.)

mary amino acids. DL-Amino acids have also been separated in fruit juices as fluorinated esters on chiral GC phases (*167*).

Amino Acid Resolution Techniques

Today, enantiomeric purity of amino acids can be easily and routinely analyzed by liquid chromatography (LC), thin-layer chromatography (TLC), gas chromatography (GC), or capillary electrophoresis (CE) because of the increasing number of chiral stationary phases (CSPs) and derivatizing agents developed in the past decade.

High-Performance Liquid Chromatography

HPLC resolution of amino acid enantiomers can be achieved with CSPs (*55, 78, 110–113, 164, 180–237*), with chiral mobile-phase additives (metal chelates, albumin, crown ethers, counterions, cyclodextrins) (*181, 238–249*), by derivatization with achiral reagents prior to separation on CSPs, or by derivatization with chiral agents and separation of the diastereomers on conventional stationary phases (*46, 49, 53, 78, 110–115, 131, 164, 168–177, 179–189, 247–314*). For the last approach, optically pure derivatizing agents are required as well as reaction conditions under which no racemization takes place (*295*). Both direct and indirect methods for HPLC amino acid resolutions have been reviewed (*221–223*). Direct enantiomeric separations in HPLC are based on the formation of labile diastereomeric complexes with an immobilized chiral selector. No single stationary phase can be used for all enantiomeric resolutions, since many different simultaneous interactions between the enantiomer and the chiral selector are required (i.e., the three-point interaction model). However, amino acids are structurally similar enough that many can be resolved by a single technique.

The four ways to resolve native underivatized amino acids chromatographically are as follows: (1) ligand-exchange methods (*235, 245, 315–321*), (2) chiral crown ether stationary phases (*227*), (3) a new macrocyclic antibiotic (teicoplanin) stationary phase (*322*), and (4) an α-cyclodextrin bonded phase for aromatic amino acids (*232*).

Ligand-exchange chromatography was introduced by Davankov and co-workers (*323, 324*). An optically active bidentate ligand is grafted onto a support surface and loaded with Cu^{2+}, which is usually a component of the mobile phase (*315–317*). Ternary complexes are formed with the amino acid analytes, and if the free energies of the diastereomeric complexes are sufficiently different, separation occurs. This system is very selective and can also be used in TLC. An alternative approach for amino acid separations uses a chiral crown ether complexing agent (*224–227*). No metal ions are needed when using crown-ether-based chiral selectors (*224–228*). For a compound to be resolved (by inclusion complexation) on a CSP based on an 18-crown-6-ether, it must contain a primary amine functional group. The initial work in this area was carried out by Cram and co-workers (226, 325–328).

Macrocyclic antibiotics were introduced by Armstrong and co-workers (322) as highly efficient chiral selectors. Teicoplanin effectively resolves all native amino acids and many peptides in the reversed-phase mode (322). Typical mobile phases are methanol–water or ethanol–water.

HPLC methods are often combined with precolumn or postcolumn derivatization to achieve high-sensitivity fluorimetric or photometric detection. The sensitivity and selectivity for amino acids can be enhanced by the use of achiral derivatizing agents. Many reports exist on precolumn derivatizing reactions. Dansyl chloride (DNS-Cl) (185, 186, 229, 230, 233, 244, 245, 250–254), dinitrobenzoyl chloride (DNB-Cl) and related derivatives (55, 78, 110–113, 164, 183, 185, 228, 255), phenylisothiocyanate (PITC) (256–259), O-phthalaldehyde (OPA) plus a thiol reagent (46, 49, 114, 115, 131, 168–177, 180, 248, 249, 260–280), and 7-chloro-4-nitrobenzo-2-oxa-1,3-diazole (NBD-Cl) (246, 296, 329), as well as the newer derivatizing agents 9-fluorenylmethyl chloroformate (FMOC-Cl), 9-fluorenylmethoxy-carbonylglycine chloride (FMOC-Gly-Cl), and 6-aminoquinolyl-N-hydroxysuccinimidyl carbamate (AQC) (110–113, 164, 182, 183, 189, 203, 281–292) have all proved to be highly versatile fluorescent agents for amino acid derivatizations. These derivatizing agents have been successfully used for many chiral separations as well as for quantitative achiral separations. Sanger's reagent can be used when detection is mostly dependent on the dinitrophenyl chromophore (293, 294). Quantitation of amino acids also can be achieved by precolumn derivatization with naphthylethylisocyanate (330).

There are relatively few reports on the chromatographic resolution of N-tert-butoxycarbonyl (N-t-Boc) amino acids on CSPs (187, 204–206). Most N-t-Boc amino acids can be resolved by HPLC with a hydroxypropyl-derivatized β-cyclodextrin bonded-phase column in the reversed-phase mode (187). The tert-butoxycarbonyl (t-Boc) group is one of the most important amine-protecting groups in peptide synthesis. N-t-Boc-amino acids are resistant to racemization during peptide synthesis, and the t-Boc group can be easily removed by acid-catalyzed hydrolysis (331). Enantiomers derivatized with this reagent have been efficiently resolved by reversed-phase HPLC (187).

Phenylthiohydantoin (PTH) amino acids are produced via the Edman degradation (295, 332). PITC reacts with the free α-amino acids to form the phenylthiocarbamyl derivatives of the peptide. The terminal amino acid is released as a thiazolinone and converted to the more stable isomeric PTH derivative. The PTH amino acids are easily separated by HPLC and detected by ultraviolet absorbance at 254 nm.

NBD-Cl can be used for the analysis of proline and hydroxyproline as well as other secondary amines (295). The reaction between NBD-Cl reacts much faster with secondary amino acids than with primary amino acids.

Some amino acids can be derivatized with an N-3,5-dinitrobenzoyl group or related derivatives, followed by enantiomeric separations on a π-complex hydrogen-bonding stationary phase (*219, 224, 255*) or via chiral inclusion mechanisms with cyclodextrin stationary phases (*185, 219*). A variety of CSPs can also be used to resolve aromatic derivatized amino acids (*78, 112, 113, 180–184, 231–234, 333*).

Several of these derivatizing agents suffer from some disadvantages (such as limited stability over time) when used to enhance sensitivity of amino acids. PITC requires evaporation of excess reagent before HPLC separation. When the analyte is present at trace levels in a complex mixture (as in foods and beverages), DNS-Cl derivatization is difficult to accomplish (*334*). OPA reacts only with primary amines, in the presence of a suitable thiol compound, to produce fluorescent and electrochemically active 1-alkylthio-2-alkyl-substituted isoindoles (*278, 302*). This reagent works faster and is more sensitive than the DNS-Cl reagent. Even though OPA and DNS-Cl have been around for many years, they are not stable in solution for an extended period (OPA suffers from instability of the isoindoles). The instability of OPA can be reduced by the use of organic solvents.

Three classes of derivatizing agents seem to be superior for the detection and resolution of amino acids in biological or "real-world" samples: FMOC, FMOC-Gly-Cl, and AQC (*110–113, 164, 182, 183, 189, 203, 281–292*). These "tagging" reagents are easily used with aqueous samples and have been shown not to cause racemization with proper use. Furthermore, they tend to stabilize the amino acid during workup and analysis. Also, enantioresolutions with these derivatives can be carried out on a large scale, and the retention order can be reversed by choice of reagent.

Both primary and secondary amino acids have been successfully separated by HPLC on cyclodextrin bonded columns after precolumn derivatization with FMOC and FMOC-Gly-Cl. These highly fluorogenic reagents avoid the disadvantages of other reagents. They are easy to use and have enabled analysis of ultratrace impurities in (supposedly) optically pure amino acid standards (*164*). They also eliminate the need for postcolumn derivatization, which leads to longer analysis times, sample dilution, and less efficient workup for the native amino acids (*335*).

A common problem with precolumn derivatization methods for amino acids is interference from excess reagent that tends to absorb or fluoresce at a similar wavelength to the analyte. A relatively new fluorescent tagging agent, AQC, does not suffer from this problem. It has been used for successful enantiomeric resolutions of primary and secondary amino acids on different bonded cyclodextrin stationary phases (*183, 281–283*). This reagent is especially useful because the by-products of the derivatization reaction do not seem to interfere chromatographically with the analytes of interest. Figure 5 shows the reaction of AQC with an amino acid (*183*). As can be seen, the excess reagent is hydrolyzed into *N*-hydroxysuccinimide and 6-aminoquinoline, which have significantly different fluorescence maxima. The retention behavior of AQC-derivatized amino acids is very similar to that of FMOC derivatives. In polar organic systems, AQC amino acids can be easily separated on native cyclodextrin bonded stationary phases

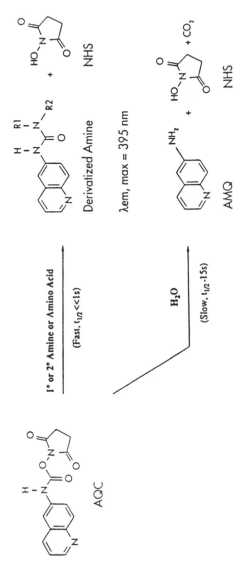

Figure 5. Derivatization chemistry. AQC stands for 6-aminoquinolyl-N-hydroxysuccinimidyl carbamate, NHS stands for N-hydroxysuccin-imide, and AMQ stands for 6-aminoquinoline. (Reproduced with permission from reference 183. Copyright 1993 Elsevier Science.)

(*183*). In the reversed-phase mode, the detection limit for amino acids ranges from 40 to 300 fmol (*283*). The use of cyclodextrin stationary phases and polar-organic (or "magic") mobile phases leads to faster equilibration, a more stable baseline, and better sensitivity, since no quenching effects are present as in reversed-phase cyclodextrin system. The detection limit was 10 fmol for many of the AQC amino acids in these systems (*183*).

Another important factor is that compatible achiral–chiral coupled column-switching procedures (columns that use compatible mobile phases) are most easily accomplished with cyclodextrin columns used in either the reversed-phase mode or the polar-organic mode (*78, 110–113, 180, 201–203*). For biological or other real-world samples, the derivatized amino acid is frequently much easier to isolate (by solid-phase extraction, liquid–liquid extraction, precipitation of proteins, etc.) and detection of low levels is straightforward.

Recently, Armstrong and co-workers described a new class of highly effective chiral selectors for HPLC (*204–206, 322, 336*), CE (*336–339*), TLC (*346*), GC (*207*), and SFC (*207*). These all belong to a family of compounds known as macrocyclic antibiotics and exist as acidic, basic, or neutral molecules with molecular weights generally between 600 and 2200. Five macrocyclic chiral selectors (vancomycin, rifamycin B, thiostrepton, teicoplanin, and ristocetin A) have been covalently linked to silica gel and evaluated by HPLC for their ability to resolve racemic mixtures (*204–207*). These CSPs can be used with great enantioselectivity in both reversed-phase and normal-phase modes for a variety of underivatized compounds and have allowed excellent resolution of *N*-blocked amino acids (*218, 219*). Generally, racemic analytes need not be derivatized before separation on macrocyclic antibiotic stationary phases. Figure 6 shows the structures of three of the macrocyclic antibiotics used in these studies.

(+)-1-(9-Fluorenyl)ethyl chloroformate (FLEC) is a fairly new chiral reagent used for the separation of amino acid enantiomers and optically active amines (*285, 295*). This reagent is related to FMOC, a precolumn derivatization reagent for primary and secondary amino acids. The reaction of FLEC with amino acids occurs at room temperature under basic conditions in aqueous solution. Highly fluorescent diastereomers of amino acid derivatives are obtained within 4 min.

Marfey's reagent, (1-fluoro-2,4-dinitrophenyl)-5-L-alanine amide, has been used to make diastereomeric derivatives of optically active amino acids that can be separated by HPLC, with the L-isomers eluting first (*293, 297, 298*). Another recently developed derivatizing reagent for enantiomeric analysis of amino acids is benzyloxycarbonyl-L-α-aminoisobutyric acid–Gly–*N*-oxysuccinimide (*299*). Mosher's acid, α-methoxy-α-trifluoromethylphenylacetic acid, is commercially available in both its (*R*)- and (*S*)-isomeric forms (*341, 342*). Before use, Mosher's acid must be activated by conversion to the acid chloride. This is followed by reaction with primary and secondary chiral alcohols and amines to produce diastereomeric esters and amides (*341, 343–345*).

There are a number of chiral ion interaction agents that are added to the mobile phase and used to resolve oppositely charged racemates (*341–351*). The

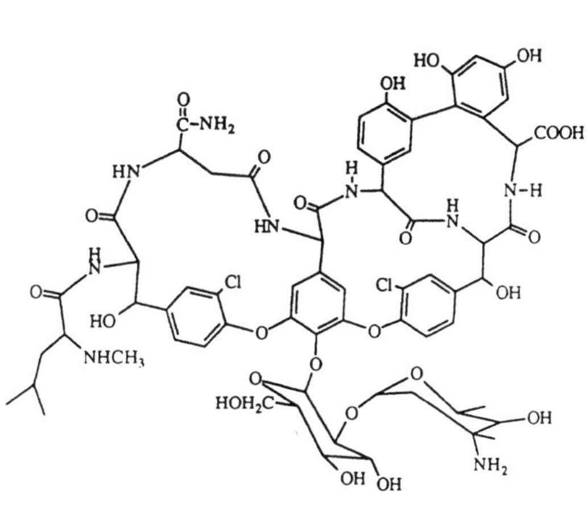

VANCOMYCIN

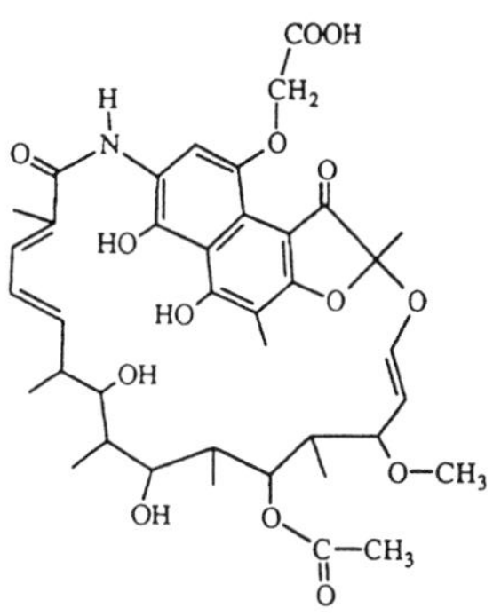

RIFAMYCIN B

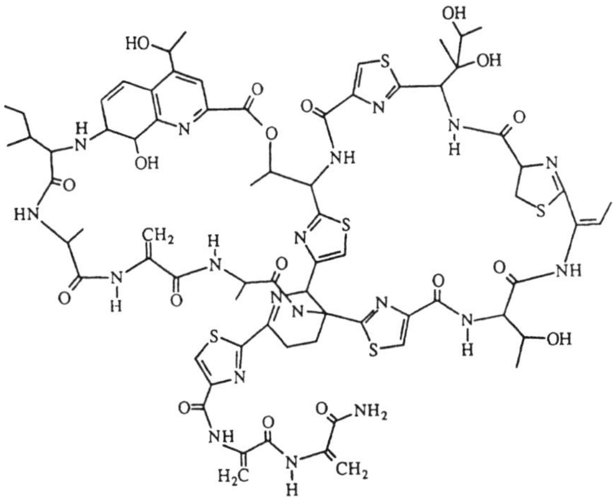

THIOSTREPTON

Figure 6. Structures of the macrocyclic antibiotics vancomycin, thiostrepton, and rifamycin B used as bonded stationary-phase chiral selectors. Vancomycin consists of three fused macrocyclic rings, thiostrepton has a large macrocyclic ring, and rifamycin B is a smaller macrocyclic compound. (Reproduced from reference 206. Copyright 1994 American Chemical Society.)

two oppositely charged chiral ions are retained on the stationary phase as an overall neutral diastereomeric pair. At least two reports use chiral mobile-phase additives to resolve the aromatic amino acid *N-t*-Boc-phenylalanine (*343, 344*). Other acidic reagents, such as mandelic acid (*350*), *O*-acetylmandelic acid (*351*), and *N*-(3,5-dinitrobenzoyl)glycine (*350*), are used for chiral amines, with which they form pairs of ions in nonpolar solvents such as trichloroethylene and benzene.

Thin-Layer Chromatography

TLC is a popular separation method because of its simplicity, low cost, low solvent consumption, and wide versatility and applicability. TLC is used extensively in amino acid, peptide, and protein research, since its reproducibility makes it ideal for microscale analysis and separation of amino acids (*352*). The enantiomeric resolution of biological molecules is influenced by the composition of the mobile phase, the type of stationary phase used, the enantioselectivity of the chiral selector used, and the efficiency of the separation (which, in turn, depends on the spot size as the plate is developed).

In TLC there are two basic ways to separate enantiomers. The most common way is through the use of chiral mobile-phase additives, especially cyclodextrins and their derivatives (*220, 353–358*). For example, Armstrong and co-workers separated many racemic dansyl amino acids using hydroxypropyl-β-cyclodextrin as the mobile-phase additive (*353, 354*). They also reported the first use of urea to increase the solubility of β-cyclodextrin in the aqueous mobile phase (*354*). Another way to separate enantiomers by TLC is by the use of CSPs (*359–366*). CSPs are used less in TLC than in other chromatographic methods, since only certain types of ligand-exchange stationary phases (ternary complex ligand-exchange plates) are commercially available. Cyclodextrin bonded stationary phases containing no nitrogen-based linkages have been shown to be very effective in high-performance TLC (*367*). Alak and Armstrong reported separation of several racemic dansyl amino acids on these CSPs (*361*). Weinstein (*362*), Grinberg and Weinstein (*368*), and Günther et al. (*363*) separated dansyl amino acids on reversed-phase plates impregnated with copper(II) complexes of chiral alkyl α-amino acid derivatives. In 1980, Yuasa et al. (*364*) reported the first enantiomeric separation of some amino acids (tryptophan, histidine, phenylalanine, tyrosine, and β-3,4-dihydroxyphenylalanine) on crystalline cellulose TLC plates. Weinstein (*362*) separated all dansyl-protected amino acids (except proline) on reversed-phase plates treated with copper(II) complexes of *N,N*-dipropyl-L-alanine. This work was later improved by Grinberg and Weinstein (*368*) using two-dimensional techniques. Brinkman and Kamminga (*365*) used commercially available chiral plates to separate underivatized chiral amino acids. Marchelli et al. (*353*) separated dansyl amino acids by one- and two-dimensional TLC with or without chiral additives in the eluent (water–acetonitrile) under isocratic conditions. If the same type of solvent system is used, TLC can (in most cases) use the same derivatizing agents, CSPs, and chiral mobile-phase additives used for HPLC.

Armstrong and Zhou (340) reported the first attempts to separate enantiomeric amino acids and racemic drugs using vancomycin as a new chiral mobile-phase additive for TLC. Vancomycin is a macrocyclic antibiotic produced by the bacterium *Streptomyces orientalis* (369). The amino acids in this study were derivatized with AQC or DNS-Cl, while the racemic drugs were underivatized. The reversed-phase mode (with a mobile phase consisting of water, acetonitrile, vancomycin, and 0.6 M NaCl) was shown to separate almost all of these compounds with ease using diphenyl-F stationary-phase plates. Salt was added to the mobile phase to stabilize the binder on the stationary-phase plates. All of the AQC amino acids had the same retention order as the dansyl amino acids (Figure 7).

Gas Chromatography

There was a tremendous expansion in the development and commercialization of CSPs for HPLC starting in the 1980s. In contrast, there appear to be at least two important periods in the development and commercialization of CSPs for GC.

The first period occurred before the HPLC expansion and is generally considered to have begun in 1966 with Gil-Av's seminal publication on the use of amino acid and peptide derivatives as GC stationary phases for the resolution of volatile amino acid esters and other compounds (370). The eventual result of all the work in this area was a single amino acid based commercialized CSP known as Chirasil-Val (371). This stationary phase, which consisted of a polysiloxane backbone with side chains containing L-valine-*tert*-butylamide residues, was developed by Frank, Nicholson, and Bayer (371–373).

The second and possibly most significant period of development for chiral GC began at the end of the 1980s and lasted into the early 1990s (32, 39, 43, 44, 51–53, 56, 86, 98, 99, 116–119, 181, 223, 236, 314, 367, 370, 373–490). It was found that derivatized cyclodextrins produced exceptional GC stationary phases for the resolution of a wide range of chiral compounds. Most of the compounds that could not be resolved by HPLC (e.g., nonaromatic and nonderivatizable compounds) were easily resolved by GC on these CSPs. The cyclodextrin-based CSPs for GC were rapidly commercialized and found to be thermally stable, widely applicable, and easy to use. They now dominate GC chiral separations. Schurig recently published an excellent review article on GC CPSs (382).

Enantioselective GC offers several other advantages over HPLC (491): speed of analysis, separation of nonderivatizable compounds, low detection limits for enantiomeric impurities, universal detectors, easy resolution of most volatile compounds, durability, and high absolute sensitivity that is independent of functional groups. The three main types of GC CSPs used in routine chiral amino acid analyses are (383) (1) chiral amino acid derivatives (Chirasil-Val) (370, 371,

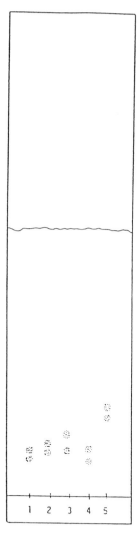

*Figure 7.   TLC chromatogram showing the separation of racemic (1) AQC ethionine, (2) AQC
α-amino-2-thiopheneacetic acid, (3) AQC α-amino-3-thiopheneacetic acid, (4) AQC α-amino-
phenylacetic acid, and (5) AQC 3-aminopiperidine dihydrochloride. The mobile phase consisted of
0.025 M vancomycin in 1:5 (by volume) acetonitrile–0.6 M NaCl (aq.). (Reproduced with permission from reference 340. Copyright 1994 Marcel Dekker Inc.)*

373–381, 492), (2) chiral metal coordination compounds (*56, 493–495*), and (3) cyclodextrin derivatives (*384–388*). Chirasil-Val (*370, 373–378, 492*) and XE-60-L-valine-(*S*)-α-phenylethylamide (*378–381*) are two complementary methods of GC enantiomeric separations that act via hydrogen bonding on chiral polysiloxane stationary phases. These CSPs are now both commercially available (*496*). All of the enantiomeric amino acids commonly present in proteins have been separated on Chirasil-Val in one chromatographic run (*371, 373*).

Many chiral food analyses have been performed with GC CSPs. Brückner and co-workers analyzed several amino acids in fermented foods and beverages on Chirasil-Val and XE-60-L-valine-(*S*)-α-phenylethylamide (*47–52, 86, 117, 118*). All fermented foods were shown to contain significant amounts of D-amino acids. Enantiomeric separations on hydrogen-bonding CSPs require derivatization of the amino acid to increase volatility or to introduce suitable functionality for additional hydrogen bonding (*382*). Since native amino acids are not volatile, all amino acids must be derivatized twice before they can be separated by GC. First, the carboxylic acid moiety must be esterified. Second, an amide must be formed by reacting the amine group with a desired agent (for example, trifluoroacetic anhydride).

The most commonly used achiral derivatization for enantiomeric amino acids in GC is perfluoroacylation using trifluoroacetyl derivatives (*56, 182, 367, 381, 497*) and pentafluoropropionyl derivatives (*371, 495, 498*). The acylation can be performed by substituting trifluoroacetyl, pentafluoropropionyl, or heptafluorobutyryl residues for the amino or hydroxyl groups (*56, 367*). The use of isopropyl esters was suggested by Gil-Av. Other functional groups, such as methyl or *n*-propyl esters, have been used as well (*387*). For quantitative amino acid analysis, methyl esters are not recommended because of their high volatility after trifluoroacetylation (*387*).

Isocyanates are also highly versatile derivatizing agents for enantiomeric analysis of amino acids by GC (*27, 499*). They react readily with chiral amines to form ureido derivatives. The reaction can be easily performed at room temperature, and the derivatized compounds can be separated with high separation factors (*500*). In the presence of a trace of pyridine, α-amino acids can also be separated after esterification and reaction with isocyanates at room temperature. The *tert*-butyl isocyanate worked better than the isopropyl isocyanate (*501*). These types of derivatives may be advantageous for amino acids such as proline and pipecolic acid, which are not easily separated as trifluoroacetyl derivatives.

Phosgene is another highly versatile derivatizing agent for GC; it forms *N*-methyloxazolidine-2,5-diones from *N*-methylamino acids (*27, 502*). However, phosgene is highly toxic and is generally avoided today. Cyclodextrin-based GC CSPs for separating volatile amino acid derivatives are the dipentylpropionyl and dipentylbutyryl derivatives (Chiraldex G-PN and Chiraldex G-BP, respectively) (*388*). One advantage of using GC and volatile amino acid derivatives is that many amino acids can be resolved in one chromatographic run. The disadvantage is that the analyte of interest must be derivatized twice to be sufficiently volatile and selective.

Capillary Electrophoresis

CE is a powerful analytical approach for the analysis of trace amounts of amino acids (*503*). CE can be used as an alternative method for the separation of many compounds (both polar and nonpolar). The short separation time (minutes) is convenient, but the small amount of solvent used (nanoliters) is considered a disadvantage for preparative-scale separations and detection sensitivity. Types of CE include capillary zone electrophoresis, capillary electrokinetic chromatography, capillary gel electrophoresis, capillary isotachophoresis, and capillary isoelectric focusing (*504*). Some require derivatization, while others separate underivatized amino acids. Only the capillary electrokinetic chromatography is able to separate chiral nonionic or neutral analytes (*504*).

The most common chiral additives for enantiomeric separations are cyclodextrins and their derivatives (*476, 477, 505–516*), surfactants (*516–522*), ligand-exchange compounds (metal complexes) (*520, 523*), proteins (*524–527*), crown ethers (*528, 529*), oligosaccharides (*530*), and most recently macrocyclic antibiotics (*207, 336–339*). Except for the antibiotics, the chiral selectors used in CE were simply borrowed from previous HPLC work (*504*).

CE is useful for the direct analysis of underivatized amino acids (*531, 532*) as well as for charged enantiomeric compounds such as dansyl amino acids (*477, 504–506, 521, 533*). CE using micelles, sometimes misnamed "micellar electrokinetic chromatography," can separate optical isomers of amino acids by using chiral surfactants and, sometimes, metal chelate complexes. Tran et al. used free solution CE and micellar CE in combination with L-Marfey's reagent and D-Marfey's reagent to separate amino acids (*534*). Sensitive phenylthiocarbamyl amino acids have been separated in a single run with UV detection at femtomole levels (*503, 535*). Cohen et al. reported the first separation of dansyl amino acids by using copper(II) complexes and N,N-didecyl-L-alanine as a chelating detergent (*536*). Because of the poor solubility of the chiral surfactant, sodium dodecyl sulfate was added to form mixed micelles.

Other reported resolutions of amino acids include N-(3,5-dinitrobenzoyl)-O-isopropyl ester derivatives separated by CE using N-dodecynoyl-L-valinate (an anionic chiral surfactant) (*522, 537*) and PTH amino acids separated with the same chiral surfactant (*538–540*). Dansyl amino acids have been separated by CE using bile salts (common biological surfactants synthesized in the liver) such as sodium taurocholate, sodium taurodeoxycholate, or sodium deoxycholate (*504, 521*). The hydroxyl-substituted steroid backbone of the bile salt contains both hydrophobic and hydrophilic parts (*504*). Digitonin has been used in CE to separate PTH amino acids (*516, 541*), while saponins have been used for separation of both dansyl and PTH amino acids (*533*).

Cyclodextrins are well suited for enantiomeric separation of amino acids by GC and HPLC and as chiral running-buffer additives for open-tube CE (*506, 515*). Some dansyl amino acids and some chiral drugs (thiopental, barbital, pentobarbital, and phenobarbital) have been separated using a combination of β-cyclodextrins, γ-cyclodextrins, and micellar solutions (*513, 542*). Enantiomeric separation of dansyl amino acids has also been reported by CE using taurodeoxycholic acid and β-cyclodextrin (*514*). Some enantioseparations of dansyl amino acids by polyacrylamide gel zone electrophoresis with the aid of β-cyclodextrin have also been reported (*505*).

Recently, Armstrong et al. demonstrated that macrocyclic antibiotics may be the most effective chiral selectors for the resolution of virtually any *N*-blocked amino acid (*207, 336–338*). This new class of chiral selectors may have both greater selectivity and wider applicability than any currently available chiral selectors used for CE. Vancomycin has been evaluated as an excellent chiral mobile-phase additive for both CE (*336*) and TLC (*340*) resolution of racemic amino acids and other compounds containing free carboxylate groups. More than 100 racemates have been resolved, including many *N*-derivatized amino acids, nonsteroidal anti-inflammatory drugs, and antineoplastic compounds. The amino acids in the CE study were derivatized with AQC, DNS-Cl, or PHTH (highly fluorescent moieties) or with *N*-3,5-dinitrobenzoyl or *N*-3,5-dinitropyridyl groups (strongly absorbing moieties). Other amino acids have simple blocking groups that did not enhance detectability (such as *N*-benzoyl, *N*-acetyl, or *N*-formyl groups). Many *N*-blocked amino acids have also been separated by using the macrocyclic antibiotics ristocetin A and teicoplanin as chiral running-buffer additives (*337, 338*). Enantiomeric resolutions (Rs) over 20 have been reported in some cases.

Sugars

Carbohydrates include many naturally occurring chiral polyhydroxy compounds, both monomers and polymers (*543*). Many are conjugated with lipids or proteins (*544*). Some sugars, especially aldohexoses and aldopentoses (*545*), occur naturally as both enantiomers (*546, 547*). Although many publications cover the topic of sugar separations, few concern the separation of epimeric, anomeric, or enantiomeric sugars. Interest in the separation of carbohydrates by HPLC increased in the 1970s and 1980s, although there was not much concern about sugar stereoisomers (*56, 202, 237, 311–313, 367, 371, 380, 452–455, 543–567*).

One way to resolve sugar enantiomers is the indirect method, which involves converting them into diastereomers by reacting each enantiomer with a chiral resolving agent (*202, 311–313, 543, 548–559, 568, 569*). The resolving agents are other enantiomerically pure compounds (amino acids, chiral thiols, chiral hydroxy compounds, etc.) (*311–313, 548–559, 568, 569*). The resulting diastereomers can then be separated by either HPLC or gas–liquid chromatography (GLC).

Oshima et al. (*551, 558, 559*) used (–)-methylbenzylamine as a chiral resolving agent for carbohydrates in HPLC. The reductive amination of the sugar enantiomers is carried out in the presence of the chiral amine and sodium cyanoborohydrate. The diastereomers can then be either selectively acetylated and separated by HPLC or trimethylsilylated and separated by GLC on polar capillary columns. Little et al. (*202, 543*) separated three pairs of sugars as diastereomeric dithioacetals. In the presence of an acid catalyst, sugars form dithioacetals when treated with thiols.

Some of the earlier attempts to separate chiral carbohydrates by GC involved formation of diastereomers (*312, 313, 548–554*). In 1968, Pollock and Jermany (*311, 556*) oxidized aldohexoses to aldonic acids and then esterified them with an optically active alcohol and peracetylated the hydroxyl groups. The volatile diastereomeric esters were then separated by GLC. Vliegenthart et al. (*312, 313*) formed glycoside diastereomers using (–)-2-butanol and acetylation. Zablocki et al. (*550*) produced acyclic sugar diastereomers of the two enantiomeric sugar pairs fucose and galactose using ethyl L-lactate as a chiral resolving agent. In this method, the sugars were first converted into acyclic diethyl dithioacetals and then acetylated and converted into the corresponding acyclic *O*-acetals using ethyl L-lactate. The resulting bis(ethyl L-lactate) acetal peracetates were resolved by GC.

All of these earlier methods suffered from problems and limitations such as low volatility of the derivatives and kinetic differentiation of the diastereomers. Another way to separate sugar enantiomers is by direct GC or HPLC separation using differential interactions with CSPs (*56, 237, 313, 367, 371, 380, 452–455, 557, 560–565*). When possible, these direct approaches are generally preferred today. They are discussed in the following two sections.

Resolution by HPLC

Large open-chain monosaccharides, consisting of five or more carbon atoms, can form cyclic structures in aqueous solution. When the sugars cyclize, two diastereomers are obtained because an additional chiral center is formed. These diastereomers are referred to as anomers. Figure 8 shows the open-chain and cyclic structures as well as the stereochemistry of D-(+)-glucose. Cyclodextrin stationary phases have proved to be very stable, efficient, and successful for saccharide analysis by HPLC (*237*). Armstrong et al. have separated both anomeric and general sugars by LC (*560*). Both α- and β-cyclodextrin columns can be used to separate many anomeric sugars, regardless of whether they show mutarotation. Figures 9 and 10 show pairs of anomers separated on α- and β-cyclodextrin columns, respectively. Both stationary phases are very effective, although the β-cyclodextrin column separated anomers somewhat better than the α-cyclodextrin column. Under the same chromatographic conditions, the retention time on the β-cyclodextrin column is slightly longer. Mutarotation can complicate a separation, since it can occur during the chromatographic process. By adjusting flow rates, separation temperatures, composition of the mobile phases, and type of

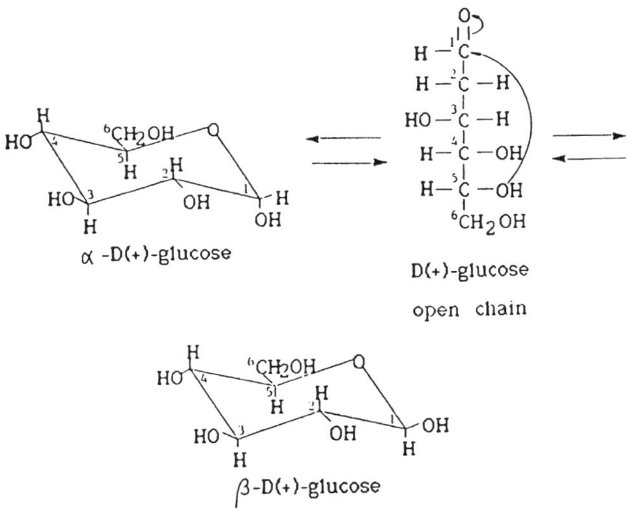

Figure 8. Open-chain and cyclic structures and stereochemistry of D-(+)-glucose. Note that upon ring closure two different diastereoisomers (α- and β-anomers) can be formed. Carbon 1 becomes an additional chiral center upon formation of the cyclic hemiacetals. The α- and β-anomers have the opposite configuration at this center. α-D-(+)-Glucose has a specific rotation of +112° and a melting point of 146 °C, while β-D-(+)-glucose has a specific rotation of +19° and a melting point of 115 °C. If either pure anomer is dissolved in water, the specific rotation will gradually reach the equilibrium value of +52.7°. (Reproduced with permission from reference 560. Copyright 1989 Wiley–Liss.)

stationary phases, mutarotation can be controlled. Table III lists the effects of some generalized separation conditions for anomers (*560*).

Resolution by GC

Sugars are found in both enantiomeric configurations in nature. In the past, the method of choice for quantitative determinations of isolated sugars has been polarimetry (*566*). A more recent method is gas–liquid chromatography (GLC), using many different types of CSPs (*56*). Examples are the XE-60-L-valine-(*S*)-phenylethylamide CSP developed by König et al. for the separation of enantiomeric pairs of aldopentoses, aldohexoses, and polyols (*561–563*) and the L-valine-*tert*-butylamide CSP coupled to a siloxane copolymer (Chirasil-Val) for the separation of sugars (*313, 371, 564*). Sherman et al. have also reported separations of carbohydrates by GC, in which pentoses and hexoses were made into their per(heptafluorobutyryl) derivatives and then separated on Chirasil-Val (*564, 565*).

Separation of enantiomeric sugars by capillary GLC on derivatized cyclodextrin stationary phases is currently the method of choice in carbohydrate

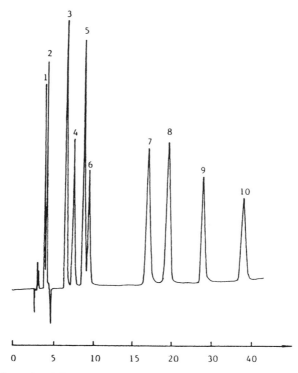

Figure 9. Separation of five pairs of anomers on a 250 × 4.6-mm (inner diameter) α-cyclodex-trin column. The peaks are (1) p-nitrophenyl-α-L-thiofucopyranoside; (2) p-nitrophenyl-β-L-thiofucopyranoside; (3) o-nitrophenyl-α-D-galactopyranoside; (4), o-nitrophenyl-β-D-galactopy-ranoside; (5) p-nitrophenyl-α-D-mannopyranoside; (6) p-nitrophenyl-β-D-mannopyranoside; (7) p-aminophenyl-α-D-glucopyranoside; (8) p-aminophenyl-β-D-glucopyranoside; (9) p-nitro-phenyl-β-D-maltoside; (10) p-nitrophenyl-α-D-maltoside. The mobile phase consisted of 91:7:2 (v/v/v) ethyl acetate–methanol–water. The flow rate was 1.5 mL/min. UV detection at 254 nm was employed. (Reproduced with permission from reference 560. Copyright 1989 Wiley–Liss.)

research. Because cyclodextrins are crystalline solids with high melting points, it is difficult to use them directly as stationary coatings for GLC (*570, 571*), but a number of suitable derivatized cyclodextrins have been developed. Examples are the liquid, lipophilic alkyl cyclodextrin derivatives developed by König et al. (*468*), used to separate sugars (*452*), and the liquid, hydrophilic cyclodextrin de-rivatives developed by Armstrong and co-workers (*367, 453, 454*). The three most common cyclodextrins are α-cyclodextrin (cyclohexaamylose), β-cyclodex-trin (cycloheptaamylose), and γ-cyclodextrin (cyclooctaamylose). Certain deriva-tives of β-cyclodextrin seem to have high enantioselectivity for sugars. The 2,3,6-tri-*O*-pentyl-derivatized and 2,6-di-*O*-pentyl-derivatized cyclodextrins are quite nonpolar, while the (*S*)-hydroxypropyl-per-*O*-methyl-derivatized cyclodextrins are

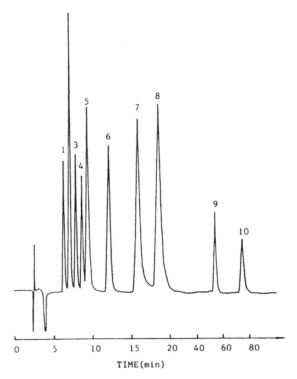

Figure 10. Separation of five pairs of anomers on a 250 × 4.6-mm (inner diameter) β-cyclodextrin column. The peaks are (1) α-naphthyl-α-D-galactopyranoside; (2) phenyl-β-D-glucopyranoside; (3) α-naphthyl-β-D-galactopyranoside; (4) phenyl-α-D-glucopyranoside; (5) p-nitrophenyl-α-D-mannopyranoside; (6) p-nitrophenyl-β-D-mannopyranoside; (7) p-aminophenyl-β-D-glucopyranoside; (8) p-aminophenyl-α-D-glucopyranoside; (9) p-nitrophenyl-β-D-maltoside; (10) p-nitrophenyl-α-D-maltoside. The mobile phase consisted of 96:2:2 (v/v/v) ethyl acetate–methanol–water. The flow rate was 1.5 mL/min. UV detection at 254 nm was employed. (Reproduced with permission from reference 560. Copyright 1989 Wiley–Liss.)

relatively polar. Most derivatized cyclodextrin CSPs are stable over a wide range of operating temperatures, as long as the carrier gas (nitrogen, hydrogen, or helium) is free of oxygen and water. König and co-workers (452) reported separations of trifluoroacetylated carbohydrate enantiomers, including more than 20 sugars, on fully pentylated α-cyclodextrin stationary phases coated on Pyrex glass capillaries. Thirteen sugars were enantiomerically separated on a variety of derivatized cyclodextrin CSPs (454). To enhance solute volatility, all sugars were trifluoroacetylated. This study also showed that only the α- and β-cyclodextrin stationary phases resolved the derivatized enantiomeric sugars. Figure 11 shows the separations of trifluoroacetylated DL-ribose on three different β-cyclodextrin stationary phases at various temperatures (454). Figure 12 shows the separation of

Table III. Chromatographic Behavior of Anomers (Mutarotation versus Separation Time)

Relative Rates	Effects on Chromatograms	Typical Chromatographic Conditions
Mutarotation $t_{1/2} \approx$ retention time	Broad peak or peak doubling	Average flow rates (0.5–1.5 mL/min), average retention times (from 10 to 90 min), combination of mobile phase and stationary phase that allows discrimination between anomers, room temperature
Mutarotation $t_{1/2} \ll$ retention time	Single sharp peak containing both anomers	Average flow rates, elevated column temperature, combination of mobile phase and stationary phase that gives poor discrimination between anomers
Mutarotation $t_{1/2} \gg$ retention time	Two peaks representing the separated anomers	High flow rates (2–5 mL/min), small retention times, temperature less than room temperature, combination of mobile phase and stationary phase that allows discrimination between anomers

SOURCE: Data taken from reference 560.

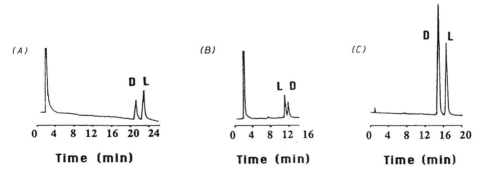

Figure 11. Enantiomeric separation of DL-ribose after trifluoroacetylation: (A) 9-m fused-silica column coated with hydroxypropylpermethyl-derivatized β-cyclodextrin, temperature 100 °C; (B) 9-m fused-silica column coated with dipentyl-derivatized β-cyclodextrin, temperature 80 °C; (C) 9-m fused-silica column coated with dipentyltrifluoroacetyl-derivatized β-cyclodextrin, temperature 120 °C. (Reproduced with permission from reference 454. Copyright 1990 Elsevier Science.)

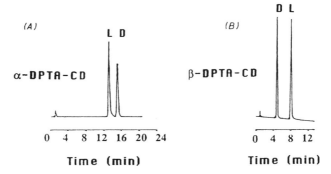

Figure 12. Enantiomeric separation of DL-*gulose after trifluoroacetylation on a 9-m fused-silica column coated with 2,6-di-O-pentyl-3-O-trifluoroacetyl-derivatized α-cyclodextrin (A) and the analogous derivatized β-cyclodextrin (B), temperature 120 °C. (Reproduced with permission from reference 454. Copyright 1990 Elsevier Science.)*

trifluoroacetylated DL-gulose on dipentyltrifluoroacetyl-derivatized α- and β-cyclodextrin stationary phases (*454*). Ten trifluoroacetylated sugars have been separated on pentylated β-cyclodextrin coated capillary columns (*455*), while several trifluoroacetylated sugars were separated on capillary columns coated with permethyl-*O*-(*S*)-2-hydroxypropyl-β-cyclodextrin (*367, 453*) and dipentyl-β-cyclodextrin (*367*). For sugars separated on two different cyclodextrins, the retention (*k'*) increases with the polarity of the cyclodextrin substituent. With CSPs like these, it is frequently possible to reverse the elution order of the enantiomeric solute by using cyclodextrin derivatives with different polarities or cyclodextrin stationary phases of different sizes (*367*).

Other Chiral Compounds

Lactones

Nearly all major food classes contain lactones, which are formed by various chemical and enzymatic reactions (*572, 573*). Both the absolute amount and the optical purity of chiral lactones are important in understanding the aroma profiles of fruits and other foods and beverages, since enantiomeric lactones may have different aromas and different characteristic intensities (*16*). Chiral aroma compounds from fruits and other natural sources have characteristic distributions of enantiomers that are formed via specific enzymatic pathways (*457, 458*). The enzyme activities and chemical pathways, as well as the enantiomeric purity of chiral components, may change during maturation and ripening, and these changes can alter the aroma, flavor, and texture of fruits and vegetables. Lactones make a key contribution to flavor in many fermentation products.

Many research papers have been published on γ- and δ-lactones in food and beverage products (*15, 17, 31, 39, 456–462, 574–579*). δ-Lactones are found mainly in animal products (as flavor components in heated meats, in fats, and in dairy products such as milk, butter, cream, and ghee) and fermented foods. γ-Lactones are considered key aroma components in fruits and are also found in vegetables and other plants (*18, 572, 580, 581*). The enantiomeric composition and chiroptical and sensory properties of naturally occurring γ- and δ-lactones have been evaluated, and their 4- and 5-alkyl homologues have been analyzed with respect to fruit-specific enantiomeric distribution (*15–17, 23, 119, 379, 463–466, 478, 575, 578, 579, 582–587*). γ-Lactones have been found in strawberries and other fruits, wines, juices, and nectars (*459*). γ-Butyrolactones, which are found in sherry, wine, and beer, can be formed during yeast fermentation.

In 1962, Muys et al. showed that yeasts could convert 4- and 5-oxo acids into optically active γ- and δ-lactones (*584*). The optically pure stereoisomers exhibit characteristic sensory differences. The γ- and δ-lactones can have very pleasant and intense fruit aromas (*16*), especially the 4- and 5-alkyl-substituted forms (*18, 572*). The 4-alkyl-substituted γ-lactones (especially 4-alkyl-substituted γ-decalactones) are known to give the distinct characteristic notes of strawberries, peaches, apricots, nectarines, and coconuts (*32, 457, 460, 479, 572, 585, 588–590*). Even though the (*R*)-enantiomers of γ- and δ-lactones predominate in most fruits in nature, the (*S*)-enantiomers may be found in amounts of 1% to 15% (compared with total γ- and δ-lactones). γ-Decalactone is known to have a fruity-peach tone, δ-decalactone a coconut tone, and γ-nonalactone a licorice or anise fragrance (*480, 591, 592*). Figure 13 shows two of these lactones separated on a 9-m 2,6-di-*O*-pentyl-3-*O*-trifluoroacetyl-γ-cyclodextrin column (*480*). Nectarines (*Prunus persica* var. *nucipersica*) contain both γ- and δ-decalactones as major aroma constituents (*460, 593*). These lactones are also characteristic aroma components of pineapple (*594*). δ-Octalactone and δ-decalactone are natural components of co-

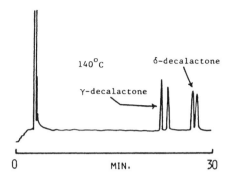

Figure 13.   Chromatogram showing the enantioseparation of γ-decalactone and δ-decalactone on a 9-m 2,6-di-O-pentyl-3-O-trifluoroacetyl-γ-cyclodextrin column, temperature 140 °C. (Reproduced from reference 480. Copyright 1990 American Chemical Society.)

conuts, while γ-lactones are absent (*469, 595*). The enantiomeric composition of
δ-octalactone and δ-decalactone in coconut cream has also been determined
(*379*). Fresh strawberries contain optically pure (4*R*)-γ-decalactone and (4*R*)-γ-
dodecalactone, while the (4*R*)-γ-undecalactone exists in the enantiomeric ratio of
70:30 (*457, 470, 582, 586, 596*). The prevalence of the (4*R*)-enantiomer corresponds to earlier findings in the literature (*586, 596*). Also, the γ-decalactone and
γ-dodecalactone is the distillate of an apricot brandy showed optical purities
similar to those in strawberries. Tang and Jennings were the first to identify γ-lactones in apricots (*597, 598*). Guichard investigated enantiomeric γ-lactones in
apricots and found that odd-numbered γ-lactones occur only at trace levels (*460*).
The even-numbered γ-lactones (C_6 to C_{12}) were the most abundant, and the proportions of the (4*R*)-enantiomer were shown to increase with increasing length of
the alkyl side chain, as was also seen for the lactones in fresh strawberries (*456,
460, 461, 587*). The most abundant γ-lactone in apricots is the γ-decalactone. It
has the most apricotlike aroma, sometimes describes as a fatty, sweet, fruity, and
coconut tone. The γ-octalactone in apricots is found in amounts of 7% to 25%
(compared with all γ-lactones in apricots) and is known to have aroma characteristics of the spicy-green, coconut, and almond (*460*). γ-Heptalactone and γ-nonalactone were found in amounts less than 2.5%, while γ-undecalactone was present in only trace amounts (compared with all γ-lactones in apricots). Again, as
for strawberries, the (4*R*)-enantiomer was always found to be predominant for γ-
lactones (*14, 16–18, 460, 461, 481, 576, 599, 600*). In apricot wines, the γ-decalactone and γ-dodecalactone contain a much higher level of the (4*R*)-enan-
tiomers (*458*). The flavors of these fruits and other food products come not from
the lactones alone but from a combination of a variety of aroma components
(*601*). Some examples are 1,4-undecanolide in peaches (*602, 603*) and the characteristic aroma compounds of mangos (*604*) and strawberries (*596, 605, 606*).

    Since natural racemic γ-lactones have not yet been observed in fruits and
fruit-containing foods and beverages, the detection of racemic γ-lactones in
commercial fruit products indicates the addition of synthetic flavoring (*30, 458,
470*). Racemic γ-deca-, γ-undeca-, and γ-dodecalactones have been found in
many commercial fruit-containing beverages (*457*). Yogurts with strawberries
were found to contain γ-decalactone and γ-dodecalactone in enantiomeric
amounts that differ from their distribution in fresh strawberries (*457, 459*). Commercial fruit nectars and other fruit-containing beverages (multivitamin fruit nectar, diet fruit nectar, and alcoholic beverages) also showed the addition of
racemic lactones (γ-nonalactone and γ-undecalactone) (*119, 457–459, 466, 467,
607*). These results prove that "nature-identical" racemic γ-lactones have been
added to the beverages. Racemic γ-lactones have also been found in
orange–maracuja nectar and passion fruit drink by multidimensional GC. The
addition of these compounds to fruit juices violates German food regulations
(*458*). Artho et al. found chiral γ-lactones in fruit-containing yogurt, coconut aroma, and apricot beverages by coupled LC–GC (*479*). Analyses revealed that
some artificial flavors are added or used in direct contradiction to the law and the

content declaration. Since standard (synthetic) lactones are required for sensory evaluation, a very reliable method for generating optically pure γ- and δ-lactones must be available. All aroma-relevant chiral γ- and δ-lactones have been synthesized in their optically pure forms and their sensory properties evaluated (*576, 586*).

Other important lactones include solerone (5-oxo-4-hydroxyhexanoic acid γ-lactone) and solerole (4,5-dihydroxyhexanoic acid γ-lactone), which have been found in sherry by HPLC and capillary GC analyses (*472, 608, 609*). 3-Methyl-4-octanolide (β-methyl-γ-octalactone), often called "whiskey lactone" or "*Quercus* lactone," was isolated from oak wood and separated into stereoisomers by GC and HPLC (*15, 470, 579*). The naturally occurring stereoisomers were found to be 77% (3*S*,4*S*)-configuration of the *cis*-lactone and 23% (3*R*,4*R*)-configuration of the *trans*-lactone (*314, 464, 470, 579*). Detection and separation were done by GC on a hexakis(3-*O*-acetyl-2,6-di-*O*-pentyl)-α-cyclodextrin stationary phase.

Analytical methods for separation of chiral γ- and δ-lactones utilize either derivatization to diastereomeric esters (*30, 31, 66, 466, 577, 610*) or direct resolution (mostly using modified cyclodextrin stationary phases) (*119, 236, 314, 457–475, 587, 611–615*), followed by analysis by complexation GC (*457, 460, 463, 616*), multidimensional GC (a very sensitive method used for differentiating γ-lactone isomers from complex flavor matrices) (*457, 461, 467*), or ordinary GC. Although both direct and indirect resolution methods are reviewed in this section, the preferred method today is direct separation on specific cyclodextrin-based CSPs.

Some derivatizing agents used for the GC separation of lactones are (*S*)-*O*-acetyllactic acid chloride (*31*), (–)-2,3-butanediol (for δ-lactones) (*379*), and (–)-(2*R*,3*R*)-butanediol–*p*-toluenesulfonic acid (for γ-lactones) (*379*). Ring opening followed by derivatization is one way to form diastereomers from lactones. (*R*)-(+)-Phenylethylisocyanate is a good derivatizing agent for 4- and 5-hydroxy acid esters (obtained from chiral γ- and δ-lactones through ring opening) (*345*). Mosher's acid, (*R*)-(+)-α-methoxy-α-trifluoromethylphenylacetic acid, is useful for converting chiral 4-hydroxy compounds, such as γ-lactones (but not δ-lactones), into diastereomeric esters (*341, 343, 344, 379*).

At least one functional group is required for the conversion of enantiomers into diastereomeric derivatives. Some examples are γ-lactones of peaches and strawberries, which can be analyzed after derivatization into 4-alkyl-substituted diastereomeric esters (*585, 596*). Diastereomeric esters of 4-hydroxy-3-methyloctanoic acid isopropyl ester can be separated by HPLC (*18*). These derivatized aroma-relevant γ-lactones are stable, since the hydrolysis and relactonization take place without racemization (*15, 16, 579*). For the separation of δ-lactones, the alkyl-substituted δ-lactones are derivatized to form diastereomers of 5-hydroxyalkanoic acid isopropyl esters (*18*). The enantiodiscrimination of chiral 1,4-diol acetonides (*616*) by complexation GC has proved to be a further possibility for differentiating chiral γ-lactones after LiAlH$_4$ reduction with subsequent cyclization to chiral 1,3-dioxepanes and stereoanalysis by GC using CSPs (*463*).

The first direct chirospecific analyses of chiral γ-lactones in food, beverages, and other fruit-containing commercial preparations have been achieved by off-line coupling of HPLC with GC on modified chiral α- and β-cyclodextrin stationary phases (*119, 462, 470, 614*). Armstrong et al. separated β-butyrolactone and α-methyl-β-butyrolactone on a 9-m fused-silica capillary column (0.25-mm inner diameter) coated with permethyl-*O*-(S)-2-hydroxypropyl-β-cyclodextrin (Figure 14) (*453*). They also separated 10 chiral lactones using 2,6-di-*O*-pentyl-3-*O*-trifluoroacetyl α-, β-, and γ-cyclodextrins as highly selective CSPs for capillary GC (*471*).

Alcohols

1-Octen-3-ol

The chiral allyl alcohol 1-octen-3-ol is sometimes called the "mushroom alcohol" because its (*R*)-(–)-enantiomer is the main compound responsible for the flavor of mushrooms (*17, 44, 617–620*). The unsaturated (*R*)-(–)-1-octen-3-ol occurs in many edible mushroom species, including *Psalliota campestris* and *P. bispora*, in which it exhibits the fruity, soft odor of fresh mushrooms (*621*). It has also been found in several molds (*622*). The (*S*)-(+)-enantiomer, on the other hand, is responsible for the unpleasant, moldy, grassy, and artificial mushroom aroma (*30*).

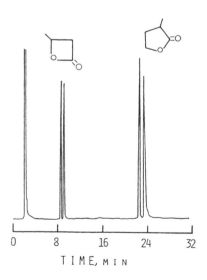

Figure 14. Enantiomeric separation of lactones on a 9-m fused-silica capillary column (0.25-mm inner diameter) coated with permethyl-O-(S)-2-hydroxypropyl-β-cyclodextrin. The column temperature was held at 60 °C for 10 min and then increased to 70 °C at 3 °C/min. The carrier gas was N₂, 3 psi. (Reproduced from reference 453. Copyright 1990 American Chemical Society.)

1-Octen-3-ol has also been found in cattle odors as a strong olfactory stimulant and attractant for the tsetse fly (*23*).

1-Octen-3-ol is known to be an autoxidation product of linoleic acid found in butter oil (*236*). Tressl et al. showed that 1-octen-3-ol is formed from linolenic acid and that the 1-octen-3-one intermediate is subsequently reduced to 1-octen-3-ol by an alcohol oxidoreductase (*619, 623*). An increase in 1-octen-3-ol during storage is proof of degradation of the butter oil due to peroxidation (*624*). Varoquaux et al. showed that this alcohol is formed enzymatically during the aerobic oxidation of linoleic acid (a major fatty acid) from lipids of *P. bispora* (*625*). Schurig reported difficulties with GC separations of both derivatized and underivatized 1-octen-3-ol (*44*). However, enantiomeric separations of trifluoroacetylated 1-octen-3-ol by GC were reported by Stalcup et al., using three different cyclodextrin CSPs: heptakis (2,6-di-*O*-pentyl-3-trifluoroacetyl)-β-cyclodextrin, octakis(2,6-di-*O*-pentyl-3-trifluoroacetyl)-γ-cyclodextrin, and permethyl-*O*-[(*S*)-2-hydroxypropyl]-β-cyclodextrin (*483*).

The first attempts to characterize the sensory properties of 1-octen-3-ol were reported by Dijkstra and Wiken (*626*). Ullrich and Grosch also indicated the importance of 1-octen-3-ol for sensory evaluation of mushrooms (*627*). In the late 1980s the optical purity of 1-octen-3-ol as diastereomeric esters in mushrooms was investigated (*31*). This analysis showed that there is a fruit-specific distribution of enantiomeric flavor compounds in nature. 1-Octen-3-ol in mushrooms can also be separated by GC via diastereomer formation with Mosher's acid (*379*), via (*S*)-*O*-acetyllactic acid (*31, 578*), or by enantiomeric resolution of 1-octen-3-yl acetate (*23*). Also, Mosandl et al. (*17*) used ^{1}H NMR spectroscopy to determine the final configuration of 1-octen-3-ol and to evaluate the optical purity of the enantiomers. König and co-workers (*628*) separated 1-octen-3-ol in essential oils using capillary GC with modified cyclodextrins. Other separations of 1-octen-3-ol enantiomers have been reported (*629*).

Alkan-2-ols

Alkan-2-ols are important aroma constituents of fruits and vegetables (*23, 30–32, 236, 344, 463, 483, 630–634*). One of the major volatile aroma constituents of blackberries (*Rubus laciniata* L.) is heptan-2-ol, which occurs mainly as the (*S*)-enantiomer (*630*), although traces of the (*R*)-enantiomer have also been detected (*32*). In corn (*Zea mays*), the major constituent of heptan-2-ol is the (*R*)-enantiomer (*32*). As the chain length of the secondary alcohol increases, so does the proportion of the (*S*)-enantiomer. This pattern points toward two different pathways or enzymes in corn, which seem to compete and lead to different enantiomeric ratios of secondary alcohols. Octan-2-ol, nonan-2-ol, decan-2-ol, and undecan-2-ol are all present in corn. The same types of enantiomeric secondary alcohols and trends were also found in coconuts (*Cocos nucifera*) (*32*). The major alcohol is heptan-2-ol, which again shows the highest amount of the (*R*)-enantiomer. 1-Octen-3-ol, pentan-2-ol, and heptan-2-ol are all characteristic flavor compounds of certain fruits and vegetables (*30, 463, 483, 631, 632*). The last two are also well-known components of chiral fruit esters in bananas, where they exist as the optically pure (*S*)-(+)-enantiomers (*30, 31, 631, 632*). Secondary alcohols have been investigated as different diastereomeric derivatives (*635, 636*). Odd-numbered secondary alcohols (pentan-2-ol, heptan-2-ol, nonan-2-ol, etc.) are present as aroma components in yellow and purple passion fruits, while the corresponding esters are typical only for the purple variety (*344, 633*). Pentan-2-ol and heptan-2-ol have been separated in the yellow passion fruit (*Passiflora edulis* f. *flavicarpa*) by GC as diastereomeric ester derivatives of (*R*)-(+)-Mosher's acid and were shown to have a higher amount of the (*S*)-(+)-enantiomers (*344*). This is in contrast to the findings from the purple passion fruit (*Passiflora edulis* sims), which contains mostly the (*R*)-(–)-enantiomer (92%) of heptan-2-ol (*344*). Obviously, different enzyme pathways must exist in the production of these chiral alcohols. Several chiral alcohols (pentan-2-ol, heptan-2-ol, octan-2-ol, nonan-2-ol, and undecan-2-ol) were separated as diastereomeric esters of Mosher's acid and as diastereomeric urethanes of (*R*)-(+)-phenylethylisocyanate (*341, 343–345*). Armstrong et al. separated enantiomers of 19 alcohols known to be present in coffee, tea, or cocoa using coated capillary GC columns (*483*). Previously they enantioresolved alcohols by GC using various derivatives of α-, β-, and γ-cyclodextrins as CSPs (*231, 471*). Reiher and Hamann separated several chiral alcohols by GC using hexakis(2,3,6-tri-*O*-pentyl)-α-cyclodextrin as the CSP (*482*).

   Although two general approaches for the resolution of these enantiomers have been reviewed (direct resolution on a CSP and reaction with a chiral derivatizing agent followed by separation of diastereomers), the preferred approach today is direct resolution on a CSP.

Menthol

Menthol, a chiral monoterpene alcohol, is an important component of many commercial products such as mentha oil, other essential oils, tobacco, cosmetics, pharmaceuticals, alcoholic beverages, other beverages, and foods (*39, 236, 637–640*). The eight stereoisomers of menthol (an aliphatic pheromone alcohol) have been separated by complexation GC on manganese(II) and nickel(II) bis[3-(heptafluorobutanoyl)-(1*R*)-camphorate] (*28, 44*). They have also been separated by GC as esters of (*R*)-(+)-Mosher's acid (*344*) and as urethanes of (*R*)-(+)-phenylethyl isocyanate (*27, 345*). Although eight stereoisomers of menthol exist, only the (1*R*,3*R*,4*S*)-isomer has the well-known minty odor (*23, 24, 27–29*). Multidimensional GC with derivatized β-cyclodextrin as the CSP was used to separate the menthol stereoisomers in peppermint oils (*639*). This technique allows the direct enantiomeric resolution of menthol from the essential peppermint oils without previous purification and isolation. Enantiomeric ratios can be used to distinguish between natural and synthetic flavors and fragrances. In the previous analysis it was found that commercial mint oils contained (+)-menthol, which can be regarded as evidence of adulteration (*639*). Recently, menthol enantiomers were directly resolved by capillary GC columns coated with permethyl-O-[(S)-2-hydroxypropyl]-β-cyclodextrin (*483*). Additional separations of menthol enantiomers have been published by several research groups (*29, 39, 629, 640–642*).

Other Alcohols

The three stereoisomers of butane-2,3-diol found in wine and sherry have been directly separated by Mosandl and co-workers using GC and the heptakis(2,3,6-tri-O-methyl)-β-cyclodextrin CSP (*643*).

Armstrong and co-workers reported the first known separation of both pairs of enantiomers of 3-methyl-2-pentanol by GC (*483*). Figure 15 shows the separation of all four stereoisomers of trifluoroacetylated 3-methyl-2-pentanol. Since no pure enantiomeric standard existed for the individual isomers at the time of separation, it was not possible to determine the absolute configuration of each peak.

Mosandl and co-workers investigated the enantiomeric distributions of 2-alkyl branched alcohols in apples and pineapples (*644*). The analysis was per-

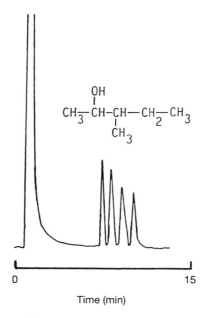

Figure 15. GC chromatogram showing the resolution of all four stereoisomers (two pairs of enantiomers) of 3-methyl-2-pentanol as the trifluoroacetyl derivative. The separation was done at 25 °C on a 10 m × 0.25 mm (inner diameter) Chiraldex β-trifluoroacetyldipentyl cyclodextrin (B-TA) column. The split ratio was 100:1, and flame ionization detection was used. (Reproduced from reference 483. Copyright 1993 American Chemical Society.)

formed with capillary GC using permethylated β-cyclodextrin as the CSP. They also analyzed apples, apple juices, and apple wines for the enantiomeric composition of 2-methylbutan-1-ol using multidimensional GC and permethylated β-cyclodextrin as the CSP (*484, 485, 645, 646*). LC was also used to analyze chiral 2-alkyl branched alcohols (*644, 647*).

Esters, Acids, Aldehydes, and Ketones

Either enantiomeric form of a molecule can predominate in different plants and animals. One example is 3-hydroxybutyrate. The (*S*)-enantiomer predominates in the yellow passion fruit, while in the mango and purple passion fruit the (*R*)-enantiomer is in excess (*648*). The same trend can be found for ethyl-3-hydroxybutyrate: In yellow passion fruit the (*S*)-(+)-isomer predominates (82%), while in purple passion fruit and mango the (*R*)-(−)-enantiomer constitutes 69% and 78% of this compound, respectively (*344*). Other 3-hydroxyacid esters have been found in pineapple (*644*), passion fruit (*650*), and mango (*651*). 3-Hydroxyacids can be separated on CSPs after formation of carbamate or amide derivatives (*501*) or 3-pentylesters (*652*). The corresponding diastereomers were resolved by derivatization with (*S*)-(−)-phenylethylisocyanate or (−)-camphanic acid chloride (*653*).

Goodall et al. used reversed-phase ion-exchange LC methods with a diode-laser-based dual polarimetric ultraviolet detector combination for analyzing (*S*)-(−)-malic acid in fruit juices and ciders (*654, 655*). Only one achiral HPLC separation was necessary instead of the usual HPLC plus enzyme assay. Because malic acid occurs only as the (*S*)-enantiomer in nature, adulteration with racemic malic acid is relatively easy to detect. Almost all fruit juices and ciders assayed were found to contain only the natural (*S*)-malic acid, while a pear drink was shown to contain some synthetic (*RS*)-malic acid. This finding is acceptable according to the food standards, which require that all juices be natural, whereas drinks need not be (*655*). The enantiomers of malic acid, together with various other acids and esters, has been analyzed and separated by GC on a derivatized β-cyclodextrin CSP (*480*).

Mosandl and co-workers analyzed apples and pineapples for enantiomeric distributions of acids and esters (*644*). The analysis was performed with capillary GC using permethylated β-cyclodextrin as the CSP. They also analyzed apples, apple juices, and apple wines for the enantiomeric composition of methyl- and ethyl-2-methylbutyrates using multidimensional GC and permethylated β-cyclodextrin as the CSP (*484, 485, 640, 646*). Chiral acids and esters have also been separated by LC (*647*). The enantiomers of 2-methylbutyric acid were analyzed in alcoholic beverages, fruits, fruit-containing yogurts, cheeses, and jams (*486*). Multidimensional GC, using heptakis(2,3,6-tri-*O*-methyl)-β-cyclodextrin as the CSP, was used to analyze all samples. The (*S*)-enantiomer predominated (>80%) in all food samples (*486*). Several ketones have been separated by various methods in the past (*28, 615, 656*). König and co-workers separated several ketones on chiral polysiloxane stationary phases (*657*) and, later on, using cyclodextrin CSPs in GC (*465*).

Olefinic Monoterpene Hydrocarbons

The most volatile and abundant components of essential oils are monoterpene hydrocarbons (*487*). They contribute significantly to flavor and taste. Monoterpenoids and sesquiterpenoids are primary plant products and are well-known characteristic components of essential oils (*573*). Limonene, α-pinene, and β-pinene are examples of chiral monoterpene hydrocarbons.

Limonene is known to occur in several ethereal oils, including those of lemon, orange, mandarin, grapefruit, bergamot, caraway, and dill (*23, 658–660*). The (*S*)-enantiomer of limonene smells like lemons, while (*R*)-limonene smells like oranges. Limonene has also been identified as a metabolite of fungi (*573*).

α-Pinene and β-pinene are constituents of many volatile oils and have the characteristic odor of turpentine. Generally, α-pinene accounts for 58% to 65% and β-pinene for 30% of the total pinenes (*661*). GC utilizing direct resolution via cyclodextrin-based stationary phases seems to be the most successful technique for resolving chiral hydrocarbons (*662*). The separation efficiency for open tubular capillary columns is remarkably better than that obtained with packed columns. Smolková-Keulemansová and co-workers separated structural hydrocarbon isomers by packed-column GC using native, methylated, and acylated cyclodextrins (*620, 663*). Kocielski and co-workers separated α- and β-pinene on a similar GC system using α-cyclodextrin as a selective agent in GLC (*571, 664*). They also separated chiral hydrocarbons by GLC using zeolite coated with α-cyclodextrin in formamide solution (*665*).

Lindström et al. (*666*) used packed columns (with solutions of cyclodextrins as CSPs on a solid support) to separate the enantiomers of limonene, α-pinene, and β-pinene in an extract from Norway spruce (*Picea abies*). Schurig and Nowotny (*393*) separated several monoterpenes using open tubular columns coated with permethylated β-cyclodextrin and separated several unfunctionalized cycloalkanes using capillary GC with β-cyclodextrin dissolved in silicone oil (*489*). König and co-workers (*628*) separated α-pinene and limonene in different pine needle oils using GC and heptakis(3-*O*-methyl-2,6-di-*O*-pentyl)-β-cyclodextrin as the CSP. They also separated limonene, α-pinene, β-pinene, and other olefinic monoterpene hydrocarbons occurring in essential oils by GC using heptakis(6-*O*-methyl-2,3-di-*O*-pentyl)-β-cyclodextrin and octakis(6-*O*-methyl-2,3-di-*O*-pentyl)-γ-cyclodextrin as CSPs (*487*). Additional GC separations of monoterpene hydrocarbons have been published by König and co-workers (*490, 667, 668*).

Recently, Armstrong et al. (*483*) analyzed chiral components known to be present in coffee, tea, or cocoa. α-Pinene and limonene were enantioresolved on capillary GC columns coated with octakis(2,6-di-*O*-pentyl-3-trifluoroacetyl)-γ-cyclodextrin and permethyl-*O*-[(*S*)-2-hydroxypropyl]-β-cyclodextrin. Limonene enantiomers were also resolved by GC using various *O*-[(*S*)-2-hydroxypropyl] derivatives of α- and β-cyclodextrins (453). Mosandl and co-workers (669, 670) investigated the enantiomeric composition of limonene, α-pinene, and β-pinene in different plant sources using GC and capillary columns with permethyl-β-cyclodextrin diluted in polysiloxane OV 1701. They also investigated the enantiomeric distribution of limonene, α-pinene, and β-pinene in several foods, drugs, and essential oils (671–673). Armstrong et al. (674) reported separations of aromatic and aliphatic hydrocarbon biomarkers (tetralins, indans, isoprenoids, etc.) on capillary GC columns coated with alkylacylcyclodextrin derivatives and hydroxypropylated cyclodextrins. Sybilska and co-workers (675) have previously separated α- and β-pinene by GC with α-cyclodextrin as a selective agent. They have also investigated the enantiomeric composition of limonene, α-pinene, and β-pinene in juniper oil and in juniper berries, where the main characteristic component is known to be α-pinene (676).

Previously, there have been few successful separations of enantiomeric hydrocarbons by LC. GC is better than LC at directly resolving molecules with limited functionality. In the early 1980s, two reports were published on the indirect resolution of olefins as diastereomeric platinum(II) complexes (677, 678). Zukowski (679) recently reported the direct HPLC resolution of α-pinene using α-cyclodextrin as a mobile-phase additive. Also, Armstrong and Zukowski (662) reported successful separations of commercial standards of α- and β-pinene by reversed-phase HPLC using a native α-cyclodextrin bonded-phase column. All commercial samples were shown to contain significant amounts of enantiomeric impurities. Additional reported separations of monoterpene hydrocarbons (either standard compounds or components of essential oils, foods, and beverages) have been published by several researchers (488, 629, 634, 680–682). Menthol, a chiral monoterpene alcohol, is discussed in the previous alcohol section.

Vitamins

Many vitamins are important industrial chemicals and belong to a broad class of compounds that generally contain one or more chiral centers. Some examples are vitamin A, with four double bonds in the side chain and 16 possible geometric isomers; folic acid (vitamin Bc or vitamin M), with one chiral center; folinic acid, with two chiral centers; pantothenic acid (vitamin B_3 or vitamin B_5), with one chiral center; vitamin C (L-ascorbic acid), with two chiral centers; vitamin D_2 (calciferol or ergocalciferol), with six chiral centers; vitamin D_3 (cholecalciferol or colecalciferol), with five chiral centers; vitamin D_4, with six chiral centers; vitamin E (α-tocopherol), with three chiral centers; biotin (vitamin H), with three chiral centers; vitamin K_1 (koagulationsvitamin), with two chiral centers; vitamin L, with four chiral centers; and vitamin U (S-methylmethionine), with one chiral center. Vitamins can occur naturally or be used as food additives (fortifiers, stabilizing agents, etc.). In either case, an evaluation of their stereochemical composition could provide information on their effectiveness, age, storage conditions, origin, and so on. Vitamins L and U are called vitamins in the literature, even though they do not fit the broad definition of vitamins (*683*).

Unlike most of the previous classes of compounds in this review, the stereochemical composition of many vitamins is either assumed to be purely the "active" form or ignored completely. Consequently, there have been relatively few reports on the stereochemical analysis of vitamins or on the biological effects and biochemical pathways of vitamin isomers. This will undoubtedly change in the future.

Vitamin A

In vertebrates, plant dietary carotenoids (especially β-carotene) are converted into vitamin A (retinol or vitamin A_1), which exists in the *all-trans*-configuration. Since practically all vitamin A found in nature is derived from carotenoids, they are called provitamin A. Vitamin A isomers have been separated by TLC, GC, and HPLC (*684–687*).

Vitamin A is an important component of the visual cycle. In the blood, *all-trans*-retinol is oxidized by alcohol dehydrogenase (with NADP) to retinal, followed by isomerization in the retina to 11-*cis*-retinal (*688–694*). 11-*cis*-Retinal will then combine with opsin (*all-trans*-retinol will not combine with opsin) to form rhodopsin. After exposure to light, several changes occur in rhodopsin, and the retinal eventually splits off as the *all-trans*-isomer and can presumably be reused in the retina.

Stereospecific synthesis of the *all*-(E)-form of vitamin A has been achieved (*695–697*). Natural vitamin A, which can be used as a therapeutic agent in certain types of cancers, also causes toxic liver cancer (*698*). On the other hand, 13-*cis*-retinoic acid (the unnatural geometric isomer) has a similar proven therapeutic ability with far less liver toxicity.

The *trans*-isomers of vitamin A were detected in 10 sample lots of fresh tomatoes, whereas the *cis*-isomers were not detected (*699*). In the same study, 52 tomato products showed different amounts of *cis*-isomers of β-carotene. (Carotenoids are converted into vitamin A by the liver.) A recent HPLC investigation showed that β-carotene, the only provitamin A carotenoid present in spinach, accounted for 28% of the total amount of carotenoids (*700*). When the β-carotene isomers were not separated, the vitamin A values were overestimated by 14% and 16% for fresh and dehydrated samples, respectively.

Another study demonstrated the utilization of a natural β-carotene stereoisomeric mixture from dried mycelium of the fungus *Phycomyces blakesleeanus* as a natural source of vitamin A in the diet of rats (*701*). The β-carotene in the fungus is composed of 10% to 20% of the 9-*cis*-stereoisomer, while the rest is the *all-trans*-stereoisomer. The *all-trans*-β-carotene was shown to have a higher vitamin A biopotency than the 9-*cis*-β-carotene.

Stalcup et al. (*687*) separated 15 different compounds related to carotene (including *trans*- and *cis*-β-carotenes, α-carotene, substituted carotene, lycopene, xanthopyll, zeaxanthin, retinal, and retinol) by HPLC on a Cyclobond I column.

Folic Acid

Folic acid (vitamin Bc or vitamin M) is known to reduce the incidence of all forms of congenital abnormalities as well as guarding against heart disease in middle-aged men (*702*). To our knowledge there are no reports on its stereochemical analysis and no studies on the physiologic effects of different isomers.

Folinic Acid

(–)-Folinic acid is the metabolic intermediate of folic acid and the active physiologic form that is incorporated into the body (the (+)-form is inactive) (*703, 704*). The chiral center is located at the carbon 6 of the pteridine ring (*705, 706*). High doses of the (6S)-stereoisomer of folinic acid have been used in combination with fluorouracil for antitumor treatments of patients with advanced colorectal carcinoma (*705, 706*). Some commercial folinic acid consists of equal amounts of the (6R)- and (6S)-stereoisomers. The (6S)-stereoisomer is readily transformed into active folate cofactors, while the (6R)-stereoisomer is not. One disadvantage of this treatment is the possible interference from the (6R)-stereoisomer at the cellular level (*705, 706*).

Pantothenic Acid

Pantothenic acid (vitamin B_3 or vitamin B_5) belongs to the B-complex vitamins. The starting material for pantothenic acid is isobutyraldehyde (*707*). The physiologically active form is the (+)-enantiomer, while the (–)-enantiomer is inactive (*708*).

Vitamin C

The active form of vitamin C is the L-enantiomer of ascorbic acid, while the D-enantiomer is an inactive analogue (*709*). The L-enantiomer functions as an electron carrier and is converted to dehydro-L-ascorbic acid by giving up two electrons (*709*). In nature, L-ascorbic acid in animals and plants is biosynthesized from D-glucose, D-galactose, and other carbohydrates. In the laboratory, L-ascorbic acid can be produced by the use of recombinant "metabolically engineered" bacterial strains able to synthesize 2-keto-L-gulonic acid, a key intermediate in the manufacture of L-ascorbic acid (*710*).

Vitamin D

Several substances are capable of performing the nutritional function of vitamin D (promoting growth and preventing rickets in young animals) (*711*). Vitamin D_2

(calciferol or ergocalciferol) and vitamin D_3 (cholecalciferol or colecalciferol) are the best known (and economically most important) examples of this group of vitamins. The D vitamins are not nutritionally required by organisms exposed to sunlight (ultraviolet light). The D vitamins are steroids and take part in the hormonal system that regulates calcium homeostasis. Various stereoselective novel synthetic approaches to vitamin D have been reviewed (*712*). As with most of the vitamins mentioned previously, there are few, if any, reports on stereoselective separations or on the physiologic effect of different isomers.

Vitamin E

The vitamin E group of compounds consists of many naturally occurring, lipophilic tocopherols, tocotrienols, and tocopherol esters (*713*). Natural vitamin E consists of two types of compounds: the tocotrienols, with an unsaturated side chain, and the tocopherols, with a saturated side chain. The major components of vegetable fats with vitamin activity are the tocopherols. Four of the vitamin E tocopherols (α-, β-, γ-, and δ-tocopherol, in order of descending biological activity) have been detected in plant oils (including cooking oil) by HPLC (*714, 715*). The three chiral centers of each tocopherol allow for a total of eight stereoisomers (*713*). The most biologically active tocopherol isomer is α-tocopherol, which exists in the chloroplasts of young plants, fruits, and berries (*716*). As the plants and fruits mature, other tocopherols begin to appear (*717*). The naturally occurring stereoisomer of α-tocopherol found in nature is *d*-α-tocopherol, with all three chiral centers in the (*R*)-configuration (*713*). Naturally occurring tocotrienols have *all-trans*-configurations. While the best method for analyzing the eight isomers of each tocopherol appears to be HPLC, much of the literature still refers only to the analysis and separation of the four different tocopherols (α-, β-, γ-, and δ-tocopherol) (*718–720*).

Vitamin E isomers are important natural antioxidants (*720*). In a recent study, 13 types of amaranth (*Amaranthus cruentus*) were investigated for the composition of vitamin E compounds. The most common tocopherols found were α-tocopherol, β-tocotrienol, and γ-tocotrienol. Vitamin E is required for normal growth in animals and humans. The lack of this vitamin may cause infertility and abnormalities of the central nervous system (*721*). The stereospecificity of the uptake of vitamin E isomers in different body tissues has been discussed by Wayner (*722*). The biological activities of the natural vitamin E isomer (*d*-α-tocopherol) and of synthetic stereoisomers have also been reviewed by Wallet (*723*).

In nutritional supplements, α-tocopherol acetate is the most available form of vitamin E (*724*). The available (*R,R,R*)-stereoisomer and the other stereoisomers have been analyzed for relative bioavailability using deuterium-labeled isotopes. The analyses were performed by HPLC and GC–MS. The results showed that the bioavailability ratio of (*R,R,R*)- to *all-rac*-α-tocopherol acetate is greater than the currently accepted ratio of 1.36. Another study concluded that (*R,R,R*)-α-tocopherol acetate is hydrolyzed more slowly than the (*S,S,R*)- and (*S,S,S*)-isomers (725). In a human study, the serum level of α-tocopherol was found to vary with both the stereochemical purity of the tocopherol and the matrix of the administered preparation (726). In another study, vitamin E deficiencies in dogs did not seem to alter the preferential incorporation of (R,R,R)-α-tocopherol compared with all-rac-α-tocopherol in plasma (727).

All eight stereoisomers of all-rac-α-tocopherol acetate have been analyzed and characterized by GLC, followed by biological biopotency assays (728). The individual diastereomeric purity was shown to be close to 100%. The (2R)-diastereomers had higher activities than the corresponding (2S)-form in all diastereomeric pairs analyzed: (R,R,R)–(S,R,R), (R,R,S)–(S,R,S), (R,S,R)–(S,S,R), and (R,S,S)–(S,S,S). The trend was similar for the (4R)- and (4S)-diastereomers, except for the (S,R,S)–(S,S,S) diastereomeric pair (728). NMR studies have also been done on the uptake, transport, and elimination of the stereoisomers of α-tocopherol in vivo (729).

Biotin

Biotin is also known as vitamin H. The physiologically active form is (+)-β-biotin, with the β-anomer side chains in the cis-form (all-cis-configuration) (730).

Vitamin K_1

Vitamin K_1 (koagulationsvitamin), found in alfalfa, was the first of the vitamins discovered belonging to the K group. Vitamin K_1 is essential for blood coagulation. It exerts a procoagulation effect in vivo (it acts as a cofactor in the synthesis of four clotting proteins) (731). All four stereoisomers of (R)-vitamin K_1 have been synthesized with high enantiomeric purity and separated by HPLC using poly(trityl methacrylate) as the CSP (supported on Nucleosil) (732).

Vitamin L

Vitamin L (o-aminobenzoic acid) is thought to be a lactation factor. It contains the active component (L_2) and an adenine derivative (683).

Vitamin U

The role of vitamin U (S-methylmethionine) has been studied in humans and rodents (733). After absorption and accumulation in the liver and kidneys, vitamin U is incorporated into the bodies of humans and rats predominantly as the L-enantiomer.

Clearly, of all the vitamins listed in this section, only vitamins A and E have had even cursory examination of their stereochemistry reported. This is surprising, given the importance of these compounds. Future studies in this area should produce information that is relevant not only to food and agricultural science but also to studies involving metabolism, enzymology, and pharmacology.

Summary

The determination of stereochemical compositions of compounds in foods and beverages is an area of growing importance. Most of the major components of nutritional value are chiral (e.g., proteins, amino acids, carbohydrates, fats, and vitamins). In addition, many of the components responsible for the taste and aroma of foods and beverages are chiral. Furthermore, different enantiomers are known to have different sensory effects. It is now clear that if we are to properly understand the nutritional value of a food or beverage or why it tastes and smells the way it does, enantiomeric compositions must be known. Enantiomeric analyses of foods and beverages can also be used to identify adulterated products, to control or monitor fermentation processes, to evaluate age and storage effects, to fingerprint complex mixtures, and so on (*480*).

The revolution in enantiomeric separation techniques (e.g., chiral HPLC, GC, and CE) over the past decade is largely responsible for the increased interest in this area of food and beverage analysis. Indeed, it was the ready availability and reliability of these methods, as well as the pointed questions of scientists in this area of research, that forced the Food and Drug Administration to finally issue guidelines on the development of new stereoisomeric drugs (*734*). The regulatory aspects of this science and technology will inevitably expand to other areas.

Today the preferred method for analyzing enantiomeric compositions is direct analysis of chiral components either on a CSP or with a chiral additive to the mobile phase (or running buffer for CE). Derivatization with a chiral reagent to form diastereomers is no longer the only or best option. Diastereomer formation is frequently considered a method of last resort, to be used when all else fails or is impractical. Regulatory agencies also seem to prefer direct methods because problems with impure chiral derivatizing agents, kinetic discrimination, side reactions, and so forth are avoided. The achiral derivatization of enantiomeric compounds is still acceptable and even preferable for enhancing sensitivity (e.g., fluorescent tagging agents in HPLC and CE), enhancing volatility (in GC), stabilizing an easily racemized or degraded component, and improving the concentration or cleanup of a compound from a real-world sample or a complex matrix.

One area of enantiomeric analysis that was not addressed in this chapter involves pesticides, herbicides, and other environmentally prevalent compounds that can be closely associated with foods and beverages (*735*). These compounds and their residues are often chiral. Enantiomeric analysis will certainly play a role in the study and regulation of these substances. There undoubtedly will be useful aspects of these studies that extend to and overlap with those involving foods and beverages (*735*).

Studies on the enantiomeric and diastereomeric effects and analyses of food and agricultural products will continue to expand. With them will come a greater and more exact understanding of nutrition, aroma, and taste as well as the biochemical pathways in which they are involved. In addition, we will be able to more exactly control and understand the production, storage, and possibly forensic aspects of our foods and beverages.

References

1. Takeoka, G. R.; Flath, R. A.; Güntert, M.; Jennings, W. *J. Agric. Food Chem.* **1988,** *36,* 553–560.
2. Wyllie, S. G.; Cook, D.; Brophy, J. J.; Richter, K. M. *J. Agric. Food Chem.* **1987,** *35,* 768–770.
3. Moshonas, M. G.; Shaw, P. E. *J. Agric. Food Chem.* **1987,** *35,* 161–165.
4. Buttery, R. G.; Takeoka, G.; Teranishi, R.; Ling, L. C. *J. Agric. Food Chem.* **1990,** *38,* 2050–2053.
5. Laleye, L. C.; Simard, R. E.; Gosselin, C.; Lee, B. H.; Giroux, R. N. *J. Food Sci.* **1987,** *52,* 303–307, 311.
6. Shi, J.; Hu, Y.; Yue, F.; Mang, L. *Fenixi Ceshi Tongbao* **1991,** *10(4),* 81–83.
7. Assa, K.; Tomov, E.; Lichev, V. *Izv. Khim.* **1991,** *24,* 191–198.
8. Vasconcelos, A. M. P.; Chaves das Neves, H. J. *J. High Resolut. Chromatogr.* **1990,** *13,* 494–498.
9. Vasconcelos, A. M. P.; Chaves das Neves, H. J. *J. Agric. Food Chem.* **1989,** *37,* 931–937.
10. Ohloff, G. In *Riechstoffe und Geruchssinn—Die molekulare Welt der Düfte;* Springer: Berlin, Germany, 1990.
11. Russell, G. F.; Hills, J. I. *Science (Washington, D.C.)* **1971,** *172,* 1043–1044.
12. Friedman, L.; Miller, J. G. *Science (Washington, D.C.)* **1971,** *172,* 1044–1046.
13. Ohloff, G. *Experientia* **1986,** *42,* 271–279.
14. Mosandl, A.; Heusinger, G. *Liebigs Ann. Chem.* **1985,** *6,* 1185–1191.
15. Günther, C.; Mosandl, A. *Liebigs Ann. Chem.* **1986,** *12,* 2112–2122.
16. Mosandl, A.; Günther, C. *J. Agric. Food Chem.* **1989,** *37,* 413–418.
17. Mosandl, A.; Heusinger, G.; Gessner, M. *J. Agric. Food Chem.* **1986,** *34,* 119–122.
18. Mosandl, A.; Gessner, M. *Z. Lebensm. Unters. Forsch.* **1988,** *187,* 40–44.
19. Pickenhagen, W. In *Flavor Chemistry—Trends and Developments;* Teranishi, R.; Buttery, R. G.; Shahidi, F., Eds.; ACS Symposium Series 388; American Chemical Society: Washington, DC, 1989; pp 151–157.
20. Acree, T. E.; Nishida, R.; Fukami, H. *J. Agric. Food Chem.* **1985,** *33,* 425–427.
21. Mosandl, A. Habilitation Thesis, University of Würzburg, 1982.
22. Demole, E.; Enggist, P.; Ohloff, G. *Helv. Chim. Acta* **1982,** *65,* 1785–1794.
23. Mosandl, A. *Food Rev. Int.* **1988,** *4,* 1–43.
24. Mosandl, A.; Heusinger, G. In *Analysis of Volatiles—Methods and Applications;* Schreier, P., Ed.; de Gruyter: Berlin, Germany, 1984; pp 343–356.
25. Ohloff, G. In *Olfaction and Taste 4;* D. Schneider, Ed.; Wissenschaftliche Verlagsgesellschaft MBH: Stuttgart, Germany, 1972; p 156.
26. Langeneau, E. E. In *American Society of Testing and Materials;* Special Technical Publications; Fritzsche Brothers: New York, 1968; pp 71–86.
27. Benecke, I.; König, W. A. *Angew. Chem.* **1982,** *94,* 709; *Angew. Chem. Int. Ed. Engl.* **1982,** *21,* 709; *Angew. Chem. Suppl.* **1982,** 1605–1613.
28. Schurig, V.; Weber, R. *Angew. Chem.* **1983,** *95,* 797–798; *Angew. Chem. Int. Ed. Engl.* **1983,** *22,* 772–773; *Angew. Chem. Suppl.* **1983,** 1130.
29. Gerlach, D.; Schneider, S.; Göllner, T.; Kim, K. S.; Schreier, P. In *Bioflavour '87, Analysis—Biochemistry Biotechnology;* Schreier, P., Ed.; de Gruyter: Berlin, Germany, 1988; pp 543–554.

30. Mosandl, A.; Günther, C.; Gessner, M.; Deger, W.; Singer, G.; Heusinger, G. In *Bioflavour '87, Analysis—Biochemistry Biotechnology;* Schreier, P., Ed.; de Gruyter: Berlin, Germany, 1988; pp 55–74.

31. Gessner, M.; Deger, W.; Mosandl, A. *Z. Lebensm. Unters. Forsch.* **1988,** *186,* 417–421.

32. Engel, K. H. In *Bioflavour '87, Analysis—Biochemistry Biotechnology;* Schreier, P., Ed.; de Gruyter: Berlin, Germany, 1988; pp 75–88.

33. Berger, R. G.; Drawert, F.; Hädrich, S. In *Bioflavour '87, Analysis—Biochemistry Biotechnology;* Schreier, P., Ed.; de Gruyter: Berlin, Germany, 1988; pp 415–434.

34. Eriksson, C. E. In *Flavor of Foods and Beverages;* Charalambous, G.; Inglett, G. E., Eds.; Academic: Orlando, FL, 1978; pp 1–14.

35. Blumenthal, A.; Cerny, M. *Mitt. Geb. Lebensmittelunters. Hyg.* **1976,** *67,* 515–520.

36. Ribas, C.; Pigati, P.; Ferreira, M. S.; Dias Netto, N. *Biologico* **1977,** *43,* 208–212.

37. Werkhoff, P.; Bretschneider, W.; Herrmann, H.-J.; Schreiber, K. *LaborPraxis* **1989,** *13,* 306–1121.

38. Werkhoff, P.; Bretschneider, W.; Herrmann, H.-J.; Schreiber, K. *LaborPraxis* **1990,** *14,* 51–352.

39. Werkhoff, P.; Brennecke, S.; Bretschneider, W. *Chem. Mikrobiol. Technol. Lebensm.* **1991,** *13,* 129–152.

40. Schneider, M. P.; Laumen, K. In *Bioflavour '87, Analysis—Biochemistry Biotechnology;* Schreier, P., Ed.; de Gruyter: Berlin, Germany, 1988; pp 483–529.

41. Schildknecht, H. *Angew. Chem.* **1981,** *93,* 164; *Angew. Chem. Int. Ed. Engl.* **1981,** *20,* 164–184.

42. Bernays, E.; De Luca, C. *Experientia* **1981,** *37,* 1289–1290.

43. Kubo, I.; Nakanishi, K. In *Host Plant Resistance to Pests;* Hedin, P. A., Ed.; ACS Symposium Series 62; American Chemical Society: Washington, DC, 1977; pp 165–178.

44. Schurig, V. In *Bioflavour '87, Analysis—Biochemistry Biotechnology;* Schreier, P., Ed.; de Gruyter: Berlin, Germany, 1988; pp 35–54.

45. Casy, A. F. *Trends Anal. Chem.* **1993,** *12(4),* 185–189.

46. Brückner, H.; Wittner, R.; Godel, H. *J. Chromatogr.* **1989,** *476,* 73–82.

47. Brückner, H.; Hausch, M. *Molk. Ztg. Hildesheim Ger.* **1987,** *41,* 1536–1537.

48. Brückner, H.; Schäfer, S.; Bahnmüller, D.; Hausch, M. *Fresenius' Z. Anal. Chem.* **1987,** *327,* 30–31.

49. Brückner, H.; Wittner, R.; Hausch, M.; Godel, H. *Fresenius' Z. Anal. Chem.* **1989,** *333,* 775–776.

50. Brückner, H.; Hausch, M. In *Chirality and Biological Activity;* Frank, H.; Holmstedt, B.; Testa, B., Eds.; Liss: New York, 1989; pp 129–136.

51. Brückner, H.; Hausch, M. *J. High Resolut. Chromatogr.* **1989,** *12,* 680–684.

52. Brückner, H.; Hausch, M. In *Amino Acids: Chemistry, Biology and Medicine;* Lubec, G.; Rosenthal, G. A., Eds.; ESCOM Scientific: Leiden, Netherlands, 1990; pp 1172–1182.

53. Palla, G.; Marchelli, R.; Dossena, A.; Casnati, G. *J. Chromatogr. A* **1989,** *475,* 45–53.

54. Blaschke, G. *Angew. Chem.* **1980,** *92,* 14–25; *Angew. Chem. Int. Ed. Engl.* **1980,** *19,* 13.

55. Armstrong, D. W. *J. Liq. Chromatogr.* **1984,** *7(S-2),* 353–376.

56. König, W. A. In *The Practice of Enantiomeric Separation by Capillary Gas Chromatography;* Bertsch, W.; Jennings, W. G.; Kaiser, R. E., Eds.; Hüthig: Heidelberg, Germany, 1987.

57. *Umami: A Basic Taste: Biochemistry, Nutrition, Food Science;* Kawamura, Y.; Kare, M. R., Eds.; Marcel Dekker: New York, 1987; p 649.

58. Teranishi, R. In *Flavor Chemistry—Trends and Developments;* Teranishi, R.; Buttery, R. G.; Shahidi, F., Eds.; ACS Symposium Series 388; American Chemical Society: Washington, DC, 1989; pp 1–6.

59. Kirimura, J.; Shimizu, A.; Kimizuka, A.; Ninomiya, T.; Katsuya, N. *J. Agric. Food Chem.* **1969**, *17*, 689–695.
60. Solms, J. *J. Agric. Food Chem.* **1969**, *17*, 686–688.
61. Strominger, J. L.; Izaki, K.; Matsuhashi, M.; Tipper, D. J. *Fed. Proc.* **1967**, *26*, 9–22.
62. Meister, A. In *Biochemistry of the Amino Acids*, 2nd ed.; Academic: Orlando, FL, 1965; Vol. 1, Chapter 1, pp 113–118.
63. Schleifer, K. H.; Kandler, O. *Bacteriol. Rev.* **1972**, *36*, 407–477.
64. Bada, J. L.; Cronin, J. R.; Ho, M.-S.; Kvenvolden, K. A.; Lawless, J. G.; Miller, S. L.; Oro, J.; Steinberg, S. *Nature (London)* **1983**, *301*, 494–496.
65. Reaveley, D. A.; Burge, R. E. *Adv. Microb. Physiol.* **1972**, *7*, 1–81.
66. Davies, J. S. In *Chemistry and Biochemistry of Amino Acids, Peptides, and Proteins*; Weinstein, B., Ed.; Marcel Dekker: Basel, Switzerland, 1977; pp 1–27.
67. Preston, R. L. *Comp. Biochem. Physiol.* **1987**, *87B*, 55–62.
68. Corrigan, J. J. *Science (Washington, D.C.)* **1969**, *164*, 142–149.
69. D'Aniello, A.; Giuditta, A. *J. Neurochem.* **1978**, *31*, 1107–1108.
70. Felbeck, H. *J. Exp. Zool.* **1985**, *234*, 145–149.
71. Matsushima, O.; Katayama, H.; Yamada, K.; Kado, Y. *Mar. Biol. Lett.* **1984**, *5*, 217–225.
72. Robinson, T. *Life Sci.* **1976**, *19*, 1097–1102.
73. Wieser, H.; Jugel, H.; Belitz, H.-D. *Z. Lebensm. Unters. Forsch.* **1977**, *164*, 277–282.
74. Kato, H.; Rhue, M. R.; Nishimura, T. In *Flavor Chemistry—Trends and Developments*; Teranishi, R.; Buttery, R. G.; Shahidi, F., Eds.; ACS Symposium Series 388; American Chemical Society: Washington, DC, 1989; pp 158–174.
75. Nagashima, Z.; Nakagawa, M.; Tokumura, H.; Toriumi, Y. *Nippon Nogeikagaku Kaishi* **1957**, *31*, 169–175.
76. Sakato, Y. *J. Agric. Chem. Soc. Jpn.* **1950**, *23*, 262–267.
77. Sakato, Y.; Hashizume, T. *J. Agric. Chem. Soc. Jpn.* **1950**, *23*, 269–271.
78. Armstrong, D. W.; Gasper, M. P.; Lee, S. H.; Ercal, N.; Zukowski, J. *Amino Acids* **1993**, *5*, 299–315.
79. Nagata, Y.; Akino, T. *Experientia* **1990**, *465*, 466–468.
80. Konno, R.; Niwa, A.; Yasumura, Y. *Biochem. J.* **1990**, *268*, 263–265.
81. Liardon, R.; Hurrell, R. F. *J. Agric. Food Chem.* **1983**, *31*, 432–437.
82. Bunjapamai, S.; Mahoney, R. R.; Fagerson, I. S. *J. Food Sci.* **1982**, *47*, 1229–1234.
83. Friedman, M.; Zahnley, J. C.; Masters, P. M. *J. Food Sci.* **1981**, *46*, 127–131, 134.
84. Hayashi, R.; Kemada, I. *Agric. Biol. Chem.* **1980**, *44*, 891–895.
85. Gandolfi, I.; Palla, G.; Delprato, L.; De Nisco, F.; Marchelli, R.; Salvadori, C. *J. Food Sci.* **1992**, *57*, 377–379.
86. Brückner, H.; Hausch, M. *Chromatographia* **1989**, *28*, 487–492.
87. Friedman, M. In *Nutritional and Toxicological Consequences of Food Processing;* Friedman, M., Ed.; Plenum: New York, 1991; pp 447–481.
88. Whitaker, J. R.; Feeney, R. E. *CRC Crit. Rev. Food Sci. Nutr.* **1983**, *19*, 173–212.
89. Masters, P. M.; Friedman, M. *J. Agric. Food Chem.* **1979**, *27*, 507–511.
90. Friedman, M.; Liardon, R. *J. Agric. Food Chem.* **1985**, *33*, 666–672.
91. Blackburn, S. In *Amino Acid Determinations: Methods and Techniques*, 2nd ed.; Blackburn, S., Ed.; Marcel Dekker: New York, 1978; pp 7–37.
92. Liardon, R.; Ledermann, S. *J. Agric. Food Chem.* **1986**, *34*, 557–565.
93. Friedman, M.; Gumbmann, M. R.; Masters, P. M. In *Nutritional and Toxicological Aspects of Food Safety;* Friedman, M., Ed.; Plenum: New York, 1984; Vol. 177, Chapter 18, pp 367–412.
94. Tovar, L. R.; Schwass, D. E. In *D-Amino Acids in Processed Proteins: Their Nutritional Consequences;* ACS Symposium Series 234; American Chemical Society: Washington, DC, 1983; pp 169–185.
95. Man, E. H.; Bada, J. L. *Ann. Rev. Nutr.* **1987**, *7*, 209–225.

96. Calendar, R.; Berg, P. *J. Mol. Biol.* **1967**, *26*, 39–54.
97. Cavallini, D.; DeMarco, D.; Mondovi, B. *J. Biol. Chem.* **1958**, *230*, 25–30.
98. Brückner, H.; Hausch, M. *Milchwissenschaft* **1990**, *45*, 421–425.
99. Frank, H. In *Amino Acid Analysis by Gas Chromatography;* Gehrke, C. W.; Kuo, K. C. T.; Zumwalt, R. W., Eds.; CRC: Boca Raton, FL, 1987; Vol. 2, p 55.
100. Neuberger, A. *Biochem. Soc. Symp.* **1948**, *1*, 20–32.
101. Berg, C. P. In *Protein and Amino Acid Nutrition;* Albanese, A. A., Ed.; Academic: Orlando, FL, 1959; pp 57–96.
102. Friedman, M.; Gumbmann, M. R. *J. Nutr.* **1984**, *114*, 2089–2096.
103. Friedman, M.; Gumbmann, M. R. *J. Nutr.* **1984**, *114*, 2301–2310.
104. Kies, C.; Fox, H.; Aprahamian, S. *J. Nutr.* **1975**, *105*, 809–814.
105. Stegink, L. D.; Bell, E. F.; Filer Jr., L. J.; Ziegler, E. E.; Andersen, D. W.; Seligson, F. H. *J. Nutr.* **1986**, *116*, 1185–1192.
106. Bender, D. A. In *Amino Acid Metabolism*, 2nd ed.; Wiley: Chichester, England, 1985.
107. Fuse, M.; Hayase, F.; Kato, H. *J. Jpn. Soc. Nutr. Food Sci.* **1984**, *37*, 349–354.
108. Jenkins, W. L.; Tovar, L. R.; Schwass, D. E.; Liardon, R.; Carpenter, K. J. *J. Agric. Food Chem.* **1984**, *32*, 1035–1041.
109. Masters, P. M.; Friedman, M. In *Chemical Deterioration of Proteins;* Whitaker, J. R.; Fujimaki, M., Eds.; ACS Symposium Series 123; American Chemical Society: Washington, DC, 1980; Chapter 8, pp 165–194.
110. Ekborg-Ott, K. H.; Armstrong, D. W.; Taylor, A. D. *J. Agric. Food Chem.*, in press.
111. Rundlett, K. L.; Armstrong, D. W. *Chirality* **1994**, *6*, 277–282.
112. Ekborg-Ott, K. H.; Armstrong, D. W. *Chirality* **1996**, *8*, 49–57.
113. Pawlowska, M.; Armstrong, D. W. *Chirality* **1994**, *6*, 270–276.
114. Brückner, H.; Wittner, R.; Godel, H. *Chromatographia* **1991**, *32*, 383–388.
115. Brückner, H.; Jaek, P.; Langer, M.; Godel, H. *Amino Acids* **1992**, *2*, 271–284.
116. Brückner, H.; Lüpke, M. *Chromatographia* **1991**, *31*, 123–128.
117. Brückner, H.; Hausch, M. *Milchwissenschaft* **1990**, *45*, 356–360.
118. Brückner, H.; Becker, D.; Lüpke, M. *Chirality* **1993**, *5*, 385–392.
119. Mosandl, A.; Kustermann, A. *Z. Lebensm. Unters. Forsch.* **1989**, *189*, 212–215.
120. Liardon, R.; Ledermann, S.; Ott, U. *J. Chromatogr.* **1981**, *203*, 385–395.
121. Finley, J. W. In *Chemical Changes in Food During Processing;* Richardson, T. R.; Finley, J. W., Eds.; Institute of Food Technologists Basic Symposium Series, 8th; AVI: Westport, CT, 1985; pp 443–482.
122. Lawrence, R. C.; Thomas, T. D.; Terzaghi, B. E. *J. Dairy Res.* **1976**, *43*, 141–193.
123. Hammes, W. P. In *Biochemical Engineering—A Challange for Interdisciplinary Cooperation;* Chmiel, H.; Hammes, W. P.; Bailey, J. E., Eds.; Fischer: Stuttgart, Germany, 1987; p 11.
124. Hammes, W. P. In *Enzyes in der Lebensmitteltechnologie;* Kroner, K. H.; Lösche, K.; Schmid, R. D., Eds.; VCH: Weinheim, Germany, 1989; p 3.
125. Johnstone, M. M.; Diven, W. F. *J. Biol. Chem.* **1969**, *244*, 5414–5420.
126. Boebel, K.; Baker, D. H. *J. Nutr.* **1982**, *112*, 367–376.
127. Borg, B. S.; Wahlstrom, R. C. In *Absorption and Utilization of Amino Acids;* Friedman, M., Ed.; CRC: Boca Raton, FL, 1989; Vol. 2, pp 155–172.
128. Friedman, M.; Gumbmann, M. R. In *Absorption and Utilization of Amino Acids;* Friedman, M., Ed.; CRC: Boca Raton, FL, 1989; Vol. 2, pp 117–132.
129. Friedman, M.; Gumbmann, M. R. In *Absorption and Utilization of Amino Acids;* Friedman, M., Ed.; CRC: Boca Raton, FL, 1989; Vol. 2, pp 173–190.
130. Friedman, M.; Gumbmann, M. R.; Brandon, D. L. *Front. Gastrointest. Res.* **1988**, *14*, 79–90.
131. Pierce Chemical Company. In *DAA Marfey's Reagent;* Technical Bulletin; Pierce Chemical Company: Rockford, IL, 1985.
132. Roberts, E. A. H. *J. Sci. Food Agric.* **1952**, *3*, 193–198.

133. Wickremasinghe, R. L. *Adv. Food Res.* **1978,** *24*, 229–286.
134. Finger, A.; Kuhr, S.; Engelhardt, U. H. *J. Chromatogr.* **1992,** *624*, 293–315.
135. Saijo, R.; Takeo, T. *Agric. Biol. Chem.* **1970,** *34*, 227–233.
136. Co, H.; Sanderson, G. W. *J. Food Sci.* **1970,** *35*, 160–164.
137. Millin, D. J.; Crispin, D. J.; Swaine, D. *J. Agric. Food Chem.* **1969,** *17*, 717–722.
138. Selvendran, R. R. *Ann. Bot.* **1970,** *34*, 825–833.
139. Neumann, K.; Montag, A. *Dtsch. Lebensm. Rundsch.* **1983,** *79*, 160–164.
140. Cartwright, R. A.; Roberts, E. A. H.; Wood, D. J. *J. Sci. Food Agric.* **1954,** *5*, 597–599.
141. Sakato, Y. *J. Agric. Chem. Soc. Jpn.* **1950,** *23*, 262–267.
142. Nakagawa, M. *Jpn. Agric. Res. Quest.* **1970,** *5*, 43–47.
143. Nakagawa, M. *Jpn. Agric. Res. Quest.* **1975,** *9*, 156–160.
144. Casimir, J.; Jadot, J.; Renard, M. *Biochim. Biophys. Acta* **1960,** *39*, 462–468.
145. Stag, G. V. *J. Sci. Food Agric.* **1974,** *25*, 1015–1034.
146. Kimura, R.; Murata, T. *Chem. Pharm. Bull.* **1980,** *28*, 664–666.
147. Kimura, R.; Murata, T. *Chem. Pharm. Bull.* **1971,** *19*, 1257–1261.
148. Kimura, R.; Murata, T. *Chem. Pharm. Bull.* **1971,** *19*, 1301–1307.
149. Kimura, R.; Kurita, M.; Murata, T. *Yakugaku Zasshi* **1975,** *95*, 892–895.
150. Pearson, D. In *The Chemical Analysis of Food*, 7th ed.; Chemical Publishing Company: New York, 1976; Chapter 10; pp 324–364.
151. Kaneda, H.; Kimura, T.; Kano, Y.; Koshino, S.; Osawa, T.; Kawakishi, S. *J. Ferment. Bioeng.* **1991,** *72*, 26–30.
152. García-Sánchez, F.; Carnero, C.; Heredia, A. *J. Agric. Food Chem.* **1988,** *36*, 80–82.
153. Kavanagh, T. E.; Clarke, B. J.; Gee, P. S.; Miles, M.; Nicholson, B. N. *Tech. Q. Master Brew. Assoc. Am.* **1991,** *28*, 111–118.
154. Atkinson, B. *Chem. Ind. (London)* **1991,** 304–307.
155. von Frisch, K. In *The Dance Language and Orientation of Bees*; Harvard University: Cambridge, MA, 1967.
156. Davies, A. M. C. *J. Food Technol.* **1976,** *11*, 515–523.
157. Gilliam, M.; McCaughey, W. F.; Wintermute, B. *J. Apic. Res.* **1980,** *19*, 64–72.
158. Gilbert, J.; Shepherd, M. J.; Wallwork, M. A.; Harris, R. G. *J. Apic. Res.* **1981,** *20*, 125–135.
159. Poncini, L.; Wimmer, F. L.; Vakamoce, V. *Fiji Agric. J.* **1983,** *45*, 65–70.
160. Davies, A. M. C. *J. Apic. Res.* **1975,** *14*, 29–39.
161. White Jr., J. W.; Rudyj, O. N. *J. Apic. Res.* **1978,** *17*, 89–93.
162. Deifel, A. *Chem. Unser Zeit.* **1989,** *23*, 25–33.
163. Perez, C.; Conchello, P.; Arino, A.; Yanguela, J.; Herrera, A. *Sci. Aliments* **1989**, *9*, 203–207.
164. Zukowski, J.; Pawlowska, M.; Armstrong, D. W. *J. Chromatogr.* **1992,** *623*, 33–41.
165. Ikeda, K. *J. Tokyo Chem. Soc.* **1908,** *30*, 820.
166. Yoshida, T. In *Kirk–Othmer Encyclopedia of Chemical Technology*, 3rd ed.; Wiley: New York, 1978; Vol. 2, pp 410–421.
167. Ooghe, W.; Kasteleyn, H.; Temmerman, I.; Sandra, P. *HRCCC J. High Resolut. Chromatogr. Chromatogr. Commun.* **1984,** *7*, 284–285.
168. Kuneman, D. W.; Braddock, J. K.; McChesney, L. L. *J. Agric. Food Chem.* **1988,** *36*, 6–9.
169. Buck, R. H.; Krummen, K. *J. Chromatogr.* **1987,** *387*, 255–265.
170. Euerby, M. R.; Partridge, L. Z.; Gibbons, W. A. *J. Chromatogr.* **1989**, *483*, 239–252.
171. Buck, R. H.; Krummen, K. *J. Chromatogr.* **1984,** *315*, 279–285.
172. Nimura, N.; Kinoshita, T. *J. Chromatogr.* **1986,** *352*, 169–177.
173. Brückner, H.; Wittner, R.; Godel, H. In *Amino Acids: Chemistry, Biology and Medicine*;

Lubec, G.; Rosenthal, G. A., Eds.; ESCOM Scientific: Leiden, Netherlands, 1990; pp 143–151.

174. Einarsson, S.; Folestad, S.; Josefsson, B. *J. Liq. Chromatogr.* **1987**, *10*, 1589–1601.
175. Jegorov, A.; Tríska, J.; Trnka, T.; Cerny, M. *J. Chromatogr.* **1988**, *434*, 417–422.
176. Jegorov, A.; Matha, V.; Trnka, T.; Cerny, M. *J. High Resolut. Chromatogr.* **1990**, *13*, 718–720.
177. Höfle, G.; Bedorf, N. In *Wissenschaftlicher Ergebnisbericht der Gesellschaft für Biotechnologische Forschung;* Gesellschaft für Biotechnologische Forschung mbH: Braunschweig, Germany, 1987; pp 90–92.
178. Boehm, M. F.; Bada, J. L. *Proc. Natl. Acad. Sci. U.S.A.* **1984**, *81*, 5263–5266.
179. Carisano, A. *J. Chromatogr.* **1985**, *318*, 132–138.
180. Armstrong, D. W.; Duncan, J. D.; Lee, S. H. *Amino Acids* **1991**, *1*, 97–106.
181. Armstrong, D. W.; Han, S. M. *CRC Crit. Rev. Anal. Chem.* **1988**, *19*, 175–224.
182. König, W. A. In *Analysis of Volatiles—Methods and Applications;* Schreier, P., Ed.; de Gruyter: Berlin, Germany, 1984; pp 77–91.
183. Pawlowska, M.; Chen, S.; Armstrong, D. W. *J. Chromatogr.* **1993**, *641*, 257–265.
184. Armstrong, D. W.; Gasper, M.; Lee, S. H.; Zukowski, J.; Ercal, N. *Chirality* **1993**, *5*, 375–378.
185. Hilton, M. L.; Chang, S.-C.; Gasper, M. P.; Pawlowska, M.; Armstrong, D. W.; Stalcup, A. M. *J. Liq. Chromatogr.* **1993**, *16*, 127–147.
186. Allenmark, S.; Andersson, S. *J. Chromatogr.* **1986**, *351*, 231–238.
187. Chang, S. C.; Wang, L. R.; Armstrong, D. W. *J. Liq. Chromatogr.* **1992**, *15*, 1411–1429.
188. Chen, S.; Pawlowska, M.; Armstrong, D. W. *J. Liq. Chromatogr.* **1994**, *17*, 483–497.
189. Spöndlin, C.; Küsters, E. *Chromatographia* **1994**, *35*, 325–329.
190. Roumeliotis, P.; Unger, K. K.; Kurganov, A. A.; Davankov, V. A. *J. Chromatogr.* **1983**, *255*, 51–66.
191. Allenmark, S.; Bomgren, B.; Borén, H. *J. Chromatogr.* **1983**, *264*, 63–68.
192. Ôi, N.; Nagase, M.; Doi, T. *J. Chromatogr.* **1983**, *257*, 111–117.
193. Pirkle, W. H.; Pochapsky, T. C. *J. Chromatogr.* **1986**, *369*, 175–177.
194. Pirkle, W. H.; House, D. W.; Finn, J. M. *J. Chromatogr.* **1980**, *192*, 143–148.
195. Domenici, E.; Bertucci, C.; Salvadori, P.; Felix, G.; Cahagne, I.; Motellier, S.; Wainer, I. *Chromatographia* **1990**, *29*, 170–176.
196. Jadaud, P.; Wainer, I. W. *J. Chromatogr.* **1989**, *476*, 165–174.
197. Lindner, W.; Hirschboeck, I. *J. Pharm. Biomed. Anal.* **1984**, *2*, 183–189.
198. Armstrong, D. W. *Anal. Chem.* **1987**, *59*(2), 84A–91A.
199. Liu, J.; Xu, X.; Huang, J. *Sepuer* **1990**, *8*(4), 229–232.
200. Lee, S. H.; Berthod, A.; Armstrong, D. W. *J. Chromatogr.* **1992**, *603*, 83–93.
201. Armstrong, D. W.; Zukowski, J.; Ercal, N.; Gasper, M. *J. Pharm. Biomed. Anal.* **1993**, *11*, 881–886.
202. Stalcup, A. M.; Faulkner Jr., J. R.; Tang, Y.; Armstrong, D. W.; Levy, L. W.; Regalado, E. *Biomed. Chromatogr.* **1991**, *5*, 3–7.
203. Zukowski, J.; Pawlowska, M.; Nagatkina, M.; Armstrong, D. W. *J. Chromatogr.* **1993**, *629*, 169–179.
204. Armstrong, D. W.; Liu, Y.; Ekborg-Ott, K. H. *Chirality* **1995**, *7*, 474–497.
205. Armstrong, D. W.; Ekborg-Ott, K. H.; Liu, Y., submitted for publication in *J. Chromatogr. A.*
206. Armstrong, D. W.; Tang, Y.; Chen, S.; Zhou, Y.; Bagwill, C.; Chen, J.-R. *Anal. Chem.* **1994**, *66*, 1473–1484.
207. Armstrong, D. W. *Pittsburg Conference Technical Program Book;* Pittsburgh, PA, 1994; Abstract 572.
208. Armstrong, D. W.; Han, Y. I.; Han, S. M. *Anal. Chim. Acta* **1988**, *208*, 275–281.

209. Han, S. M.; Han, Y. I.; Armstrong, D. W. *J. Chromatogr.* **1988,** *441,* 376–381.
210. Berthod, A.; Chang, S.-C.; Armstrong, D. W. *Anal. Chem.* **1992,** *64,* 395–404.
211. Armstrong, D. W.; Stalcup, A. M.; Hilton, M. L.; Duncan, J. D.; Faulkner J. R., Jr.; Chang, S.-C. *Anal. Chem.* **1990,** *62,* 1610–1615.
212. Lindner, W. F.; Petersson, C. In *Liquid Chromatography in Pharmaceutical Development*; Wainer, I., Ed.; Aster: Springfield, OR, 1985; pp 63–131.
213. Pirkle, W. H.; Finn, J. M. In *Asymmetric Synthesis*; Mosher, J. D., Ed.; Academic: Orlando, FL, 1983; Vol. 1, Chapter 6, pp 87–124.
214. Berthod, A.; Jin, H. L.; Stalcup, A. M.; Armstrong, D. W. *Chirality* **1990,** *2,* 38–42.
215. Roumeliotis, P.; Kurganov, A. A.; Davankov, V. A. *J. Chromatogr.* **1983,** *266,* 439–450.
216. Allenmark, S.; Bomgren, B.; Borén, H. *J. Chromatogr.* **1982,** *237,* 473–477.
217. Armstrong, D. W.; DeMond, W. *J. Chromatogr. Sci.* **1984,** *22,* 411–415.
218. Armstrong, D. W.; Chen, S.; Chang, C.; Chang, S. *J. Liq. Chromatogr.* **1992,** *15,* 545–556.
219. Armstrong, D. W.; Hilton, M.; Coffin, L. *LC-GC* **1992,** *9,* 646–651.
220. Duncan, J. D.; Armstrong, D. W.; Stalcup, A. M. *J. Liq. Chromatogr.* **1990,** *13,* 1091–1103.
221. Ward, T. J.; Armstrong, D. W. *J. Liq. Chromatogr.* **1986,** *9,* 407–423.
222. Han, S. M.; Armstrong, D. W. In *Chiral Separations by HPLC*; Krstulovic, A. M., Ed.; Ellis Horwood: Chichester, England, 1989; pp 208–284.
223. *Chiral Separations by LC*; Ahuja, S., Ed.; ACS Symposium Series 471; American Chemical Society: Washington DC, 1991.
224. Hilton, M.; Armstrong, D. W. *J. Liq. Chromatogr.* **1991,** *14,* 3673–3683.
225. Mitsubishi, K. *R & D Rev.* **1989,** *3,* 5.
226. Shinbo, T.; Yamaguchi, T.; Nishimura, K.; Sugiura, M. *J. Chromatogr.* **1987,** *405,* 145–153.
227. Hilton, M. L.; Armstrong, D. W. *J. Liq. Chromatogr.* **1991,** *14,* 9–28.
228. Ward, T. J.; Armstrong, D. W. In *Chromatographic Chiral Separations*; Zeif, M.; Crane, L. J., Eds.; Chromatographic Science Series; Marcel Dekker: New York, 1988; Vol. 40, Chapter 5, pp 131–163.
229. Fujimura, K.; Suzuki, S.; Hayashi, K.; Masuda, S. *Anal. Chem.* **1990,** *62,* 2198–2205.
230. Stalcup, A. M.; Jin, H. L.; Armstrong, D. W. *J. Liq. Chromatogr.* **1990,** *13,* 473–484.
231. Stalcup, A. M.; Chang, S.-C.; Armstrong, D. W. *J. Chromatogr.* **1991,** *540,* 113–128.
232. Armstrong, D. W.; Yang, X.; Han, S. M.; Menges, R. A. *Anal. Chem.* **1987,** *59,* 2594–2596.
233. Hinze, W. L.; Riehl, T. E.; Armstrong, D. W.; DeMond, W.; Alak, A.; Ward, T. *Anal. Chem.* **1985,** *57,* 237–242.
234. Pirkle, W. H.; Pochapsky, T. C.; Mahler, G. S.; Field, R. E. *J. Chromatogr.* **1985,** *348,* 89–96.
235. Gübitz, G.; Juffmann, F.; Jellenz, W. *Chromatographia* **1982,** *16,* 103–106.
236. Mosandl, A. *J. Chromatogr.* **1992,** *624,* 267–292.
237. Armstrong, D. W.; Jin, H. L. *J. Chromatogr.* **1989,** *462,* 219–232.
238. Nimura, N.; Toyama, A.; Kasahara, Y.; Kinoshita, T. *J. Chromatogr.* **1982,** *239,* 671–675.
239. Shimada, K.; Hirakata, K. *J. Liq. Chromatogr.* **1992,** *15,* 1763–1771.
240. Pettersson, C. *Trends Anal. Chem.* **1988,** *7,* 209–217.
241. Davankov, V. A.; Kurganov, A. A.; Bochkov, A. S. *Adv. Chromatogr.* **1983,** *22,* 71–116.
242. Pettersson, C. *J. Chromatogr.* **1984,** *316,* 553–567.
243. Pettersson, C.; No, K. *J. Chromatogr.* **1983,** *282,* 671–684.
244. Weinstein, S.; Weiner, S. *J. Chromatogr.* **1984,** *303,* 244–250.
245. Lindner, W.; LePage, J. N.; Davies, G.; Seitz, D. E.; Karger, B. L. *J. Chromatogr.* **1979,** *185,* 323–344.

246. Nimura, N.; Toyama, A.; Kinoshita, T. *J. Chromatogr.* **1982,** *234,* 482–484.
247. Nimura, N.; Toyama, A.; Kinoshita, T. *J. Chromatogr.* **1984,** *316,* 547–552.
248. Weinstein, S.; Engel, M. H.; Hare, P. E. *Anal. Biochem.* **1982,** *121,* 370–377.
249. Nimura, N.; Suzuki, T.; Kasahara, Y.; Kinoshita, T. *Anal. Chem.* **1981,** *53,* 1380–1383.
250. Bayer, E.; Grom, E.; Kaltenegger, B.; Uhmann, R. *Anal. Chem.* **1976,** *48,* 1106–1109.
251. Yu, J.; El Rassi, Z. *J. Liq. Chromatogr.* **1993,** *16,* 2931–2959.
252. Lin, J.-K.; Wang, C.-H. *Clin. Chem.* **1980,** *26,* 579–583.
253. Armani, E.; Barazzoni, L.; Dossena, A.; Marchelli, R. *J. Chromatogr.* **1988,** *441,* 287–298.
254. Armani, E.; Dossena, A.; Marchelli, R.; Virgili, R. *J. Chromatogr.* **1988,** *441,* 275–286.
255. Griffith, O. W.; Campbell, E. B.; Pirkle, W. H.; Tsipouras, A.; Hyun, M. H. *J. Chromatogr.* **1986,** *362,* 345–352.
256. Marcus, A.; Feeley, J. *Biochim. Biophys. Acta* **1961,** *46,* 603–604.
257. Cohen, S. A.; Strydom, D. J. *Anal. Biochem.* **1988,** *174,* 1–16.
258. Heinrikson, R. L.; Meredith, S. C. *Anal. Biochem.* **1984,** *136,* 65–74.
259. Gunawan, S.; Walton, N. Y.; Treiman, D. M. *J. Chromatogr.* **1990,** *503,* 177–187.
260. Brückner, H.; Bosch, I.; Graser, Y.; Fürst, P. *J. Chromatogr.* **1987,** *395,* 569–590.
261. Simons, S. S., Jr.; Johnson, D. F. *Anal. Biochem.* **1977,** *82,* 250–254.
262. Chen, R. F.; Scott, C.; Trepman, E. *Biochim. Biophys. Acta* **1979,** *576,* 440–455.
263. Alvarez-Coque, M. C. G.; Hernández, M. J. M.; Camañas, R. M. V.; Fernández, C. M. *Anal. Biochem.* **1989,** *180,* 172–176.
264. de Montigny, P.; Stobaugh, J. F.; Givens, R. S.; Carlson, R. G.; Srinivasachar, K.; Sternson, L. A.; Higuchi, T. *Anal. Chem.* **1987,** *59,* 1096–1101.
265. Donzanti, B. A.; Yamamoto, B. K. *Life Sci.* **1988,** *43,* 913–922.
266. Smith, S.; Sharp, T. *Br. J. Pharmacol. Proc. Suppl.* **1992,** *107S,* 210P.
267. Pearson, S. J.; Czudek, C.; Mercer, K.; Reynolds, G. P. *J. Neural Transm. Gen. Sect.* **1991,** *86,* 151–157.
268. Jacobs, W. A. *J. Chromatogr.* **1987,** *392,* 435–441.
269. Turiák, G.; Volicer, L. *J. Chromatogr. A* **1994,** *668,* 323–329.
270. May, M. E.; Brown, L. L. *Anal. Biochem.* **1989,** *181,* 135–139.
271. Aswad, D. W. *Anal. Biochem.* **1984,** *137,* 405–409.
272. Gil-Av, E.; Nurok, D. *Adv. Chromatogr.* **1974,** *10,* 99–172.
273. Brückner, H.; Langer, M.; Godel, H. In *Techniques in Protein Chemistry 3;* Angeletti, R. H., Ed.; Academic: Orlando, FL, 1992; pp 243–247.
274. Graser, T. A.; Godel, H.; Albers, S.; Földi, P.; Fürst, P. *Anal. Biochem.* **1985,** *151,* 142–152.
275. Alvarez-Coque, M. C. G.; Hernández, M. J. M.; Camañas, R. M. V.; Fernández, C. M. *Anal. Biochem.* **1989,** *178,* 1–7.
276. Moretti, F.; Birarelli, M.; Carducci, C.; Pontecorvi, A.; Antonozzi, I. *J. Chromatogr.* **1990,** *511,* 131–136.
277. Gil-Av, E.; Tishbee, A.; Hare, P. E. *J. Am. Chem. Soc.* **1980,** *102,* 5115–5117.
278. Roth, M. *Anal. Chem.* **1971,** *43,* 880–882.
279. Fürst, P.; Pollack, L.; Graser, T. A.; Godel, H.; Stehle, P. *J. Chromatogr.* **1990,** *499,* 557–569.
280. Betnér, I.; Földi, P. *Chromatographia* **1986,** *22,* 381–387.
281. Cohen, S. A.; Michaud, D. P. *Anal. Biochem.* **1993,** *211,* 279–287.
282. Cohen, S. A.; DeAntonis, K.; Michaud, D. M. In *Techniques in Protein Chemistry 4;* Angeletti, R. H., Ed.; Academic: Orlando, FL, 1993; pp 289–298.
283. Strydom, D. J.; Cohen, S. A. *Bioforum;* Millipore: Bedford, MA, 1992; No. 3.
284. Schuster, R. *J. Chromatogr.* **1988,** *431,* 271–284.

285. Einarsson, S.; Josefsson, B.; Möller, P.; Sanchez, D. *Anal. Chem.* **1987,** *59,* 1191–1195.
286. Carpino, L. A.; Han, G. Y. *J. Org. Chem.* **1972,** *37,* 3404–3409.
287. Carpino, L. A. *Acc. Chem. Res.* **1987,** *20,* 401–407.
288. Einarsson, S.; Folestad, S.; Josefsson, B.; Lagerquist, S. *Anal. Chem.* **1986,** *58,* 1638–1643.
289. Cunico, R.; Mayer, A. G.; Wehr, C. T.; Sheehan, T. L. *BioChromatography* **1986,** *1,* 6–14.
290. Näsholm, T.; Sandberg, G.; Ericsson, A. *J. Chromatogr.* **1987,** *396,* 225–236.
291. Carpino, L. A.; Cohen, B. J.; Stephens, K. E., Jr.; Sadat-Aalaee, S. Y.; Tien, J.-H.; Langridge, D. C. *J. Org. Chem.* **1986,** *51,* 3734–3736.
292. Malmer, M. F.; Schroeder, L. A. *J. Chromatogr.* **1990,** *514,* 227–239.
293. Marfey, P. *Carlsberg Res. Commun.* **1984,** *49,* 591–596.
294. Brückner, H.; Gah, C. *J. Chromatogr.* **1991,** *555,* 81–95.
295. *Food Analysis by HPLC;* Nollet, L. M. L., Ed.; Marcel Dekker: New York, 1992.
296. Ahnoff, M.; Grundevik, I.; Arfwidsson, A.; Fonselius, J.; Persson, B.-A. *Anal. Chem.* **1981,** *53,* 485–489.
297. Péter, A.; Tóth, G.; Tourwé, D. *J. Chromatogr. A* **1994,** *668,* 331–335.
298. Nimura, N.; Ogura, H.; Kinoshita, T. *J. Chromatogr.* **1980,** *202,* 375–379.
299. Yamada, Y.; Nonomura, S.; Fujiwara, H.; Miyazawa, T.; Kuwata, S. *J. Chromatogr.* **1990,** *515,* 475–482.
300. Nachtmann, F. *Int. J. Pharm.* **1980,** *4,* 337–345.
301. Brückner, H.; Strecker, B. *J. Chromatogr.* **1992,** *627,* 97–105.
302. Joseph, M. H.; Davies, P. *J. Chromatogr.* **1983,** *277,* 125–136.
303. Einarsson, S.; Hansson, G. In *High Performance Liquid Chromatography of Peptides and Proteins;* Mant, C. T.; Hodges, R. S., Eds.; CRC: Boca Raton, FL, 1991; p 369.
304. Kinoshita, T.; Kasahara, Y.; Nimura, N. *J. Chromatogr.* **1981,** *210,* 77–81.
305. Nambara, T. In *CRC Handbook of HPLC for the Separation of Amino Acids;* Hancock, W. S., Ed.; CRC: Boca Raton, FL, 1984; Vol. 1, pp 383–389.
306. Lindner, W. F. In *Chemical Derivatization in Analytical Chemistry;* Lawrence, J. F.; Frei, R. W., Eds.; Plenum: New York, 1982; Vol. 2, pp 145–190.
307. Lindner, W. In *Chromatographic Chiral Separations;* Zeif, M.; Crane, L. J., Eds.; Chromatographic Science Series; Marcel Dekker: New York, 1988; Vol. 40, Chapter 4, pp 91–130.
308. Watanabe, Y.; Imai, K. *Anal. Chem.* **1983,** *55,* 1786–1791.
309. Tsuruta, Y.; Date, Y.; Kohashi, K. *J. Chromatogr.* **1990,** *502,* 178–183.
310. Brückner, H.; Strecker, B. *Chromatographia* **1992,** *33,* 586–587.
311. Pollock, G. E.; Jermany, D. *J. Chromatogr. Sci.* **1970,** *8,* 296.
312. Gerwig, G. J.; Kamerling, J. P.; Vliegenthart, J. F. G. *Carbohydr. Res.* **1978,** *62,* 349–357.
313. Gerwig, G. J.; Kamerling, J. P.; Vliegenthart, J. F. G. *Carbohydr. Res.* **1979,** *77,* 1–7.
314. Mosandl, A.; Deger, W.; Gessner, M.; Günther, C.; Heusinger, G.; Singer, G. *Lebensmittelchem. Gerichtl. Chem.* **1987,** *41,* 35–39.
315. Charmot, D.; Audebert, R.; Quivoron, C. *J. Liq. Chromatogr.* **1985,** *8,* 1769–1781.
316. Davankov, V. A.; Bochkov, A. S.; Belov, Yu. P. *J. Chromatogr.* **1981,** *218,* 547–557.
317. Davankov, V. A.; Zolotarev, Yu. A. *J. Chromatogr.* **1978,** *155,* 303–310.
318. Gübitz, G.; Jellenz, W. *J. Chromatogr.* **1981,** *203,* 377–384.
319. Kiniwa, H.; Baba, Y.; Ishida, T.; Katoh, H. *J. Chromatogr.* **1989,** *461,* 397–405.
320. Oelrich, E.; Preusch, H.; Wilhelm, E. *HRCCC J. High Resolut. Chromatogr. Chromatogr. Commun.* **1980,** *3,* 269–272.
321. Galushko, S. V.; Shishkina, I. P.; Soloshonok, V. A. *J. Chromatogr.* **1992,** *592,* 345–348.

322. Berthod, A.; Liu, Y.; Bagwill, C.; Armstrong, D. W. *J. Chromatogr. A* **1996,** *731,* 123–137.
323. Rogozhin, S. V.; Davankov, V. A. *Dokl. Akad. Nauk SSSR* **1970,** *192,* 1288–1290.
324. Rogozhin, S. V.; Davankov, V. A. *Chem. Soc. Chem. Commun.* **1971,** 490.
325. Helgeson, R. C.; Timko, J. M.; Moreau, P.; Peacock, S. C.; Mayer, J. M.; Cram, D. J. *J. Am. Chem. Soc.* **1974,** *96,* 6762–6763.
326. Sogah, G. D. Y.; Cram, D. *J. Am. Chem. Soc.* **1979,** *101,* 3035–3042.
327. Newcomb, M.; Toner, J. L.; Helgeson, R. C.; Cram, D. J. *J. Am. Chem. Soc.* **1979,** *101,* 4941–4947.
328. Lingenfelter, D. S.; Helgeson, R. C.; Cram, D. J. *J. Org. Chem.* **1981,** *46,* 393–406.
329. Fager, R. S.; Kutina, C. B.; Abrahamson, E. W. *Anal. Biochem.* **1973,** *53,* 290–294.
330. Dunlop, D. S.; Neidle, A. *Anal. Biochem.* **1987,** *165,* 38–44.
331. Greene, T. W. In *Protective Groups in Organic Synthesis;* Wiley: New York, 1981.
332. Edman, P. *Acta. Chem. Scand.* **1950,** *4,* 283–293.
333. Pirkle, W. H.; Hyun, M. H. *J. Chromatogr.* **1985,** *322,* 287–293.
334. Neadle, D. J.; Pollitt, R. J. *Biochem. J.* **1965,** *97,* 607–608.
335. Moore, S.; Stein, W. H. *J. Biol. Chem.* **1951,** *192,* 663–681.
336. Armstrong, D. W.; Rundlett, K. L.; Chen, J.-R. *Chirality* **1994,** *6,* 496–509.
337. Armstrong, D. W.; Gasper, M. P.; Rundlett, K. L. *J. Chromatogr. A* **1994,** *689,* 285–304
338. Rundlett, K. L.; Gasper, M. P.; Zhou, E. Y.; Armstrong, D. W. *Chirality* **1996,** *8,* 88–107.
339. Armstrong, D. W.; Rundlett, K.; Reid III, G. L. *Anal. Chem.* **1994,** *66,* 1690–1695.
340. Armstrong, D. W.; Zhou, Y. *J. Liq. Chromatogr.* **1994,** *17,* 1695–1707.
341. Dale, J. A.; Dull, D. L.; Mosher, H. S. *J. Org. Chem.* **1969,** *34,* 2543–2549.
342. Maryanoff, B. E.; Vaught, J. L.; Shank, R. P.; McComsey, D. F.; Costanzo, M. J.; Nortey, S. O. *J. Med. Chem.* **1990,** *33,* 2793–2797.
343. Dale, J. A.; Mosher, H. S. *J. Am. Chem. Soc.* **1973,** *95,* 512–519.
344. Tressl, R.; Engel, K.-H. In *Progress in Flavour Research (Developments in Food Science);* Adda, J., Ed.; Elsevier: New York, 1985; Vol. 10, pp 441–455.
345. Tressl, R.; Engel, K. H.; Albrecht, W.; Bille-Abdullah, H. In *Characterization and Measurement of Flavor Compounds;* Bills, D. D.; Mussinan, C. J., Eds.; ACS Symposium Series 289; American Chemical Society: Washington, DC, 1985; pp 43–60.
346. Knox, J. H.; Jurand, J. *J. Chromatogr.* **1982,** *234,* 222–224.
347. Yoneda, H. *J. Liq. Chromatogr.* **1979,** *2,* 1157–1178.
348. Pettersson, C.; Schill, G. *J. Chromatogr.* **1981,** *204,* 179–183.
349. Pettersson, C.; Schill, G. *Chromatographia* **1982,** *16,* 192–197.
350. Benson, S. C.; Cai, P.; Colon, M.; Haiza, M. A.; Tokles, M.; Snyder, J. K. *J. Org. Chem.* **1988,** *53,* 5335–5341.
351. Parker, D.; Taylor, R. J. *Tetrahedron* **1987,** *43,* 5451–5456.
352. Jentsch, J. *Sci. Pharm.* **1986,** *54,* 195.
353. Marchelli, R.; Virgili, R.; Armani, E.; Dossena, A. *J. Chromatogr.* **1986,** *355,* 354–457.
354. Armstrong, D. W.; He, F.-Y.; Han, S. M. *J. Chromatogr.* **1988,** *448,* 345–354.
355. Armstrong, D. W.; Faulkner, J. R., Jr.; Han, S. M. *J. Chromatogr.* **1988,** *452,* 323–330.
356. Han, S. M.; Armstrong, D. W. In *Enantiomeric Separation by Thin-Layer Chromatography;* Touchstone, J. C., Ed.; Chemical Analysis 108 (Planar Chromatography Life Sciences); Wiley: New York, 1990; Chapter 7, pp 81–100.
357. Duncan, J. D.; Armstrong, D. W. *J. Planar Chromatogr.* **1990,** *3,* 65–67.
358. Duncan, J. D.; Armstrong, D. W. *J. Planar Chromatogr.* **1991,** *4,* 204–206.
359. Wainer, I. W.; Brunner, C. A.; Doyle, T. D. *J. Chromatogr.* **1983,** *264,* 154.
360. Gont, L. K.; Neuendorf, S. K. *J. Chromatogr.* **1987,** *391,* 343–345.

361. Alak, A.; Armstrong, D. W. *Anal. Chem.* **1986,** *58,* 582–584.
362. Weinstein, S. *Tetrahedron Lett.* **1984,** *25,* 985–986.
363. Günther, K.; Martens, J.; Schickedanz, M. *Angew. Chem.* **1984,** *96,* 514–515; *Angew. Chem. Int. Ed. Engl.* **1984,** *23,* 506.
364. Yuasa, S.; Shimada, A.; Kameyama, K.; Yasui, M.; Adzuma, K. *J. Chromatogr. Sci.* **1980,** *18,* 311–314.
365. Brinkman, U. A. T.; Kamminga, D. *J. Chromatogr.* **1985,** *330,* 375–378.
366. Martens, J.; Günther, K.; Schickedanz, M. *Arch. Pharm. (Weinheim Ger.)* **1986,** *319,* 572–574.
367. Armstrong, D. W.; Li, W.; Pitah, J. *Anal. Chem.* **1990,** *62,* 214–217.
368. Grinberg, N.; Weinstein, S. *J. Chromatogr.* **1984,** *303,* 251–255.
369. Higgins, H. M.; Harrison, W. H.; Wild, G. M.; Bungay, H. R.; McCormick, M. H. *Antibiot. Ann.* **1958,** *59,* 906–914.
370. Gil-Av, E.; Feibush, B.; Charles-Sigler, R. *Tetrahedron Lett.* **1966,** *10,* 1009–1015.
371. Frank, H.; Nicholson, G. J.; Bayer, E. *J. Chromatogr. Sci.* **1977,** *15,* 174–176.
372. Frank, H.; Nicholson, G. J.; Bayer, E. *J. Chromatogr.* **1978,** *146,* 197–206.
373. Frank, H.; Nicholson, G. J.; Bayer, E. *Angew. Chem.* **1978,** *90,* 396; *Angew. Chem. Int. Ed. Engl.* **1978,** *17,* 363–365.
374. Schurig, V. *Angew. Chem.* **1984,** *96,* 733–822; *Angew. Chem. Int. Ed. Engl.* **1984,** *23,* 747–765.
375. Gil-Av., E. *J. Mol. Evol.* **1975,** *6,* 131–144.
376. Bayer, E.; Frank, H. In *Modification of Polymers;* Carraher, C. E., Jr.; Tsuda, M., Eds.; ACS Symposium Series 121; American Chemical Society: Washington, DC, 1980; pp 341–358.
377. Bayer, E. *Z. Naturforsch. B.: Anorg. Chem. Org. Chem.* **1983,** *38B,* 1281–1291.
378. König, W. A. *HRCCC J. High Resolut. Chromatogr. Chromatogr. Commun.* **1982,** *5,* 588–595.
379. Tressl, R.; Engel, K. H. In *Analysis of Volatiles—Methods and Applications;* Schreier, P., Ed.; de Gruyter: Berlin, Germany, 1984; pp 323–342.
380. König, W. A.; Benecke, I.; Sievers, S. *J. Chromatogr.* **1981,** *217,* 71–79.
381. König, W. A.; Sievers, S.; Benecke, I. In *Proceedings of the 4th International Symposium on Capillary Chromatography;* Kaiser, R. E., Ed.; Institute of Chromatography: Bad Dürkheim, Germany, 1981; pp 703–725.
382. Schurig, V. *J. Chromatogr. A* **1994,** *666,* 111–129.
383. Gil-Av, E. In *The Science of Chromatography;* Bruner, F., Ed.; Elsevier: Amsterdam, Netherlands, 1985; Vol. 32, pp 111–130.
384. König, W. A. *Kontakte (Darmstadt)* **1990,** No. 2, 3–14.
385. Schurig, V.; Nowotny, H.-P. *Angew. Chem.* **1990,** *102,* 969; *Angew. Chem. Int. Ed. Engl.* **1990,** *29,* 939–957.
386. König, W. A. In *Enantioselective Gas Chromatography with Modified Cyclodextrins;* Bad Dürkheim and Hüthig: Heidelberg, Germany, 1992.
387. König, W. A. In *Modern Methods in Protein and Nucleic Acid Research;* Tschesche, H., Ed.; de Gruyter: Berlin, Germany, 1990; pp 213–230.
388. In *Chiraldex Handbook;* Advanced Separation Technologies: Whippany, NJ, 1994, p 9.
389. Armstrong, D. W.; Ward, T. J.; Armstrong, R. D.; Beesley, T. E. *Science (Washington, D.C.)* **1986,** *232,* 1132–1135.
390. König, W. A.; Lutz, S.; Mischnick-Lübbecke, P.; Brassat, B. Wenz, G. *J. Chromatogr.* **1988,** *447,* 193–197.
391. König, W. A.; Lutz, S.; Hagen, M.; Krebber, R.; Wenz, G.; Baldenius, K.; Ehlers, J.; tom Dieck, H. *J. High Resolut. Chromatogr.* **1989,** *12,* 35–39.
392. Koppenhoefer, B.; Hintzer, K.; Weber, R.; Schurig, V. *Angew. Chem.* **1980,** *92,* 473–474; *Angew. Chem. Int. Ed. Engl.* **1980,** *19,* 471.

393. Schurig, V.; Nowotny, H.-P. *J. Chromatogr.* **1988,** *441,* 155–163.
394. Ôi, N. In *Amino Acid Analysis by Gas Chromatography;* Zumwalt, R. W.; Kuo, K. C. T.; Gehrke, C. W., Eds.; CRC: Boca Raton, FL, 1987; Vol. 3, Chapter 3, pp 47–63.
395. Armstrong, D. W. *Pittsburg Conference Technical Program Book;* Pittsburgh, PA, 1989; Abstract 001.
396. Schurig, V.; Schmalzing, D.; Schleimer, M. *Angew. Chem.* **1991,** *103,* 994; *Angew. Chem. Int. Ed. Engl.* **1991,** *30,* 987–989.
397. Fischer, P.; Aichholz, R.; Bölz, U.; Juza, M.; Krimmer, S. *Angew. Chem.* **1990,** *102,* 439; *Angew. Chem. Int. Ed. Engl.* **1990,** *29,* 427–429.
398. Schurig, V.; Schmalzing, D.; Mühleck, U.; Jung, M.; Schleimer, M.; Mussche, P.; Duvekot, C.; Buyten, J. C. *J. High Resolut. Chromatogr.* **1990,** *13,* 713–717.
399. Jung, M.; Schurig, V. *J. Microcol. Sep.* **1993,** *5,* 11–22.
400. Bayer, E.; Frank, H.; Gerhardt, J.; Nicholson, G. *J. Assoc. Off. Anal. Chem.* **1987,** *70,* 234.
401. Walser, M. Ph.D. Dissertation, University of Tübingen, 1987.
402. Schmalzing, D.; Jung, M.; Mayer, S.; Rickert, J.; Schurig, V. *J. High Resolut. Chromatogr.* **1992,** *15,* 723–729.
403. Frank, H. *HRCCC J. High Resolut. Chromatogr. Chromatogr. Commun.* **1988,** *11,* 787–792.
404. Ruffing, F.-J.; Lux, J. A.; Roeder, W.; Schomburg, G. *Chromatographia* **1988,** *26,* 19–28.
405. Berthod, A.; Li, W.; Armstrong, D. W. *Anal. Chem.* **1982,** *64,* 873–879.
406. Metzger, J. W.; Jung, M.; Schmalzing, D.; Bayer, E.; Schurig, V. *Carbohydr. Res.* **1991,** *222,* 23–35.
407. Nowotny, H.-P. Ph.D. Dissertation, University of Tübingen, 1989.
408. Schmarr, H.-G.; Mosandl, A.; Neukom, H.-P.; Grob, K. *J. High Resolut. Chromatogr.* **1991,** *14,* 207–210.
409. Hardt, I.; König, W. A. In *14th International Symposium on Capillary Chromatography;* Sandra, P.; Lee, M. L., Eds.; Hüthig: Heidelberg, Germany, 1992; p 243.
410. Keim, W.; Köhnes, A.; Meltzow, W.; Römer, H. *J. High Resolut. Chromatogr.* **1991,** *14,* 507–529.
411. Jung, M.; Schmalzing, D.; Schurig, V. *J. Chromatogr.* **1991,** *552,* 43–57.
412. König, W. A.; Icheln, D.; Hardt, I. *J. High Resolut. Chromatogr.* **1991,** *14,* 694–695.
413. Schurig, V.; Juvancz, Z.; Nicholson, G. J.; Schmalzing, D. *J. High Resolut. Chromatogr.* **1991,** *14,* 58–62.
414. Schurig, V.; Glausch, A. *Naturwissenschaften* **1993,** *80,* 468–469.
415. Schurig, V.; Schleimer, M.; Jung, M.; Mayer, S.; Glausch, A. In *Progress in Flavour Precursor Studies;* Schreier, P.; Winterhalter, P., Eds.; Allured: Carol Stream, IL, 1993; pp 63–75.
416. Schreier, P. In *Progress in Flavour Precursor Studies;* Schreier, P.; Winterhalter, P., Eds.; Allured: Carol Stream, IL, 1993; pp 45–61.
417. König, W. A. *Carbohydr. Res.* **1989,** *192,* 51–60.
418. Mayer, S.; Schmalzing, D.; Jung, M.; Schleimer, M. *LC-GC Int.* **1992,** *5,* 58.
419. Blum, W.; Aichholz, R. *J. High Resolut. Chromatogr.* **1990,** *13,* 515–518.
420. König, W. A.; Icheln, D.; Runge, T.; Pforr, I.; Krebs, A. *J. High Resolut. Chromatogr.* **1990,** *13,* 702–707.
421. Bicchi, C.; Artuffo, G.; D'Amato, A.; Manzin, V.; Galli, A.; Galli, M. *J. High Resolut. Chromatogr.* **1993,** *16,* 209–214.
422. Bicchi, C.; Artuffo, G.; D'Amato, A.; Manzin, V.; Galli, A.; Galli, M. *J. High Resolut. Chromatogr.* **1992,** *15,* 710–714.
423. König, W. A.; Gehrcke, B.; Icheln, D.; Evers, P.; Dönnecke, J.; Wang, W. *J. High Resolut. Chromatogr.* **1992,** *15,* 367–372.
424. Hardt, I.; König, W. A. *J. Microcol. Sep.* **1993,** *5,* 35–40.

425. Jung, M.; Schurig, V. *J. High Resolut. Chromatogr.* **1993,** *16,* 289–298.
426. König, W. A.; Lutz, S.; Wenz, G.; Goergen, G.; Neumann, C.; Gäbler, A.; Boland, W. *Angew. Chem.* **1989,** *101,* 180–181.
427. Schurig, V.; Ossig, A.; Link, R. *Angew. Chem.* **1989,** *101,* 197–200.
428. Juvancz, Z.; Alexander, G.; Szejtli, J. *HRCCC J. High Resolut. Chromatogr. Chromatogr. Commun.* **1987,** *10,* 105–107.
429. Venema, A.; Tolsma, P. J. A. *J. High Resolut. Chromatogr.* **1989,** *12,* 32–34.
430. Mosandl, A.; Bruche, G.; Askari, C.; Schmarr, H.-G. *J. High Resolut. Chromatogr.* **1990,** *13,* 660–662.
431. Guichard, E.; Hollnagel, A.; Mosandl, A.; Schmarr, H.-G. *J. High Resolut. Chromatogr.* **1990,** *13,* 299–301.
432. Takeoka, G.; Flath, R. A.; Mon, T. R.; Buttery, R. G.; Teranishi, R.; Güntert, M.; Lautamo, R.; Szejtli, J. *J. High Resolut. Chromatogr.* **1990,** *13,* 202–206.
433. Armstrong, D. W.; Jin, H. L. *J. Chromatogr.* **1990,** *502,* 154–159.
434. Mosandl, A.; Schubert, V. *J. Ess. Oil Res.* **1990,** *2,* 121–132.
435. Köpke, T.; Mosandl, A. *Z. Lebensm. Unters. Forsch.* **1992,** *194,* 372–376.
436. König, W. A. *Clin. Pharmacol.* **1993,** *18,* 107–137.
437. Lickefett, H.; Krohn, K.; König, W. A.; Gehrcke, B.; Syldatk, C. *Tetrahedron: Asymmetry* **1993,** *4,* 1129–1135.
438. König, W. A.; Gehrcke, B. *J. High Resolut. Chromatogr.* **1993,** *16,* 175–181.
439. König, W. A.; Nippe, K. S.; Mischnick, P. *Tetrahedron Lett.* **1990,** *31,* 6867–6868.
440. Wenz, G.; Mischnick, P.; Krebber, R.; Richters, M.; König, W. A. *J. High Resolut. Chromatogr.* **1990,** *13,* 724–728.
441. König, W. A.; Lutz, S.; Evers, P.; Knabe, J. *J. Chromatogr.* **1990,** *503,* 256–259.
442. König, W. A. *Nachr. Chem. Tech. Lab.* **1989,** *37,* 471–472, 474–476.
443. König, W. A.; Lutz, S.; Mischnick-Lübbecke, P.; Brassat, B.; von der Bey, E.; Wenz, G. *Starch/Staerke* **1988,** *40,* 472–476.
444. Ernst-Cabrera, K.; König, W. A. *React. Polym. Ion Exch. Sorbents* **1987,** *6,* 267–274.
445. Schmarr, H.-G.; Maas, B.; Mosandl, A.; Bihler, S.; Neukom, H.-P.; Grob, K. *J. High Resolut. Chromatogr.* **1991,** *14,* 317–321.
446. Dietrich, A.; Maas, B.; Messer, W.; Bruche, G.; Karl, V.; Kaunzinger, A.; Mosandl, A. *J. High Resolut. Chromatogr.* **1992,** *15,* 590–593.
447. Lou, X.; Liu, Y.; Wang, Q.; Zhu, D.; Zhou, L. *Sepuer* **1992,** *10*(4), 195–198.
448. Bradshaw, J. S.; Yi, G.; Rossiter, B. E.; Reese, S. L.; Petersson, P.; Markides, K. E.; Lee, M. L. *Tetrahedron Lett.* **1993,** *34,* 79–82.
449. Yi, G.; Bradshaw, J. S.; Rossiter, B. E.; Reese, S. L.; Petersson, P.; Markides, K. E.; Lee, M. L. *J. Org. Chem.* **1993,** *58,* 2561–2565.
450. Yi, G.; Bradshaw, J. S.; Rossiter, B. E.; Malik, A.; Li, W.; Lee, M. L. *J. Org. Chem.* **1993,** *58,* 4844–4850.
451. Tang, Y.; Zhou, Y.; Armstrong, D. W. *J. Chromatogr. A* **1994,** *666,* 147–159.
452. König, W. A.; Mischnick-Lübbecke, P. M.; Brassat, B.; Lutz, S.; Wenz, G. *Carbohydr. Res.* **1988,** *183,* 11–17.
453. Armstrong, D. W.; Li, W.; Chang, C.-D.; Pitha, J. *Anal. Chem.* **1990,** *62,* 914–923.
454. Berthod, A.; Li, W. Y.; Armstrong, D. W. *Carbohydr. Res.* **1990,** *201,* 175–184.
455. Armstrong, D. W.; Li, W.; Stalcup, A. M.; Secor, H. V.; Izac, R. R.; Seeman, J. I. *Anal. Chim. Acta* **1990,** *234,* 365–380.
456. Nitz, S.; Kollmannsberger, H.; Drawert, F. *Chem. Microbiol. Technol. Lebensm.* **1989,** *12,* 75–80.
457. Mosandl, A.; Hener, U.; Hagenauer-Hener, U.; Kustermann, A. *J. High Resolut. Chromatogr.* **1989,** *12,* 532–536.
458. Mosandl, A.; Hener, U.; Hagenauer-Hener, U.; Kustermann, A. *J. Agric. Food Chem.* **1990,** *38,* 767–771.

459. Mosandl, A.; Kustermann, A.; Hener, U.; Hagenauer-Hener, U. *Dtsch. Lebensm. Rundsch.* **1989,** *85,* 205–209.

460. Guichard, E.; Kustermann, A.; Mosandl, A. *J. Chromatogr.* **1990,** *498,* 396–401.

461. Bernreuther, A.; Christoph, N.; Schreier, P. *J. Chromatogr.* **1989,** *481,* 363–367.

462. Mosandl, A. *Lebensmittelchemie* **1991,** *45,* 2–7.

463. Mosandl, A.; Palm, U.; Günther, C.; Kustermann, A. *Z. Lebensm. Unters. Forsch.* **1989,** *188,* 148–150.

464. König, W. A.; Lutz, S.; Colberg, C.; Schmidt, N.; Wenz, G.; von der Bey, E.; Mosandl, A.; Günther, C.; Kustermann, A. *HRCCC J. High Resolut. Chromatogr. Chromatogr. Commun.* **1988,** *11,* 621–625.

465. König, W. A.; Krebber, R.; Mischnick, P. *J. High Resolut. Chromatogr.* **1989,** *12,* 732–738.

466. Lehmann, D.; Askari, C.; Henn, D.; Dettmar, F.; Hener, U.; Mosandl, A. *Dtsch. Lebensm. Rundsch.* **1991,** *87,* 75–77.

467. Schomburg, G.; Husmann, H.; Hübinger, E.; König, W. A. *HRCCC J. High Resolut. Chromatogr. Chromatogr. Commun.* **1984,** *7,* 404–410.

468. König, W. A.; Lutz, S.; Wenz, G.; von der Bey, E. *HRCCC J. High Resolut. Chromatogr. Chromatogr. Commun.* **1988,** *11,* 506–509.

469. Palm, U.; Askari, C.; Hener, U.; Jakob, E.; Mandler, C.; Gessner, M.; Mosandl, A.; König, W. A.; Evers, P.; Krebber, R. *Z. Lebensm. Unters. Forsch.* **1991,** *192,* 209–213.

470. Mosandl, A.; Kustermann, A.; Palm, U.; Dorau, H.-P.; König, W. A. *Z. Lebensm. Unters. Forsch.* **1989,** *188,* 517–520.

471. Li, W.-Y.; Jin, H. L.; Armstrong, D. W. *J. Chromatogr.* **1990,** *509,* 303–324.

472. Mosandl, A.; Hollnagel, A. *Chirality* **1989,** *1,* 293–300.

473. Mosandl, A.; Hagenauer-Hener, U.; Hener, U.; Lehmann, D.; Kreis, P.; Schmarr, H.-G. In *Flavour Science and Technology (Weureman Symposium, 6th);* Bessiere, Y.; Thomas, A. F., Eds.; Wiley: Chichester, England, 1990; pp 21–24.

474. Dietrich, A.; Maas, B.; Karl, V.; Kreis, P.; Lehmann, D.; Weber, B.; Mosandl, A. *J. High Resolut. Chromatogr.* **1992,** *15,* 176–179.

475. Guichard, E.; Mosandl, A.; Hollnagel, A.; Latrasse, A.; Henry, R. *Z. Lebensm. Unters. Forsch.* **1991,** *193,* 26–31.

476. Armstrong, D. W.; Tang, Y.; Ward, T.; Nichols, M. *Anal. Chem.* **1993,** *65,* 1114–1117.

477. Nowotny, H. P.; Schmalzing, D.; Wistuba, D.; Schurig, V. *J. High Resolut. Chromatogr.* **1989,** *12,* 383–393.

478. Engel, K. H.; Flath, R. A.; Albrecht, W.; Tressl, R. *J. Chromatogr.* **1989,** *479,* 176–180.

479. Artho, A.; Grob, K. *Mitt. Geb. Lebensmittelunters. Hyg.* **1990,** *81,* 544–558.

480. Armstrong, D. W.; Chang, C.-D.; Li, W. Y. *J. Agric. Food Chem.* **1990,** *38,* 1674–1677.

481. Nitz, S.; Kollmannsberger; H.; Drawert, F. *Chem. Microbiol. Technol. Lebensm.* **1989,** *12,* 105–110.

482. Reiher, T.; Hamann, H.-J. *J. High Resolut. Chromatogr.* **1992,** *15,* 346–349.

483. Stalcup, A. M.; Ekborg, K. H.; Gasper, M. P.; Armstrong, D. W. *J. Agric. Food Chem.* **1993,** *41,* 1684–1689.

484. Rettinger, K.; Weber, B.; Mosandl, A. *Z. Lebensm. Unters. Forsch.* **1990,** *191,* 265–268.

485. Mosandl, A.; Rettinger, K.; Fischer, K.; Schubert, V.; Schmarr, H.-G.; Maas, B. *J. High Resolut. Chromatogr.* **1990,** *13,* 382–385.

486. Mosandl, A.; Rettinger, K.; Weber, B.; Henn, D. *Dtsch. Lebensm. Rundsch.* **1990,** *86,* 375–379.

487. König, W. A.; Krüger, A.; Icheln, D.; Runge, T. *J. High Resolut. Chromatogr.* **1992,** *15,* 184–189.

488. König, W. A.; Krebber, R.; Wenz, G. *J. High Resolut. Chromatogr.* **1989,** *12,* 641–644.

489. Schurig, V.; Nowotny, H.-P.; Schmalzing, D. *Angew. Chem.* **1989,** *101,* 785–786; *Angew. Chem. Int. Ed. Engl.* **1989,** *28,* 736.
490. König, W. A.; Lutz, S.; Wenz, G. *Angew. Chem.* **1988,** *100,* 989–990; *Angew. Chem. Int. Ed. Engl.* **1988,** *27,* 979–980.
491. König, W. A. *Trends Anal. Chem.* **1993,** *12,* 130–137.
492. Gil-Av, E.; Feibush, B.; Charles-Sigler, R. In *Gas Chromatography;* Littlewood, A. B., Ed.; Institute of Petroleum: London, 1967; Vol. 6, pp 227–239.
493. Schurig, V. *Chromatographia* **1980,** *13,* 263–270.
494. Weber, R.; Schurig, V. *Naturwissenschaften* **1984,** *71,* 408–413.
495. Schurig, V. *J. Chromatogr.* **1988,** *441,* 135–153.
496. Chrompack International: Middelburg, Netherlands.
497. Nakaparksin, S.; Gil-Av, E.; Oro, J. *Anal. Biochem.* **1970,** *33,* 374–382.
498. Frank, H.; Woiwode, W.; Nicholson, G.; Bayer, E. *Liebigs Ann. Chem.* **1981,** *3,* 354–365.
499. König, W. A.; Benecke, I.; Lucht, N.; Schmidt, E.; Schulze, J.; Sievers, S. *J. Chromatogr.* **1983,** *279,* 555–564.
500. König, W. A., Benecke, I.; Sievers, S. *J. Chromatogr.* **1982,** *238,* 427–432.
501. König, W. A.; Benecke, I.; Lucht, N.; Schmidt, E.; Schulze, J.; Sievers, S. In *Proceedings of the 5th International Symposium on Capillary Chromatography;* Rijks, J., Ed.; Elsevier: Amsterdam, Netherlands, 1983; pp 609–618.
502. König, W. A.; Steinbach, E.; Ernst, K. *Angew. Chem.* **1984,** *96,* 516; *Angew. Chem. Int. Ed. Engl.* **1984,** *23,* 527–528.
503. Bergman, T.; Jörnvll, H. In *Techniques in Protein Chemistry 3;* Angeletti, R. H., Ed.; Academic: Orlando, FL, 1992; pp 129–134.
504. Ward, T. *Anal. Chem.* **1994,** *66,* 632A–640A.
505. Nishizawa, H.; Nakajima, K.; Kobayashi, M.; Abe, Y. *Anal. Sci.* **1991,** *7,* 959–961.
506. Werner, A.; Nassauer, T.; Kiechle, P.; Erin, F. *J. Chromatogr. A* **1994,** *666,* 375–379.
507. Mayer, S.; Schurig, V. *J. High Resolut. Chromatogr.* **1992,** *15,* 129–131.
508. Mayer, S.; Schurig, V. *J. Liq. Chromatogr.* **1993,** *16,* 915–931.
509. Fanali, S. *J. Chromatogr.* **1991,** *545,* 437–444.
510. Shibukawa, A.; Lloyd, D. K.; Wainer, I. W. *Chromatographia* **1993,** *35,* 419–429.
511. Snopek, J.; Soini, H.; Novotny, M.; Smolková-Keulemansová, E.; Jelinek, I. *J. Chromatogr.* **1991,** *559,* 215–222.
512. Guttman, A.; Paulus, A.; Cohen, A. S.; Grinberg, N.; Karger, B. L. *J. Chromatogr.* **1988,** *448,* 41–53.
513. Nishi, H.; Fukuyama, T.; Terabe, S. *J. Chromatogr.* **1991,** *553,* 503–516.
514. Okafo, G. N.; Camilleri, P. *J. Microcol. Sep.* **1993,** *5,* 149–153.
515. Ward, T. J.; Nichols, M.; Sturdivant, L.; King, C. C. *Amino Acids* **1995,** *8,* 337–344.
516. Otsuka, K.; Terabe, S. *J. Chromatogr.* **1990,** *515,* 221–226.
517. Fanali, S. *J. Chromatogr.* **1989,** *474,* 441–446.
518. Bushey, M. M.; Jorgenson, J. W. *Anal. Chem.* **1989,** *61,* 491–493.
519. Cole, R. O.; Sepaniak, M. J.; Hinze, W. L. *J. High Resolut. Chromatogr.* **1990,** *13,* 579–582.
520. Snopek, J.; Jelínek, I.; Smolková-Keulemansová, E. *J. Chromatogr.* **1988,** *452,* 571–590.
521. Terabe, S.; Shibata, M.; Miyashita, Y. *J. Chromatogr.* **1989,** *480,* 403–411.
522. Dobashi, A.; Ono, T.; Hara, S.; Yamaguchi, J. *Anal. Chem.* **1989,** *61,* 1984–1986.
523. Gassmann, E.; Kuo, J. E.; Zare, R. N. *Science (Washington, D.C.)* **1985,** *230,* 813–814.
524. Barker, G. E.; Russo, P.; Hartwick, R. A. *Anal. Chem.* **1992,** *64,* 3024–3028.
525. Sun, P.; Wu, N.; Barker, G.; Hartwick, R. A. *J. Chromatogr.* **1993,** *648,* 475–480.
526. Sun, P.; Barker, G. E.; Hartwick, R. A.; Grinberg, N.; Kaliszan, R. *J. Chromatogr.* **1993,** *652,* 247–252.

527. Voltcheva, L.; Mohammad, J.; Pettersson, G.; Hjertén, S. *J. Chromatogr.* **1993**, *638*, 263–267.
528. Kuhn, R.; Erni, F.; Bereuter, T.; Häusler, J. *Anal. Chem.* **1992**, *64*, 2815–2820.
529. Kuhn, R.; Stoecklin, F.; Erni, F. *Chromatographia* **1992**, *33*, 32–36.
530. D'Hulst, A.; Verbeke, N. *J. Chromatogr.* **1992**, *608*, 275–287.
531. Bergman, T.; Agerberth, B.; Jörnvall, H. *FEBS Lett.* **199**, *283*, 100–103.
532. Agerberth, B.; Lee, J. Y.; Bergman, T.; Carlquist, M.; Boman, H. G.; Mutt, V.; Jörnvall, H. *Eur. J. Biochem.* **1991**, *202*, 849–854.
533. Ishihama, Y.; Terabe, S. *J. Liq. Chromatogr.* **1993**, *16*, 933–944.
534. Tran, A. D.; Blanc, T.; Leopold, E. J. *J. Chromatogr.* **1990**, *516*, 241–249.
535. Bergman, T.; Carlquist, M.; Jörnvall, H. In *Advanced Methods in Protein Microsequence Analysis;* Wittmann-Liebold, B.; Salnikow, J.; Erdmann, V. A., Eds.; Springer: Berlin, Germany, 1986; pp 45–55.
536. Cohen, A. S.; Paulus, A.; Karger, B. L. *Chromatographia* **1987**, *24*, 15–24.
537. Dobashi, A.; Ono, T.; Hara, S.; Yamaguchi, J. *J. Chromatogr.* **1989**, *480*, 413–420.
538. Terabe, S.; Otsuka, K.; Ando, T. *Anal. Chem.* **1985**, *57*, 834–841.
539. Otsuka, K.; Terabe, S. *Electrophoresis* **1990**, *11*, 982–983.
540. Otsuka, K.; Kawahara, J.; Tatekawa, K.; Terabe, S. *J. Chromatogr.* **1991**, *559*, 209–214.
541. Davies, J. S.; Mohammed, A. K. A. *J. Chem. Soc., Perkin Trans. 2* **1984**, *10*, 1723–1727.
542. Miyashita, Y.; Terabe, S. In *High Performance Capillary Electrophoresis;* Beckman, DS-767; Palo Alto, CA, 1990.
543. Little, M. R. *J. Biochem. Biophys. Methods* **1985**, *11*, 195–202.
544. Sharon, N. In *Complex Carbohydrates: Their Chemistry, Biosynthesis, and Functions;* Addison-Wesley: Reading, MA, 1975.
545. Roboz, J.; Suzuki, R.; Holland, J. F. *J. Clin. Microbiol.* **1980**, *12*, 594–602.
546. Schaffer, R. In *The Carbohydrates*, 2nd ed.; Pigman, W.; Horton, D., Eds.; Academic: Orlando, FL, 1972; Vol. 1A, pp 69–111.
547. Bell, D. J.; Baldwin, E. *J. Chem. Soc.* **1941**, 125–132.
548. Leontein, K.; Lindberg, B.; Lönngren, J. *Carbohydr. Res.* **1978**, *62*, 359–362.
549. Briggs, J.; Finch, P.; Percival, E.; Weigel, H. *Carbohydr. Res.* **1982**, *103*, 186–189.
550. Zablocki, W.; Behrman, E. J.; Barber, G. A. *J. Biochem. Biophys. Methods* **1979**, *1*, 253–256.
551. Schweer, H. *J. Chromatogr.* **1982**, *243*, 149–152.
552. Schweer, H. *J. Chromatogr.* **1983**, *259*, 164–168.
553. Schweer, H. *Carbohydr. Res.* **1983**, *116*, 139–143.
554. Little, M. R. *Carbohydr. Res.* **1982**, *105*, 1–8.
555. Halpern, B. In *Handbook of Derivatives for Chromatography;* Blau, K.; King, G. S., Eds.; Heyden: London, 1978; pp 457–499.
556. Pollock, G. E.; Jermany, D. A. *J. Gas Chromatogr.* **1968**, *6*, 412–415.
557. Oshima, R.; Kumanotani, J. *Chem. Lett.* **1981**, 943–946.
558. Oshima, R.; Yamauchi, Y.; Kumanotani, J. *Carbohydr. Res.* **1982**, *107*, 169–176.
559. Oshima, R.; Kumanotani, J.; Watanabe, C. *J. Chromatogr.* **1983**, *259*, 159–163.
560. Armstrong, D. W.; Jin, H. L. *Chirality* **1989**, *1*, 27–37.
561. König, W. A.; Benecke, I.; Bretting, H. *Angew. Chem.* **1981**, *93*, 688; *Angew. Chem. Int. Ed. Engl.* **1981**, *20*, 693–694.
562. Benecke, I.; Schmidt, E.; König, W. A. *HRCCC J. High Resolut. Chromatogr. Chromatogr. Commun.* **1981**, *4*, 553–556.
563. König, W. A.; Benecke, I. *J. Chromatogr.* **1983**, *269*, 19–21.
564. Leavitt, A. L.; Sherman, W. R. *Carbohydr. Res.* **1982**, *103*, 203–212.
565. Sherman, W. R.; Leavitt A. L.; Honchar, M. P.; Hallcher, L. M.; Phillips, B. E. *J. Neurochem.* **1981**, *36*, 1947–1951.

566. Gal, J. *LC-GC* **1987,** *5,* 106–126.
567. Kahle, V.; Tesarík, K. *J. Chromatogr.* **1980,** *191,* 121–128.
568. Little, M. R. Ph.D. Dissertation, Ohio State University, 1982.
569. Little, M. R.; Reynolds, K. A.; Rauck, C. R.; Barber, G. A.; Behrman, E. J. *Abstracts of Papers,* 179th American Chemical Society National Meeting, Houston, TX, American Chemical Society, Washington, D.C., 1980; CARB-28.
570. Krýsl, S.; Smolková-Keulemansová, E. *J. Chromatogr.* **1985,** *349,* 167–172.
571. Kocielski, T.; Sybilska, D.; Jurczak, J. *J. Chromatogr.* **1983,** *280,* 131–134.
572. Maga, J. A. *CRC Crit. Rev. Food Sci. Nutr.* **1976,** *8,* 1–56.
573. Tressl, R.; Apetz, M.; Arrieta, R.; Grünewald, K. G. In *Flavor of Foods and Beverages;* Charalambous, G.; Inglett, G. E., Eds.; Academic: Orlando, FL, 1978; pp 145–168.
574. Bernreuther, A.; Lander, V.; Huffer, M.; Schreier, P. *Flavour Fragrance J.* **1990,** *5*(2), 71–73.
575. Masuda, M.; Nishimura, K. *Chem. Lett.* **1981,** *10,* 1333–1336.
576. Gessner, M. Ph.D. Dissertation, University of Würzburg, 1988.
577. Deger, W.; Gessner, M.; Günther, C.; Singer, G.; Mosandl, A. *J. Agric. Food Chem.* **1988,** *36,* 1260–1264.
578. Mosandl, A.; Gessner, M.; Günther, C.; Deger, W.; Singer, G. *HRCCC J. High Resolut. Chromatogr. Chromatogr. Commun.* **1987,** *10,* 67–70.
579. Günther, C.; Mosandl, A. *Z. Lebensm. Unters. Forsch.* **1987,** *185,* 1–4.
580. Wadhwa, B. K.; Jain, M. K. *Indian J. Dairy Sci.* **1989,** *42,* 206–209.
581. Hagenauer-Hener, U. Ph.D. Dissertation, University of Frankfurt, 1990.
582. Tressl, R.; Albrecht, W. In *Biogeneration of Aromas;* Parliment, T. H.; Croteau, R., Eds.; ACS Symposium Series 317; American Chemical Society: Washington, DC, 1986; pp 114–133.
583. Tressl, R.; Engel, K. H.; Albrecht, W. *Food Sci. Technol.* **1988,** *30,* 67–80.
584. Muys, G. T.; van der Ven, B.; de Jonge, A. P. *Nature (London)* **1962,** *194,* 995–996.
585. Feuerbach, M.; Fröhlich, O.; Schreier, P. *J. Agric. Food Chem.* **1988,** *36,* 1236–1237.
586. Günther, C. Ph.D. Dissertation, University of Würzburg, 1988.
587. Bernreuther, A.; Koziet, J.; Brunerie, P.; Krammer, G.; Christoph, N.; Schreier, P. *Z. Lebensm. Unters. Forsch.* **1990,** *191,* 299–301.
588. Ohloff, G. *Fortschr. Chem. Org. Naturst.* **1978,** *35,* 431–527.
589. Do, J. Y.; Salunkhe, D. K.; Olson, L. E. *J. Food Sci.* **1969,** *34,* 618–621.
590. Engel, K.-H.; Ramming, D. W.; Flath, R. A. *J. Agric. Food Chem.* **1988,** *36,* 1003–1006.
591. Arctander, S. In *Perfume and Flavor Chemicals;* Arctander: Montclair, NJ, 1969.
592. Furia, T. E.; Bellanca, N. In *Feranolis Handbook of Flavor Ingredients;* Chemical Rubber: Cleveland, OH, 1981; p 803.
593. Engel, K. H.; Flath, R. A.; Buttery, R. G.; Mon, T. R.; Ramming, D. W.; Teranishi, R. *J. Agric. Food Chem.* **1988,** *36,* 549–553.
594. Engel, K. H.; Heidlas, J.; Albrecht, W.; Tressl, R. In *Flavor Chemistry—Trends and Developments;* Teranishi, R.; Buttery, R. G.; Shahidi, F., Eds.; ACS Symposium Series 388; American Chemical Society: Washington, DC, 1989; pp 8–22.
595. Eberhardt, R.; Woidich, H.; Pfannhauser, W. In *Flavor '81;* Schreier, P., Ed.; de Gruyter: Berlin, Germany, 1981; pp 377–383.
596. Krammer, G.; Fröhlich, O.; Schreier, P. In *Bioflavour '87, Analysis—Biochemistry Biotechnology;* Schreier, P., Ed.; de Gruyter: Berlin, Germany, 1988; pp 89–95.
597. Tang, C. S.; Jennings, W. G. *J. Agric. Food Chem.* **1967,** *15,* 24–28.
598. Tang, C. S.; Jennings, W. G. *J. Agric. Food Chem.* **1968,** *16,* 252–254.
599. Issanchou, S.; Schlich, P.; Guichard, E. *Sci. Aliments* **1989,** *9,* 351–369.
600. Rodriguez, F.; Seck, S.; Crouzet, J. *Lebensm. Wiss. Technol.* **1980,** *13,* 152–155.
601. Als, G. *Food Manuf.* **1973,** *48,* 31–32.
602. Jennings, W. G.; Sevenants, M. R. *J. Food Sci.* **1964,** *29,* 796–801.

603. Sevenants, M. R.; Jennings, W. G. *J. Food Sci.* **1966,** *31,* 81–86.
604. Idstein, H.; Schreier, P. *Phytochemistry* **1985,** *24,* 2313–2316.
605. Tressl, R.; Drawert, F.; Heimann, W. *Z. Naturforsch. B.: Anorg. Chem. Org. Chem.* **1969,** *24,* 1201–1202.
606. Schreier, P. *J. Sci. Food Agric.* **1980,** *31,* 487–494.
607. Nitz, S.; Kollmannsberger, H.; Drawert, F. *J. Chromatogr.* **1989,** *471,* 173–185.
608. Hollnagel, A. Ph.D. Dissertation, University of Frankfurt, 1990.
609. Hollnagel, A.; Menzel, E. M.; Mosandl, A. *Z. Lebensm. Unters. Forsch.* **1991,** *193,* 234–236.
610. Konishi, S.; Takahashi, E. *Plant Cell Physiol.* **1966,** *7,* 171–175.
611. Deans, D. R. *Chromatographia* **1968,** *1,* 18–22.
612. Schomburg, G.; Husmann, H.; Weeke, F. *J. Chromatogr.* **1975,** *112,* 205–217.
613. Guichard, E.; Etievant, P.; Henry, R.; Mosandl, A. *Z. Lebensm. Unters. Forsch.* **1992,** *195,* 540–544.
614. Mosandl, A.; Hener, U.; Fenske, H.-D. *Liebigs Ann. Chem.* **1989,** *9,* 859–862.
615. Koppenhoefer, B.; Allmendinger, H.; Nicholson, G. J. *Angew. Chem.* **1985,** *97,* 46; *Angew. Chem. Int. Ed. Engl.* **1985,** *24,* 48–49.
616. Schurig, V.; Wistuba, D. *Tetrahedron Lett.* **1984,** *25,* 5633–5636.
617. Freytag, W.; Ney, K. H. *Eur. J. Biochem.* **1968,** *4,* 315.
618. Freytag, W.; Ney, K. H. *J. Chromatogr.* **1969,** *41,* 473–474.
619. Tressl, R.; Bahri, D.; Engel, K. H. In *Quality of Selected Fruits and Vegetables of North America;* ACS Symposium Series 170; American Chemical Society: Washington, DC, 1980; pp 213–232.
620. Smolková, E.; Králová, H.; Krýsl, S.; Feltl, L. *J. Chromatogr.* **1982,** *241,* 3–8.
621. Maga, J. A. *J. Agric. Food Chem.* **1981,** *29,* 1–4.
622. Kaminski, E.; Stawicki, S.; Wasowicz, E.; Kasperek, M. *Nahrung* **1980,** *24,* 103–113.
623. Tressl, R.; Bahri, D.; Engel, K. H *J. Agric. Food Chem.* **1982,** *30,* 89–93.
624. Widder, S.; Sen, A.; Grosch, W. *Z. Lebensm. Unters. Forsch.* **1991,** *193,* 32–35.
625. Holtz, R.; Schisler, L. C. *Lipids* **1971,** *6,* 176–180.
626. Dijkstra, F. Y.; Wikén, T. O. *Z. Lebensm. Unters. Forsch.* **1976,** *160,* 255–262.
627. Ullrich, F.; Grosch, W. *Z. Lebensm. Unters. Forsch.* **1987,** *184,* 277–282.
628. König, W. A.; Krebber, R.; Evers, P.; Bruhn, G. *J. High Resolut. Chromatogr.* **1990,** *13,* 328–332.
629. Kobayashi, A. *Nippon Nogeikagaku Kaishi* **1990,** *64,* 203–207.
630. Georgilopoulos, D. N.; Gallois, A. N. *Z. Lebensm. Unters. Forsch.* **1987,** *184,* 374–380.
631. Schubert, V.; Diener, R.; Mosandl, A. *Z. Naturforsch., C: Biosci.* **1991,** *46,* 33–36.
632. Froehlich, O.; Huffer, M.; Schreier, P. *Z. Naturforsch., C: Biosci.* **1989,** *44,* 555–558.
633. Engel, K. H.; Tressl, R. *Chem. Microbiol. Technol. Lebensm.* **1983,** *8,* 33–39.
634. Mosandl, A.; Fischer, K.; Hener, U.; Kreis, P.; Rettinger, K.; Schubert, V.; Schmarr, H.-G. *J. Agric. Food Chem.* **1991,** *39,* 1131–1134.
635. Rose, H. C.; Stern, R. L.; Karger, B. L. *Anal. Chem.* **1966,** *38,* 469–472.
636. König, W. A.; Francke, W.; Benecke, I. *J. Chromatogr.* **1982,** *239,* 227–231.
637. Biyani, B. *East. Pharm.* **1982,** *25,* 37–38.
638. Maheshwari, M. L.; Sangeeta, M. J. *PAFAI J.* **1983,** *5,* 25–27.
639. Faber, B.; Dietrich, A.; Mosandl, A. *J. Chromatogr. A* **1994,** *666,* 161–165.
640. Bicchi, C.; Pisciotta, A. *J. Chromatogr.* **1990,** *508,* 341–348.
641. Koshiro, S.; Sonomoto, K.; Tanaka, A.; Fukui, S. *J. Biotechnol.* **1985,** *2,* 47–57.
642. Askari, C.; Kreis, P.; Mosandl, A.; Schmarr, H.-G. *Arch. Pharm. (Weinheim Ger).* **1992,** *325,* 35–39.
643. Hagenauer-Hener, U.; Henn, D.; Dettmar, F.; Mosandl, A.; Schmitt, A. *Dtsch. Lebensm. Rundsch.* **1990,** *86,* 273–276.
644. Rettinger, K.; Karl, V.; Schmarr, H.-G.; Dettmar, F.; Hener, U.; Mosandl, A. *Phytochem. Anal.* **1991,** *2,* 184–188.

645. Karl, V.; Rettinger, K.; Dietrich, H.; Mosandl, A. *Dtsch. Lebensm. Rundsch.* **1992,** *88,* 147–149.
646. Karl, V.; Schmarr, H.-G.; Mosandl, A. *J. Chromatogr.* **1991,** *587,* 347–350.
647. Rettinger, K.; Burschka, C.; Scheeben, P.; Fuchs, H.; Mosandl, A. *Tetrahedron: Asymmetry* **1991,** *2,* 965–968.
648. Tressl, R.; Heidlas, J.; Albrecht, W.; Engel, K. H. In *Bioflavour '87: Analysis—Biochemistry Biotechnology;* Schreier, P., Ed.; de Gruyter: Berlin, Germany, 1988; pp 221–236.
649. Creveling, R. K.; Silverstein, R. M.; Jennings, W. G. *J. Food Sci.* **1968,** *33,* 284–287.
650. von Winter, M.; Klöti, R. *Helv. Chim. Acta.* **1972,** *55,* 1916–1921.
651. Engel, K. H.; Tressl, R. *J. Agric. Food Chem.* **1983,** *31,* 796–801.
652. Koppenhoefer, B.; Allmendinger, H.; Nicholson, G. J.; Bayer, E. *J. Chromatogr.* **1983,** *260,* 63–73.
653. Wipf, B.; Kupfer, E.; Bertazzi, R.; Leuenberger, H. G. W. *Helv. Chim. Acta* **1983,** *66,* 485–488.
654. Goodall, D. M. *Trends Anal. Chem.* **1993,** *12,* 177–184.
655. Goodall, D. M.; Wu, Z.; Lisseter, S. G. *Anal. Proc.* **1992,** *29,* 14–16.
656. Bestmann, H. J.; Attygalle, A. B.; Glasbrenner, J.; Riemer, R.; Vostrowsky, O. *Angew. Chem.* **1987,** *99,* 784–785.
657. König, W. A.; Benecke, I.; Ernst, K. *J. Chromatogr.* **1982,** *253,* 267–270.
658. Buttery, R. G.; Seifert, R. M.; Guadagni, D. G.; Ling, L. C. *J. Agric. Food Chem.* **1971,** *19,* 524–529.
659. Ahmed, E. M.; Dennison, R. A.; Dougherty, R. H.; Shaw, P. E. *J. Agric. Food Chem.* **1978,** *26,* 187–191.
660. Ohloff, G.; Flament, I.; Pickenhagen, W. *Food Rev. Int.* **1985,** *1,* 99–148.
661. In *The Merck Index: An Encyclopedia of Chemicals, Drugs, and Biologicals,* 11th ed.; Budavari, S.; O'Neil, M.; Smith, A.; Heckelman, P. E., Eds.; Merck & Company: Rahway, NJ, 1989; Nos. 7414, 7415, p 1182.
662. Armstrong, D. W.; Zukowski, J. *J. Chromatogr. A* **1994,** *666,* 445–448.
663. Smolková-Keulemansová, E.; Feltl, L.; Krýsl, S. *J. Incl. Phenom.* **1985,** *3,* 183–196.
664. Kocielski, T.; Sybilska, D.; Belniak, S.; Jurczak, J. *Chromatographia* **1986,** *21,* 413–416.
665. Kocielski, T.; Sybilska, D.; Jurczak, J. *J. Chromatogr.* **1986,** *364,* 299–303.
666. Lindström, M.; Norin, T.; Roeraade, J. *J. Chromatogr.* **1990,** *513,* 315–320.
667. König, W. A.; Krebber, R.; Wenz, G. *J. High Resolut. Chromatogr.* **1989,** *12,* 790–792.
668. Ehlers, J.; König, W. A.; Lutz, S.; Wenz, G.; Dieck, H. T. *Angew. Chem. Int. Ed. Engl.* **1988,** *27,* 1556–1558.
669. Mosandl, A.; Hener, U.; Kreis, P.; Schmarr, H.-G. *Flavour Fragrance J.* **1990,** *5,* 193–199.
670. Hener, U.; Kreis, P.; Mosandl, A. *Flavour Fragrance J.* **1990,** *5,* 201–204.
671. Hener, U.; Kreis, P.; Mosandl, A. *Flavour Fragrance J.* **1991,** *6,* 109–111.
672. Kreis, P.; Hener, U.; Mosandl, A. *Dtsch. Apoth. Ztg.* **1990,** *130,* 985–988.
673. Kreis, P.; Hener, U.; Mosandl, A. *Dtsch. Lebensm. Rundsch.* **1991,** *87,* 8–11.
674. Armstrong, D. W.; Tang, Y.; Zukowski, J. *Anal. Chem.* **1991,** *63,* 2858–2861.
675. Sybilska, D.; Kowalczyk, J.; Asztemborska, M.; Stankiewicz, T.; Jurczak, J. *J. Chromatogr.* **1991,** *543,* 397–404.
676. Sybilska, D.; Asztemborska, M.; Kowalczyk, J.; Ochocka, R. J.; Ossicini, L.; Perez, G. *J. Chromatogr. A* **1994,** *659,* 389–394.
677. Köhler, J.; Schomburg, G. *Chromatographia* **1981,** *14,* 559–566.
678. Gil-Av, E. Presented at the 5th International Symposium on Column Liquid Chromatography, Avignon, France, May 1981.
679. Zukowski, J. *J. High Resolut. Chromatogr.* **1991,** *14,* 361–362.
680. Hansen, A. P.; Heinis, J. J. *J. Dairy Sci.* **1992,** *75,* 1211–1215.
681. Askari, C.; Hener, U.; Schmarr, H.-G.; Rapp, A.; Mosandl, A. *Fresenius' Z. Anal. Chem.* **1991,** *340,* 768–772.

682. Villalon Mir, M.; Lopez Garzia de la Serrana, H.; Lopez Martinez, M. del C.; Garcia Villanova, R. *An. Bromatol.* **1985**, *36*, 61–69.
683. Fryth, P. W. In *Kirk–Othmer Encyclopedia of Chemical Technology*, 3rd ed.; Grayson, M.; Eckroth, D., Eds.; Wiley: New York, 1983; Vol. 24, pp 1–7.
684. Hickman, K. C. D. *Chem. Rev.* **1944**, *34*, 51–106.
685. Bollinger, H. R.; König, A. In *Dünnschichtchromatography*, 2nd ed.; Stahl, E., Ed.; Springer: Heidelberg, Germany, 1967.
686. Vecchi, M.; Vesely, J.; Oesterhelt, G. *J. Chromatogr.* **1973**, *83*, 447–453.
687. Stalcup, A. M.; Jin, H. L.; Armstrong, D. W.; Mazur, P.; Derguini, F.; Nakanishi, K. *J. Chromatogr.* **1990**, *499*, 627–635.
688. Bieri, J. G. In *The Encyclopedia of Biochemistry*; Williams, R. J.; Lansford E. M., Jr., Eds.; Reinhold: New York, 1967; pp 834–835.
689. Wing, C. L.; Kim, S.; Rando, R. R. *J. Am. Chem. Soc.* **1989**, *111*, 793–795.
690. Law, W. C.; Rando, R. R. *Biochemistry* **1988**, *27*, 4147–4152.
691. Law, W. C.; Rando, R. R. *J. Am. Chem. Soc.* **1988**, *110*, 5915–5917.
692. Bernstein, P. S.; Law, W. C.; Rando, R. R. *Proc. Natl. Acad. Sci. U.S.A.* **1987**, *84*, 1849–1853.
693. Fulton, B. S.; Rando, R. R. *Biochemistry* **1987**, *26*, 7938–7945.
694. Bernstein, P. S.; Law, W. C.; Rando, R. R. *J. Biol. Chem.* **1987**, *262*, 16848–16857.
695. Olson, G. L.; Cheung, H.-C.; Morgan, K. D.; Borer, R.; Saucy, G. *Helv. Chim. Acta* **1976**, *59*, 567–585.
696. von Fischli, A.; Mayer, H.; Simon, W.; Stoller, H.-J. *Helv. Chim. Acta* **1976**, *59*, 397–405.
697. Cardillo, G.; Contento, M.; Sandri, S.; Panunzio, M. *J. Chem. Soc., Perkin Trans. I* **1979**, 1729–1733.
698. Chauhan, Y. S.; Chandraratna, R. A. S.; Miller, D. A.; Kondrat, R. W.; Reischl, W.; Okamura, W. H. *J. Am. Chem. Soc.* **1985**, *107*, 1028–1033.
699. Rodriguez-Amaya, D. B.; Tavares, C. A. *Food Chem.* **1992**, *45*, 297–302.
700. Ramos, D. M. R.; Rodriguez-Amaya, D. B. *Arg. Biol. Tecnol.* **1993**, *36*, 93–94.
701. Shlomai, P.; Ben-Amotz, A.; Margalith, P.; Mokady, S. *J. Nutr. Biochem.* **1992**, *3*, 415–420.
702. Douglas, K. *New Scientist* **1993**, *139*, 24–25.
703. In *The Merck Index: An Encyclopedia of Chemicals, Drugs, and Biologicals*, 11th ed.; Budavari, S.; O'Neil, M.; Smith, A.; Heckelman, P. E., Eds.; Merck & Company: Rahway, NJ, 1989; No. 4141, p 4139.
704. In *Handbook of Vitamins and Hormones;* Kutsky, R. J., Ed.; VNR: New York, 1973; Chapter 13, pp 87–95.
705. Machover, D.; Grison, X.; Goldschmidt, E.; Zittoun, J.; Lotz, J. P.; Metzger, G.; Richaud, J.; Hannoun, L.; Marquet, J.; Guillot, T. *J. Natl. Cancer Inst.* **1992**, *84*, 321–327.
706. Machover, D.; Grison, X.; Goldschmidt, E.; Zittoun, J.; Metzger, G.; Richaud, J.; Lotz, J. P.; Andre, T.; Hannoun, L.; Marquet, J. *Ann. Oncol.* **1993**, *4*(Suppl. 2), S29–S35.
707. In *Encyclopedia of Chemical Technology*, 3rd ed.; Wiley: New York, 1978; Vol. 4.
708. In *Handbook of Vitamins and Hormones*; Kutsky, R. J., Ed.; VNR: New York, 1973; Chapter 15, pp 104–111.
709. Jaffe, G. M. In *Kirk–Othmer Encyclopedia of Chemical Technology*, 3rd ed.; Grayson, M.; Eckroth, D., Eds.; Wiley: New York, 1984; Vol. 24, pp 8–40.
710. Anderson, S.; Marks, C. B.; Lazarus, R.; Miller, J.; Stafford, K.; Seymour, J.; Light, D.; Rastetter, W.; Estell, D. *Science (Washington, D.C.)* **1985**, *230*, 144–149.
711. Williams, R. J. In *The Encyclopedia of Biochemistry;* Williams, R. J.; Lansford E. M., Jr., Eds.; Reinhold: New York, 1967; pp 843–844.
712. Wilson, S. R.; Yasmin, A. *Stud. Nat. Prod. Chem.* **1992**, *10*, 43–75.

713. Tan, B.; Brzuskiewicz, L. *Anal. Biochem.* **1989,** *180,* 368–373.
714. Deng, B.; Chen, P.; Cai, X. *Gaodeng Xuexiao Huaxue Xuebao* **1988,** *9,* 333–336.
715. Yakushina, L. M.; Lykova, N. M.; Ryndakova, I. A. *Symp. Biol. Hung.* **1988,** *37,* 463–469.
716. Piironen, V.; Syväoja, E.-L.; Varo, P.; Salminen, K.; Koivistoinen, P. *J. Agric. Food Chem.* **1986,** *34,* 742–746.
717. Parrish, D. B. *CRC Crit. Rev. Food Sci. Nutr.* **1980,** *13,* 161–187.
718. Ollivier, D.; Dordonnant, J. M.; Garetier, T.; Barridas, S.; Tisse, C.; Guerere, M. *Ann. Falsif. Expert. Chim. Toxocol.* **1993,** *86,* 149–160.
719. Iakushuina, L. M.; Lykova, N. M.; Ryndakova, I. A. *Vopr. Pitan.* **1987,** *6,* 59–61.
720. Lehmann, J. W.; Putnam, D. H.; Qureshi, A. A. *Lipids* **1994,** *29,* 177–181.
721. Hodges, R. E. In *The Encyclopedia of Biochemistry;* Williams, R. J.; Lansford, E. M., Jr., Eds.; Reinhold: New York, 1967; p 844.
722. Wayner, D. D. M. *Bioelectrochem. Bioenerg.* **1987,** *18,* 219–229.
723. Wallet, S. *Fette, Seifen, Anstrichim.* **1986,** *88,* 485–490.
724. Acuff, R. V.; Thedford, S. S.; Hidiroglou, N. N.; Papas, A. M.; Odom T. A. Jr. *Am. J. Clin. Nutr.* **1994,** *60,* 397–402.
725. Zahalka, H. A.; Cheng, S. C.; Burton, G. W.; Ingold, K. U. *Biochim. Biophys. Acta* **1987,** *921,* 481–485.
726. Horwitt, M. K.; Elliott, W. H.; Kanjananggulpan, P.; Fitch, C. D. *J. Am. Clin. Nutr.* **1984,** *40,* 240–245.
727. Traber, M. G.; Pillai, S. R.; Kayden, H. J.; Steiss, J. E. *Lipids* **1993,** *28,* 1107–1112.
728. Weiser, H.; Vecchi, M. *Int. J. Vitam. Nutr. Res.* **1982,** *52,* 351–370.
729. Brownstein, S.; Burton, G. W.; Hughes, L.; Ingold, K. U. *J. Org. Chem.* **1989,** *54,* 560–569.
730. In *Handbook of Vitamins and Hormones;* Kutsky, R. J., Ed.; VNR: New York, 1973; Chapter 12, pp 79–86.
731. In *The Merck Index: An Encyclopedia of Chemicals, Drugs, and Biologicals,* 11th ed.; Budavari, S.; O'Neil, M.; Smith, A.; Heckelman, P. E., Eds.; Merck & Company: Rahway, NJ, 1989; No. 9933, p 1580.
732. Schmid, R.; Antoulas, S.; Rüttimann, A.; Schmid, M.; Vecchi, M.; Weiser, H. *Helv. Chim. Acta* **1990,** *73,* 1276–1299.
733. Gessler, N. N.; Bezzubov, A. A. *Farmakol. Toksikol.* **1987,** *50,* 49–51.
734. *Chirality* **1982,** *4,* 338–340.
735. Armstrong, D. W.; Reid, G. L., III; Hilton, M. L.; Chang, C.-D. *Environmental Pollution* **1993,** *79,* 51–58.

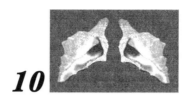

10

Chromatography as a Separation Tool for the Preparative Resolution of Racemic Compounds

Eric R. Francotte

Thanks to the concomitant development of a wide range of chiral stationary phases (CSPs) and the introduction of new chromatographic instrumentation, the preparative separation of enantiomers by chromatography is gaining increasing acceptance as a simple, rapid, and generally applicable method of preparing optically pure compounds. This chapter presents a general strategy for preparative chiral separations. Among the CSPs used, the most widely applied have been derived from polysaccharides, particularly cellulose. The wide range of application and the high loading capacity of these CSPs have contributed much to their prevalence. The chapter reviews the merits and limitations of the different chromatographic techniques. Up to now, elution batch chromatography clearly dominates in terms of the number of applications. Various approaches to improving throughput, such as close injections, peak shaving, and recycling, are discussed. However, simulated moving-bed chromatography is arousing growing interest and may soon become a very efficient tool for producing optically pure compounds.

The importance of molecular chirality to the biological activity of pharmaceuticals and agrochemicals is now widely recognized (*1–6*). In consequence of, and in accordance with, the recommendations of the regulatory authorities for

3407–8/96/0271$16.00/0
© 1996 American Chemical Society

developing single enantiomers (7–9), the demand is growing for stereoselective separation techniques and analytical assays to evaluate enantiomeric purity of chiral compounds. Among the separation techniques, chromatography on chiral stationary phases (CSPs) has become one of the most powerful tools for the analytical determination of enantiomeric purity in syntheses and in biological, pharmacologic, toxicologic, pharmacokinetic, and clinical studies (4, 10–12). In fact, this method was originally developed for preparative purposes to obtain the pure enantiomers of racemic compounds not accessible by the classical routes. After these pioneering works using lactose (13–15), cellulose (16, 17), starch (18–20), cellulose triacetate (21–23), and optically active polyacrylamides (24), emphasis has, for more than 10 years now, been placed on the analytical potential of the method. With the development of new CSPs and the introduction of more elaborate technologies, the chromatographic resolution of racemates on a preparative scale compels increasing recognition as an alternative to "preparing" pure enantiomers.

This chapter summarizes what is currently feasible and what is to be expected in the way of CSPs and technical possibilities in preparative chromatographic resolution, without listing all preparative chiral separations. Such a list is available from a recent publication (25). The usefulness of the method is illustrated with examples of separations of drugs, agrochemicals, and synthesis intermediates. Moreover, it must be emphasized that both the range of CSPs and the technologies available in this field are rapidly growing and rapidly changing.

General Strategy

Although most problems related to preparative chiral chromatography are not typical of separations on CSPs, some properties are specific to each kind of CSP and have to be considered before performing preparative chiral separations. The general strategy to be applied for preparative chiral separations can be summed up as follows (Figure 1):

1. Select an appropriate CSP by screening analytical columns filled with CSPs also available in preparative amounts.
2. Improve the separation and resolution factors by optimizing the various chromatographic parameters (mobile-phase composition, flow rate, temperature, etc.).
3. Optimize the chromatographic throughput (sample amount, column overloading).
4. Transfer the analytical separation to the preparative system, with choice of the column dimension and final adjustment of the chromatographic conditions, depending on the mode of chromatography chosen.
5. Perform the preparative separation and, if necessary, automate the process.

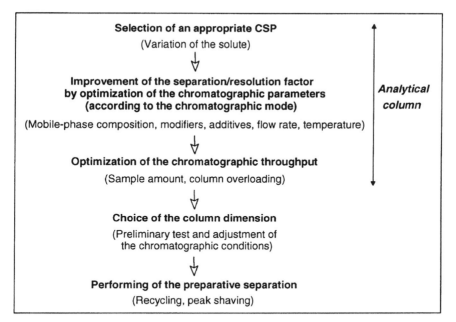

Figure 1. General strategy for preparative chiral separations.

These steps are discussed in the next sections.

As an alternative to Step 1 of this general strategy, it is sometimes advantageous to select a CSP and accommodate the solute by modification (derivatization) of the structure until it separates on that CSP. This strategy has led to successful resolutions of racemates that otherwise could not be separated on a particular CSP (*26–32*). It is especially indicated when the user is forced, for various reasons (urgency, costs, etc.), to work with a limited number of preparative CSPs. These two possible strategies for choosing the CSP are schematized in Figure 2. Finally, if the desired compound from a reaction sequence cannot be separated, it is often possible to resolve a precursor (*33–42*). In general, Strategy 1 should be evaluated first, except for design of a specific CSP, which is more elaborate and can reasonably be applied only for particularly interesting classes of racemic structures or for very important compounds that would justify such an investment.

Nevertheless, in the future it is quite possible that tailor-made CSPs will be especially designed for a defined racemic solute. This strategy is applicable to CSPs prepared by the approach developed by Pirkle and based on the concept of reciprocity (*43, 44*) or by the approach developed by Wulff (*45, 46*) and further investigated by Mosbach and co-workers (*47, 48*), which involves preparing polymers imprinted with chiral cavities that fit preferably one enantiomer of the desired compound. The first approach has already been used to prepare CSPs es-

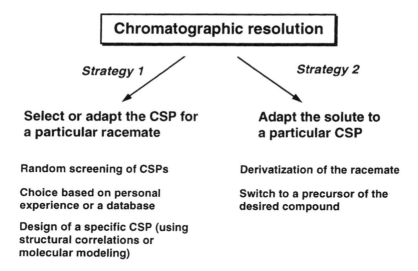

Figure 2. Options for choosing a CSP.

pecially designed for β-blockers *(49)* or naproxen *(50)*, for example. The second approach is, in principle, very elegant, but it has certain drawbacks: to imprint the polymers, it is necessary first to have the pure enantiomers of the racemate separated or of a structurally closely related compound, which implies that another method of separating the enantiomers already exists. The loading capacity of such CSPs is also not high enough for large-scale separations *(48)*.

Available CSPs

The chromatographic separation of enantiomers on CSPs was, in fact, originally developed as a preparative tool for the resolution of racemates, long before the method was recognized as useful for the analytical determination of optical purity. Indeed, when the first commercial chiral high-performance liquid chromatography (HPLC) column became available in 1980, the technique of resolving enantiomers on CSPs was already being intensively investigated and used in different research groups. At that time, preparative separations had already been carried out on lactose *(13–15)*, on starch *(18–20)*, and later more successfully on cellulose triacetate in crystalline form I (CTA-I) *(21–23)*. Moreover, the polyacrylamide CSPs introduced by Blaschke were already in use early on to "prepare" optically pure enantiomeric drugs to investigate their biological activities *(51, 52)*.

Since the introduction of the first commercial HPLC column based on the Pirkle concept *(53)*, the market has been inundated with analytical chiral HPLC columns. Not all analytical CSPs can be used for preparative separations, however, because some properties that are not essential for analytical purposes can be crucial for preparative applications. One of the major requirements for a preparative CSP is loading capacity, as discussed later. Furthermore, some general features such as wide availability, easy preparation, low preparation costs, and durability (chemical and mechanical stability) of the phase are important factors in choosing a preparative CSP.

Although the range of CSPs available on the market is rapidly changing, recent reviews have listed many existing or proposed CSPs for preparative or semi-preparative separations (*25, 54*). For some of these CSPs, the preparative potential has yet to be demonstrated or is limited to the preparation of milligram amounts for use in determining physical properties or in investigating biological activities requiring only small amounts of drugs. These CSPs are not considered here, and only chiral sorbents that could reasonably be used for separations of many grams or kilograms of racemates are discussed. A list of the CSPs meeting the preparative requirements is given in Table I. It includes the polyacrylamide CSPs, a series of variations of the Pirkle phases, some CSPs for ligand-exchange chromatography (LEC), a series of cellulose-based CSPs composed of the pure polymers or applied as coatings on silica, and some cyclodextrin-based CSPs. The structures of the most used CSPs are shown in Figures 3 through 6.

Among these preparative CSPs, the cellulose derivatives, particularly CTA-I,

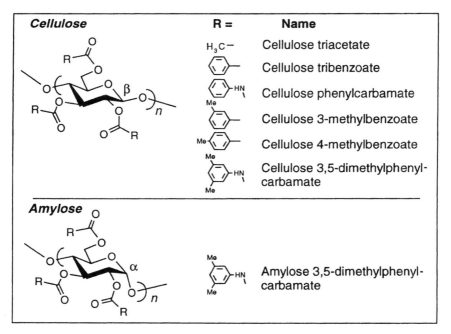

Figure 3. Structures of cellulose- and amylose-based CSPs.

**Table I. Most Used CSPs for Preparative Chromatographic
Resolution of Chiral Compounds**

Packing Name	Chiral Selector	Supplier or Ref.
Cellulose and amylose derivatives		
Chiralcel OA	Cellulose triacetate	Daicel
Chiralcel OB	Cellulose tribenzoate	Daicel
Chiralcel OC	Cellulose tris(phenylcarbamate)	Daicel
Chiralcel OD	Cellulose tris(3,5-dimethylphenylcarbamate)	Daicel
Chiralcel OJ	Cellulose tris(4-methylbenzoate)	Daicel
Chiralcel OK	Cellulose tris(cinnamate)	Daicel
Chiralpak AD	Amylose tris(3,5-dimethylphenylcarbamate)	Daicel
CTA-I	Cellulose triacetate	Daicel, Merck
CTB	Cellulose tribenzoate	Riedel–de Haen
MMBC	Cellulose tris(3-methylbenzoate beads)	66
PMBC	Cellulose tris(4-methylbenzoate beads)	66
TBC	Cellulose tribenzoate beads	65
Cyclodextrin-based CSPs		
ChiraDex	β-cyclodextrin	Merck
Cyclobond I	β-cyclodextrin	Astec
Hyd-β-CD	Hydroxypropyl-β-cyclodextrin	Merck, 154
π-acidic and π-basic CSPs		
DNBLeu	3,5-Dinitrobenzoylleucine	Regis
DNBPG-co	3,5-Dinitrobenzoylphenylglycine (covalent bonding)	Regis
DNBPG-io	3,5-Dinitrobenzoylphenylglycine (ionic bonding)	Regis
ChyRoSine-A	3,5-Dinitrobenzoyltyrosine butylamide	Sedere
DNB-Tyr-E	3,5-Dinitrobenzoyltyrosine methylester	130
NAP-AI	Naphthylalanine	Regis
Whelk-O 1	3,5-Dinitrobenzoyl tetrahydrophenanthrene amine	Regis
Polyacrylamide CSPs		
ChiraSpher	Poly[(S)-N-acryloylphenylalanine ethyl ester]	Merck
Poly-CHMA	Poly[(S-N-methacryloyl-2-cyclohexylethylamine]	56, 57
Poly-MA	Poly[N-acryloylmenthylamine]	162
Poly-PEA	Poly[(S-N-acryloylphenylalanine ethyl ester]	56, 57
CSPs for ligand-exchange chromatography		
Chirosolve-pro	L- or D-proline bound to polyacrylamide	JPS Chimie
Chirosolve-phe	L- or D-phenylalanine bound to polyacrylamide	JPS Chimie
Chirosolve-val	L-or D-valine bound to polyacrylamide	JPS Chimie
Chirosolve-hypro	L- or D-hydroxyproline bound to polyacrylamide	JPS Chimie
Chirosolve-porretine	L- or D-porretine bound to polyacrylamide	JPS Chimie
Polystyrene-Prol	L- or D-proline bound to polystyrene	176

NOTE: Suppliers are Astec (Advanced Separation Technologies, Whippany, NJ, USA); Daicel Chemical Industries, Tokyo, Japan; JPS Chimie, Bevais, Switzerland; E. Merck, Darmstadt, Germany; Regis Chemical Company, Morton Grove, IL, USA; Riedel–de Haen, Seelze, Germany; and Sedere, Alfortville, France.

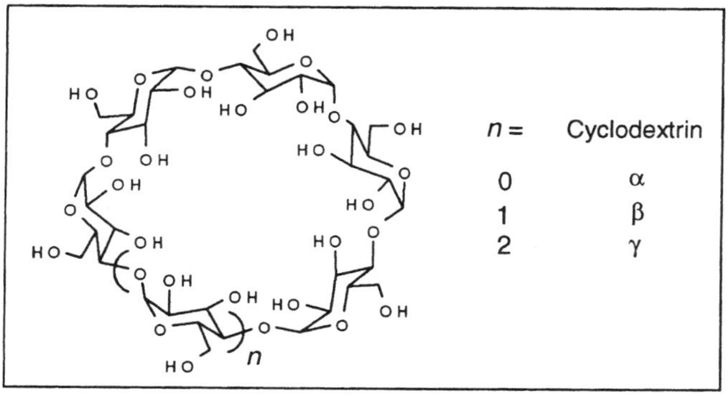

Figure 4. Structures of cyclodextrins.

were found to be especially suitable because of their versatility and high loading capacity. CTA-I is still the CSP most used for preparative separations, especially for sample sizes greater than 10 g, but the preparative potential of other CSPs has been clearly demonstrated.

Selection Criteria for the CSP

The most important criterion for choosing a CSP is obviously chiral selectivity, expressed by the separation factor α. Nevertheless, when a preparative separation is required, other parameters such as loading capacity, solubility of the solute in the mobile phase, or chemical stability of the CSP can outweigh selectivity.

Loading Capacity

Because loading capacity is a critical factor, the screening step using analytical columns should obviously be restricted to CSPs with a high enough loading capacity. The CSPs that have been, or could be, used for the chromatographic resolution of at least gram amounts of enantiomers are listed in Table I.

Numerous chiral sorbents introduced for the chromatographic resolution of racemates were obtained by immobilizing chiral compounds covalently or ionically on the modified surface of silica gel conferring mechanical stability. These compounds are small molecules, such as amino acid derivatives in the case of various Pirkle phases (*53*), or high molecular weight compounds, such as cyclodextrins (*55*) or polymethacrylamides (*56, 57*). Polymeric materials, such as cellulose and amylose derivatives, have also been used as layers coated on macroporous silica gel (*58*).

Although the mechanical properties of all these silica-based materials are

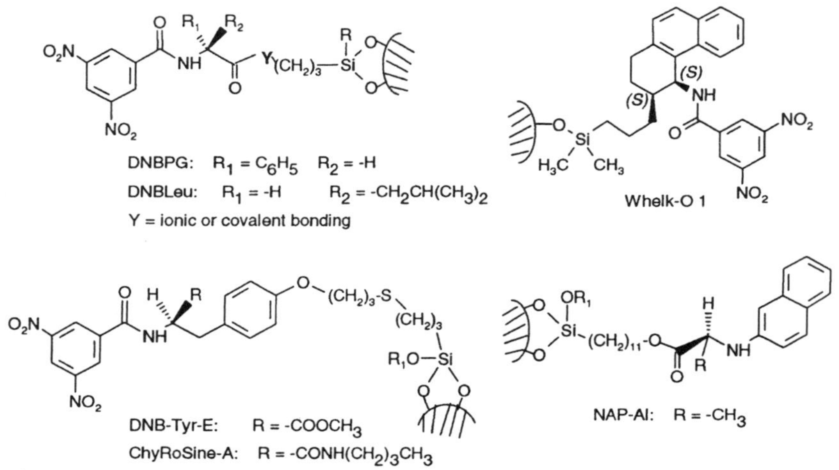

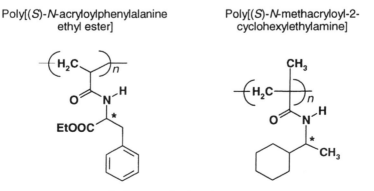

DNBPG: R₁ = C₆H₅   R₂ = -H
DNBLeu: R₁ = -H   R₂ = -CH₂CH(CH₃)₂
Y = ionic or covalent bonding

Whelk-O 1

DNB-Tyr-E: R = -COOCH₃
ChyRoSine-A: R = -CONH(CH₂)₃CH₃

NAP-AI: R = -CH₃

Figure 5. Structures of π-acidic and π-basic CSPs.

generally satisfactory, they have the disadvantage that only a part of the sorbent contains the chiral information capable of differentiating the enantiomers. Obviously, this feature affects the loading capacity of the material. A stationary phase composed mostly of silica gel with only a few chiral elements will be rapidly overloaded, and even if it has properties useful for analytical purposes, it will not be appropriate for preparative applications. The same restriction also applies when only a small part of the chiral material seems to be involved in chiral recognition. This is the case for the protein-based phases (59), which generally have very low capacities because the active site capable of chiral recognition is only a small part of the macromolecule.

Poly[(S)-N-acryloylphenylalanine ethyl ester]

Poly[(S)-N-methacryloyl-2-cyclohexylethylamine]

Figure 6. Structures of polyacrylamide-based CSPs.

The situation is completely different for other classes of CSPs, s polysaccharide derivatives and various polyacrylamide and polymetha derivatives developed as chromatographic supports for enantiomeric separations. In these materials, composed of a small chiral unit regularly repeating along the polymeric chain, the density of active sites capable of chiral recognition is very high and results in a high loading capacity. Data from the literature on sample size per unit mass of CSP correlate well with this classification, as depicted in Figure 7.

This effect is clearly demonstrated by the exceptionally high loading capacity of CTA-I (*54, 60, 61*). Resolutions of 40 to 150 g of racemate per run on a 20 × 100 cm column of CTA-I have been reported (*25, 62*). Another example, showing the separation of 1.3 kg/run of a racemate on a 50 × 45 cm column containing 32 kg of CTA-I, was also recently published (*63*). Other cellulose benzoate derivatives developed by Mannschreck and co-workers (*64*) and by Francotte and Wolf (*65, 66*) and used as pure polymeric material also have high loading capacities (*64–68*), as do the polyacrylamide and polymethacrylamide CSPs developed by Blaschke (Figure 8) (*69*).

Obviously, if the polymeric chiral sorbents are deposited on a carrier, the dilution decreases the loading capacity. This is the case for the polysaccharide-based CSPs introduced by Okamoto (*58*) and marketed under the name Chiralcel. Nevertheless, these materials have a sufficiently high capacity to be useful in preparative applications, as shown in Figure 9 for the separation of 2 g of a drug on a CSP composed of cellulose tris(phenylcarbamate) (*70*).

Furthermore, not only the number but also the availability of the potential interaction sites is important to obtaining a useful preparative sorbent. Indeed,

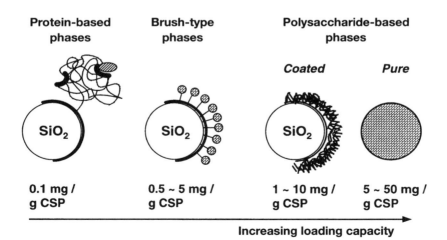

Figure 7. Loading capacity scale for various CSP types.

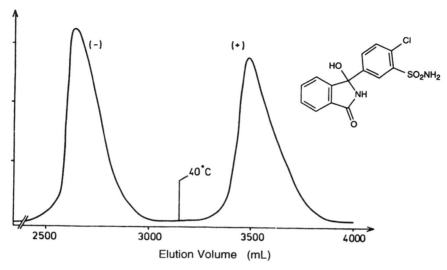

Figure 8. Chromatographic resolution of 532 mg of racemic chlorthalidone on Poly-PEA.

site availability is related to such physical properties of the material as surface area or porosity—features that must be considered for all chromatographic sorbents. Even though these properties may be changed when the silica gel surface is modified with chiral materials, they are rarely discussed in the literature.

Finally, a few cases of inversion of elution order have been observed on CTA-I CSP upon increasing the amount of injected racemate (*71, 72*). Such in-

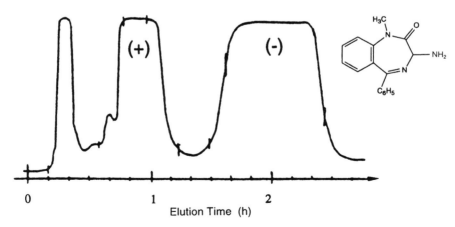

Figure 9. Chromatographic resolution of 2 g of a racemic benzodiazepine on Chiralcel OC. Column: 4 × 20 cm. Mobile phase: ethanol (anhydrous). (From reference 38. Chromatogram kindly provided by G. Dutruc-Rosset, Rhone–Poulenc Rorer.)

version is rare, but users should be aware of this phenomenon, which can be explained by the presence of different types of interaction sites in this CSP (*26, 71, 73*).

Selectivity

The most important point in ensuring the successful resolution of a racemate is to work under high-selectivity conditions—that is, conditions that give the highest separation factor. The selectivity is first determined by the recognition ability of the CSP, but it can also be influenced by the composition of the mobile phase or the presence of modifiers (*37, 41, 74–91*), the presence of additives (*89, 92, 93*), the pH (*60, 84, 88, 89, 94*), or the temperature (*41, 60, 67, 75, 76, 79, 92, 95*). These parameters can be optimized on an analytical column while maintaining good solubility of the solute and reasonable retention times. An analytical column is normally used to identify an appropriate CSP because smaller amounts are needed and the operation is less time-consuming.

Although most users still base their choice of CSP on their own experience, some predictions based on structural requirements, possible molecular interactions, or empirical rules can help. Models have also been proposed for optimization of the chromatographic parameters (*96, 97*). Wainer has proposed a selection guide (*98, 99*) but in the meantime many new chiral columns have been introduced and new applications have been established for previously existing CSPs.

A computer database developed by Roussel and co-workers (*100, 101*) also facilitates the selection of a CSP. It allows searches for separations of racemates reported in the literature on the basis of structural fragments and in general provides a means of rapidly finding the chromatographic conditions reported for a particular separation within the vigorously expanding body of published work in this field. An expert system for selecting appropriate CSPs has also recently been proposed (*102*). However, experience in my group suggests that most chiral separations can be achieved with a very limited number of CSPs.

When the desired selectivity is not achieved for the investigated racemate and when the number of CSPs available is limited, it can be useful to separate a precursor or a derivative of the racemate. For aliphatic and aromatic alcohols, for example, a good separation is generally obtained on the cellulose tribenzoate CSP after derivatization of the alcohol as a benzoyl ester (*103*). This strategy was applied for separating the enantiomers of the racemic intermediate used in the synthesis of the anticancer agent edatrexate (Figure 10) (*104*).

Chemical and Physical Stability of the CSP

When using CSPs, one must pay attention to some chromatographic properties that are usually neglected. Many CSPs are composed of organic polymeric materials that have limitations related to their chemical, physical, or mechanical

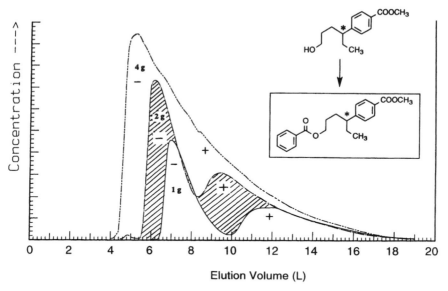

Figure 10. Preparative chromatographic separation of the enantiomers of the benzoyl ester deriva-tive of 4-(3-hexan-6-ol)benzoic acid methyl ester (drug intermediate) on TBC sample size: 1–4 g. Column: 70 × 5 cm. Mobile phase: hexane–2-butanol, 9:1.

properties. Most of these polymeric materials are not cross-linked or not immo-bilized and are therefore more or less soluble in numerous solvents. For example, cellulose triacetate, which has proved very useful for preparative purposes (*26, 54, 62, 69, 105–108*), cannot be used with chlorinated alkanes as mobile phases, because they dissolve it. Other cellulose derivatives introduced for the chromato-graphic resolution of racemates as a coating on silica or as pure polymers (*58*) generally show the same limitations regarding mobile phases, but several can be used under either normal- or reversed-phase conditions. Despite these limita-tions, cellulose derivatives are probably the most often used CSPs for preparative separations because of their broad applicability and high loading capacity.

A strong swelling has also been reported for the polyacrylamide derivatives developed in the pure polymeric form some years ago by Blaschke and Donow (*109*). Although these materials were cross-linked, they could be used only at very low pressures because of their gel structure in most organic solvents. To improve their mechanical properties, these materials were later polymerized (by graft polymerization) on the surface of silica gel, but this treatment reduced the load-ing capacity because of the lower amount of chiral material per unit of mass. Nevertheless, the feasibility of preparative separations on the resulting CSP has been demonstrated (*57, 110, 111*).

The CSPs obtained by immobilization of chiral material on silica gel are generally stable and afford greater flexibility regarding the chromatographic con-ditions, particularly the choice of mobile phase. With these classes of sorbents, the type and composition of the mobile phase can be varied widely, except that some, such as the protein-based CSPs and the crown ether CSPs, must be used under aqueous conditions. These later CSPs are not discussed here, because they are not useful for large-scale separations. The CSPs developed for LEC are also used exclusively under reversed-phase conditions, but in this case this limitation is imposed by the mechanism of interaction. Davankov (*112*), a pioneer in developing LEC, has reviewed its general features as well as recent developments. An example of a preparative separation carried out by LEC is shown in Figure 11 (*113*).

The chemical stability of the stationary phase can also be a limiting factor. Phases bearing reactive chemical functions can be altered by injection of certain solutes. For example, stationary phases bearing labile ester functions, such as cellulose triacetate or cellulose tribenzoate, and some polyacrylamides react easily with nucleophilic amines. This was also the case for the first generation of Pirkle's CSPs obtained by ionic bonding. Chemical stability must be considered before performing a separation on a particular CSP.

Solubility of the Solute

The solubility of the racemic solute in the mobile phase is frequently a limiting factor in the development of a preparative chiral separation. This factor is negligible in analytical applications, but it becomes critical in preparative applications. Fortunately, there are more and more CSPs that are compatible with a

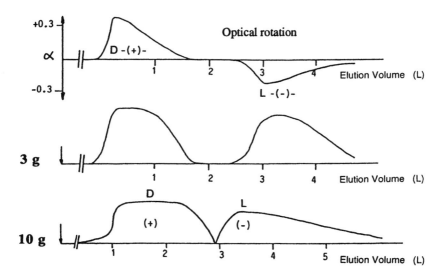

Figure 11. Chromatographic resolution of (a) 3 g and (b) 10 g of racemic threonine on Chirosolve-Pro Column: 120 × 6 cm. Mobile phase: 0.005 M copper(II) acetate, 0.05 M acetic acid. (Reproduced with permission from reference 113. Copyright 1989 Vieweg Publishing.)

wide variety of mobile phases, including nonpolar and polar solvents, under normal- and reversed-phase conditions. Among the preparative CSPs, the major limitation concerns the polysaccharide-based CSPs, which must be used under carefully chosen conditions that vary from CSP to CSP. Recommendations about conditions can usually be obtained from the supplier.

Moreover, using a better solvent can dramatically decrease or completely destroy selectivity. Indeed, better solvents show a stronger affinity for the solute and so compete more strongly with the CSP for complexation with the solute. Because the mobile phase is usually achiral, interactions that do not discriminate between the enantiomers will predominate and even suppress the chiral differentiation if the interaction with the CSP becomes negligible. In such cases, the best course is to find a compromise between solubility and selectivity or to resolve a precursor or derivative of the desired compound.

Finally, solubility can also be improved by increasing the temperature, but that usually decreases retention time and may lower selectivity.

Choice of the Chromatographic Mode

Various chromatographic modes are available for preparative resolutions. The options depend on such factors as the techniques available, the amounts desired, the solubility of the solute, and the physical properties of the CSP. At least five technologies have been applied to the preparative separation of enantiomers: batch-elution chromatography, simulated moving-bed (SMB) chromatography, supercritical or subcritical fluid chromatography (SFC or SubFC), gas chromatography (GC), and the membrane process (described in Chapter 11).

If there are no cost limitations, the choice of chromatographic mode will be governed mainly by the desired amounts of enantiomers. For amounts ranging from 1 to 50 g, the best option in terms of costs, productivity, and time is probably the classical batch-elution chromatography. For compounds that are soluble in nonpolar solvents, SFC can be considered as an alternative. Volatile racemates can be resolved into their enantiomers by GC. For amounts ranging from 100 g to several kilograms, elution chromatography can be still used, but to achieve the separation within a reasonable time, it is necessary to work with large columns and to automate the system. Working under overloaded conditions and using recycling and peak shaving (described later) are also recommended. Until SMB technology was introduced into this field, it was practically impossible to separate large amounts. Now, this technique offers a new tool for resolving racemic compounds in amounts up to production scale, provided the CSPs become available in large quantities at affordable cost.

Batch-Elution Chromatography

At present, the batch-mode process has the advantage of being a well-established, simple technology that is in principle applicable on any scale according to the needs of the user. As a rule, however, the column dimensions, and consequently the amount of CSP needed, will increase with the amount of racemate to be separated. This limits the extent to which chiral separations using batch-elution chromatography can be scaled up, because the availability of the CSPs in large quantities cannot be relied on at present, except perhaps in the case of CTA-I. Nevertheless, for modest amounts, this mode of chromatography is the fastest and the easiest to operate. Until now, most of the preparative chiral separations have been performed in this mode, as illustrated in Figure 12 for the separation of the enantiomers of 45 g of the chiral NMR solvating agent 2,2,2-trifluoro-1-(9-anthryl) ethanol on CTA-I (*25*).

Various options, such as close injections, recycling, and peak shaving, have been used to improve throughput in batch-elution chromatography. These options are reviewed in the next section. Detailed comparisons of the overloaded elution mode and the displacement mode show that the two modes give broadly similar purity and yield values (*114, 115*).

Simulated Moving-Bed Chromatography

Although most chiral separations have used the conventional batch-elution process, interest in SMB technology is growing because it saves large amounts of mobile phase and increases productivity, thus reducing production costs (*40*,

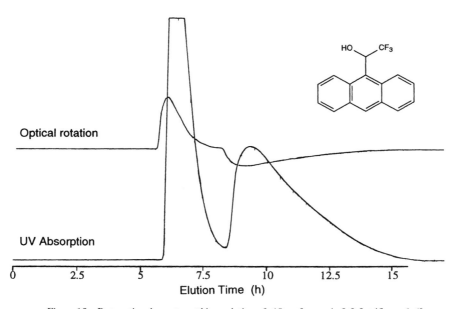

Figure 12. Preparative chromatographic resolution of 45 g of racemic 2,2,2-trifluoro-1-(9-anthryl) ethanol (TFAE) on CTA-I. Column: 100 × 20 cm. Mobile phase: ethanol–water 94:6 (5 L/h).

116–118). This technology, which was initially developed in the 1960s (*119*), is already applied in various industrial processes, even for compounds with low added value, as in the separation of *p*-xylene and other C_8 aromatics or of fructose and glucose (*120*), and it is now available on the laboratory scale. SMB chromatography allows the continuous introduction of the feed and fully utilizes the available mass-transfer area, so throughput is higher. Moreover, the technique is particularly appropriate for the separation of binary mixtures such as racemates.

Both the principle and the design of SMB systems have recently been reviewed (*117, 120–122*). Use of the method for chiral separations is just beginning, and a few applications have already been reported (Table II). However, in view of its scale-up potential, its range of applicability can be expected to increase rapidly. Besides saving mobile phase, SMB chromatography has the advantage of being highly productive with even modest amounts of CSP, thus reducing the costs for the generally very expensive chiral sorbents. These various advantages have been evidenced, for example, by the separation of the enantiomers of propranolol on Chiralcel OD reported by Gattuso and Makino (*123*): a productivity of 3.83 g/h/L of CSP (0.11 g in batch elution) with a mobile-phase consumption of less than 0.5 L to yield 1 g of racemate (25 L in batch elution). However, because of the investment required to set up this technology—which is also difficult to apply properly—its practical uses will probably be limited to compounds that are needed in amounts of at least 100 g.

Supercritical or Subcritical Fluid Chromatography

The feasibility of preparative separations using SFC or SubFC has also been demonstrated (*127–131*) up to pilot scale with 6-cm (inner diameter) axial compression columns (Table III). An excellent review summarizing the state of the art in preparative SFC has recently been published (*132*). Because of the low viscosity of the mobile phase (usually consisting mainly of carbon dioxide), SFC can be performed at high flow rates and is generally considered to be faster than HPLC (*133*), thus permitting greater throughput. Moreover, one of the main advantages of SFC is that it operates with safer (nonflammable) and considerably cheaper mobile phases, two factors that are very important for large-scale separations. Macaudière et al. have shown that a wide range of commercial chiral HPLC columns can also be used for SFC or SubFC (*133*), and an interesting recent paper has compared the performance of the two chromatographic modes and reported on the long-term stability of a cellulose-based CSP (*134*). The preparative separations performed under SFC conditions are summarized in Table III and exemplified in Figure 13, showing the resolution of the antitussive racemic drug guaifenesin on Chiralcel OD (*135*).

Nevertheless, the limited solubility of most drugs and agrochemicals in carbon dioxide considerably reduces the applicability of this technique. The solubility can be improved by adding modifiers such as alcohols, but then the modifier

Table II. Preparative Chromatographic Resolutions Performed in SMB Mode

Racemate	CSP	Productivity in SMB Mode (in batch mode)	Mobile Phase Needed to Process 1 g Enantiomer	Ref.
	Chiralcel OD	0.98 g/h/L CSP (0.016 g)	5.3 L	116
	Chirosolve L-Proline	not mentioned	not mentioned	125
	CTA-I	1.45 g/h/kg CSP (2.5 g)	400 mL	40, 126
	CTA-I	4.7 g/h/kg CSP (not mentioned)	56 mL	118
	CTA-I	119 g/h/kg CSP (2.66 g)	94 mL	124
	PMBC	4.15 g/h/kg CSP (not mentioned)	not mentioned	124
	Chiralcel OD	3.83 g/h/L CSP (0.11 g)	200 mL	123

must be evaporated after chromatography, thus diminishing the advantage of SFC over HPLC in preparative applications. Large-scale SFC devices are also available, but the large investment would be justified only for racemates that are needed in large amounts and are soluble in the SFC mobile phase.

Gas Chromatography

GC on an analytical scale is commonly used for determining the enantiomeric purity of a wide range of racemic substances (*136, 137*). By contrast, few prepar-

Table III. Preparative Chromatographic Resolutions Performed Under SFC Conditions

Racemate	CSP	Column Dimensions (cm)	Sample Size (mg)	Mobile Phase	Flow Rate (mL/min)	Ref.
	DNBPG-co	25 × 0.9	10	CO_2–2-propanol 82:18	8	127
	ChyRoSine-A	7.6 × 6	40	CO_2–ethanol 96.5:3.5	717	128
	DNB-Tyr-E	7.6 × 6	100	CO_2–ethanol 92:8	167	128 130
	DNB-Tyr-E	10 × 6	100	CO_2–ethanol 88:12	200	130
	DNB-Tyr-E	10 × 6	38	CO_2–ethanol 92:8	250	130
	Chiralcel OD	23 × 1	60	CO_2–methanol– 2-propanol 86:7:7	5	135
	Whelk O-1	25 × 2.54	200	CO_2–2-propanol– AcOH 75:25:0.5	100	131

ative applications of GC have been reported, mainly because GC is applicable only to relatively volatile compounds. For such molecules, the method undoubtedly has potential, and preparative processes based on GC could be regarded as an economical alternative.

The reported semipreparative separations of enantiomers by GC were performed on a modest scale, but they do at least demonstrate the feasibility of the method in this special field of application. Most of these separations have been achieved on cyclodextrin-based CSPs obtained by coating a dilute solution of cyclodextrin or its derivative on Chromosorb. The racemic substances include pheromones, monoterpenes, and anesthetics (Table IV). Outputs of up to 15 g/day were obtained for 1-octenyl-3-acetate on a 1 m × 2 cm column (*138*) and 5

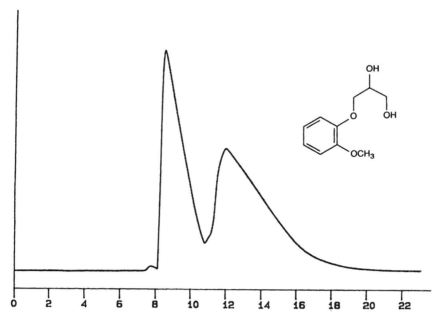

Figure 13. Preparative chromatographic separation of the enantiomers of racemic guaifenesin (60 mg) by SubFC on Chiralcel OD. Column: 23 × 1 cm. Mobile phase: carbon dioxide–methanol–2-propanol 86:7:7.

to 10 g/day for methyl 2-chloropropionate on a 1 m × 2.25 cm column (*139*). For the anesthetics enflurane and isoflurane, baseline separations were achieved by injection of 47 mg and 6 mg, respectively, on a 4 m × 0.7 cm column containing 2,6-di-*O*-pentyl-3-*O*-butanoyl-γ-cyclodextrin as the chiral selector (*140, 141*). Larger amounts of the enantiomers of these anesthetics were also separated on a trifluoroacetyl-γ-cyclodextrin CSP, and the production rate has been determined as a function of the amount of sample injected for various degrees of enantiomeric purity (*142, 143*). Using enantioselective complexation GC, Schurig and Leyer were able to separate the labile enantiomers of 1-chloro-2,2-dimethylaziridine in amounts sufficient to determine their chiroptical data and to measure their inversion barrier (*144*).

Improving the Separation

As emphasized earlier, it is always advantageous to work under the conditions giving the best selectivity. Once a CSP suitable for performing the chromatographic resolution has been identified, the separation can be further improved by adjusting parameters such as the composition of the mobile phase or the pres-

Table IV. Preparative Chromatographic Resolutions Performed by GC

Racemate	Chiral Selector	Column	Sample Size (mg)	Carrier Gas	Flow Rate	Ref.
	β-cyclodextrin	1 m × 2 cm	430	hydrogen	2.6 L/min	138
$CF_3CHClOCHF_2$	trifluoroacetyl-γ-cyclodextrin	2 m × 1 cm	75–675	helium	2.2 cm/s	142
$CF_3CHClOCHF_2$	2,6-di-O-pentyl-3-O-butanoyl-γ-cyclodextrin	4 m × 0.7 cm	6	helium	8.4 cm/s	140
$CHFClCF_2OCHF_2$	trifluoroacetyl-γ-cyclodextrin	1 m × 1 cm	50–120	hydrogen	5 cm/s	143
$CHFClCF_2OCHF_2$	2,6-di-O-pentyl-3-O-butanoyl-γ-cyclodextrin	4 m × 0.7 cm	47	helium	8.4 cm/s	140 141
	2,6-di-O-methyl-3-O-pentyl-β-cyclodextrin	1.8 m × 0.4	2	helium	400 mL/min	146
	2,6-di-O-methyl-3-O-pentyl-β-cyclodextrin	1.8 m × 0.4 cm	5	helium	400 mL/min	146
	α-cyclodextrin	2 m × 0.4 cm	—	helium	—	145
	trichloroacetyl-β-cyclodextrin	1 m × 2.25 cm	206	hydrogen	8 cm/s	139
	nickel(II) bis[(3-heptafluorobu-tanoyl)-(1R)-cam-phorate]	7.5 m × 0.3 cm	2	nitrogen	1.5 bar	144

NOTE: — means not mentioned in the reference.

ence of modifiers (*37, 41, 74–91*), the presence of additives (*89, 92, 93*), the pH (*60, 84, 88, 89, 94*), and the temperature (*41, 60, 67, 75, 76, 79, 92, 95*) to improve selectivity or resolution. The optimization is usually carried out on an analytical column. However, no general rule can be given for optimizing, because the influence of the parameters depends strongly on the CSP and the solute. In some cases, even inversions of the elution order have resulted from changing the temperature (*90, 147*) or the type of modifier (*74, 77, 80, 91*). With silica-based CSPs, an acidic or basic additive often improves the resolution. Increasing the temperature will usually reduce the retention time, but opposite effects (longer retention times) have also been observed on a CSP developed by Pirkle and Burke (*49*). Higher temperature can also improve the solubility of the solute.

Achieving Preparative Chromatographic Separations

Column Packing

Until recently, most commercial CSPs were available only as packed columns, but most CSPs suitable for preparative separations will likely soon be obtainable in bulk. The particle size of the CSPs offered for preparative purposes is generally larger than that of the material used in analytical columns. In their chromatographic properties, such as efficiency, these preparative sorbents can differ from those used for analytical determinations, and therefore it may be useful to perform the optimization on an analytical column filled with the preparative material. In SMB chromatography, the number of theoretical plates in the columns is less important than in the other chromatographic modes, and CSP materials with a large particle size are preferred.

A wide range of column hardware is currently available, and many suppliers also offer a packing service to customers. To achieve high-performance packing, there is an increasing trend toward using columns with axial compression. The column dimensions will depend on the amount of racemate to be processed and on the technology used. SMB chromatography uses shorter columns on account of the back pressure caused by setting up the columns in series.

After filling a new column, it is advisable to perform a test, which should be repeated regularly to check the performance.

Improving Throughput

To improve the separation or the throughput, techniques such as recycling and peak shaving are increasingly being used. Recycling, which is especially useful for poorly resolved racemates, involves passing the solute through the column several times (Figure 14). This technique, first applied to chiral separations by Schlögl

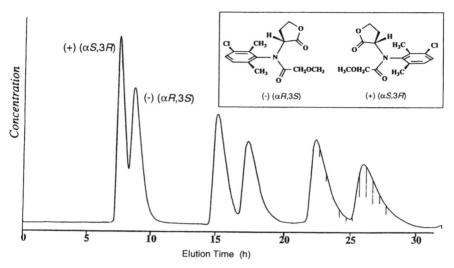

Figure 14. Resolution of 52 g of racemic clozylacon by chromatography on CTA-I with recycling. Column: 100 × 20 cm. Mobile phase: ethanol–water 94:6.

(149), has been useful in various applications *(25, 79, 149–154)*, even for large amounts (Figure 14).

In peak shaving, the first and the last fractions containing the enantiomers at the desired optical purity are collected at each pass through the column, while the eluate containing a mixture is recycled. Peak shaving is usually combined with recycling, allowing work to be performed under overload conditions to improve throughput *(79, 154)*. An example is shown in Figure 15 for the separation of the enantiomers of a nootropic (memory-enhancing agent) drug using two cycles *(104)*, but examples with up to nine cycles per injection were reported by Dingenen et al., who used this process intensively *(79)*.

The technique of peak shaving and recycling has recently been exhaustively reviewed and discussed *(154)*, and even though most of the reported separations involved only 25 to 400 mg per run, much of the review is applicable to large-scale separations. Jacobson and Guiochon have demonstrated that it is preferable to work under overloading conditions to increase the production rate *(155)*. Theoretical models for closed-loop recycling in preparative chromatography have also been proposed *(156, 157)*. Further guidelines for optimizing chromatographic yield have been discussed for specific CSPs *(37, 77–79, 110)*.

Finally, automation of the process can greatly improve the throughput and is practically indispensable for separations of large amounts of racemates. This was clearly demonstrated by the resolution of 14 kg of a racemic benzotriazole derivative on a 10 × 50 cm column filled with Chiralcel OJ CSP by automatic continuous injections of 20 g of racemate *(154)*.

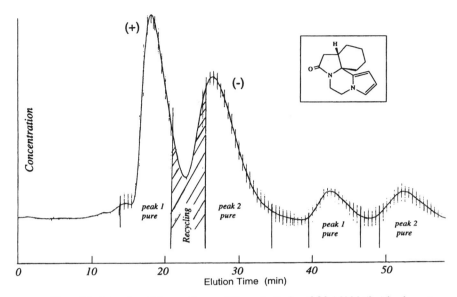

Figure 15. Separation of the enantiomers of the nootropic drug CGS 16920 (2 g) by chromatography on TBC with recycling and peak shaving. Column: 45 × 5 cm. Mobile phase: 100% methanol.

Isolating the Solute

Isolation of the resolved enantiomers is obviously a crucial step in preparative chromatographic separation. In the conventional batch-elution process, isolation is generally no difficulty. Fractionation is usually achieved in a conventional way (i.e., the fractions containing the solute at the desired optical purity are collected). As soon as automation of the process is required or techniques such as recycling and peak shaving are applied, the fractionation points have to be carefully determined. The purity of the isolated fractions should be regularly checked, and chromatographic conditions have to be readjusted when deviations are observed (*154*). In the SMB mode, there is no fractionation, but both enantiomers are collected continuously, one as raffinate and the other as extract.

After isolation, the solutes are recovered by simply evaporating the mobile phase. In SFC, however, this step, which could a priori be considered easier because of the mobile phase used, can lead to difficulties associated with pressure reduction and the formation of considerable amounts of carbon dioxide (*128*, *130*). Since this technical problem has already been solved for large-scale equipment, it can be expected that a solution will rapidly be found for laboratory-scale units. A prototype equipped with a cyclone separator for fraction collection has recently been presented (*131*).

Sometimes, the isolated solutes are apparently contaminated with substances from the CSP or from the mobile phase. In that case, recrystallization or flash chromatography on silica gel is generally enough to purify the desired material.

As a final step for compounds that have been derivatized to improve the separation, the derivatizing group must be cleaved off. This presents no difficulty if the derivatizing group has been properly chosen to ensure that no racemization occurs under the cleavage conditions. That, of course, must be checked before scale-up.

Selected Applications and Perspectives

Although the preparative potential of chromatographic resolution is clearly established, the number of published applications in which more than 1 g of racemate has been injected per run remains limited. Reported resolutions of 20 g or more per run are still rare, and most have used CTA-I as a CSP. Nevertheless, using recycling, peak shaving, and continuous injections, several hundred grams and even kilograms of racemates have been resolved on the cellulose-based CSPs Chiralcel OD and OJ, as well as on a hydroxypropyl-β-cyclodextrin CSP (*154*).

Table V summarizes the published separations of at least 1 g of injected racemate, arranged in descending order of sample size. The number of cycles is listed, which indicates whether the recycling technique was applied. Although the list is by no means exhaustive, since many preparative separations carried out in the industry have not been published and most analytical separations performed on the CSPs listed in Table V could be scaled up, it provides an overview of what has been done in this field.

The feasibility of separating enantiomers up to kilogram scale by chromatography on CSPs is now clearly established. The method is especially useful as an inexpensive means of obtaining the pure enantiomers of new drugs or pesticides in the amounts needed for preliminary biological, pharmacologic, or toxicologic testing and for field trials. It is generally applicable, rapid, and easy, and usually furnishes both enantiomers at high optical purity. Techniques such as recycling and peak shaving can greatly improve yields and throughput, but large-scale separations would require large amounts of CSP and have been considered economically unjustifiable because of the high cost of the CSPs, the high-dilution conditions, the consumption of large amounts of mobile phase, and the difficulties associated with recycling the mobile phase. Thanks to new developments, however, particularly SMB chromatography, the technical prerequisites for cost-effective large-scale separations can now be met. The SMB mode can save up to 90% of the mobile phase and achieves a much higher throughput, thus necessitating less CSP for processing a given amount of racemate. With these new possibilities, it can be expected that the technique will soon be considered a suitable alternative to the classical approach for producing optically pure compounds in large amounts. In parallel with this development, it is likely that

Table V. **Preparative Resolution of Racemic Compounds**

Structure	Sample Size per Run (g)	CSP	Column (cm)	Number of Cycle	Ref.
	1300	CTA-I	50 × 45	1	63
	150	CTA-I	100 × 20	1	25, 158
	110	CTA-I	100 × 20	1	158
	100	CTA-I	100 × 20	1	26, 158
	100	CTA-I	100 × 20	1	158
R = (CH₂)₉-CH=CH₂	100	DNBPG-co	122 × 15	1	161
	65	CTA-I	100 × 20	1	158
	52	CTA-I	100 × 20	2	25, 158–160
	50	CTA-I	100 × 20	1	158
	50	Poly-MA	not mentioned	1	162

(continued)

Table V. *Continued.* **Preparative Resolution of Racemic Compounds**

Structure	Sample Size per Run (g)	CSP	Column (cm)	Number of Cycle	Ref.
	45	CTA-I	100 × 20	1	25, 158
	45	CTA-I	100 × 20	1	158
	40	CTA-I	100 × 20	1	158
	40	CTA-I	100 × 20	1	158
	40	CTA-I	100 × 20	1	163
	25	CTA-I	100 × 20	1	158
	20	CTA-I	100 × 20	2	163, 167
	20	CTA-I	100 × 20	1	25, 62
	20	Polystyrene-Prol	not mentioned	1	176
Benztriazole Derivative	17	CTA-I	100 × 20	1	158
	15	Chiralcel OJ	50 × 10	6	177
	13.9	CTA-I	100 × 20	1	158

(continued)

Table V. *Continued.* **Preparative Resolution of Racemic Compounds**

Structure	Sample Size per Run (g)	CSP	Column (cm)	Number of Cycle	Ref.
	10	CTA-I	100 × 20	1	158
	10	Chirosolve-pro	120 × 6	1	113
	9	CTA-I	100 × 20	1	158
	8	DNBPG-io	76 × 5	1	164
Piperazine carboxamide derivative	7.5	Chiralcel OD	40 × 10	2	154
	7	DNBPG-io	76 × 5	1	168
	6	DNBPG-io	76 × 5	1	168, 169
	6	Polystyrene-Prol	not mentioned	1	176
	5.3	ChiraSpher	40 × 10	1	110, 111
	5.2	CTA-I	100 × 5	1	163

(continued)

Table V. *Continued.* **Preparative Resolution of Racemic Compounds**

Structure	Sample Size per Run (g)	CSP	Column (cm)	Number of Cycle	Ref.
	5	CTA-I	60 × 5	1	165
	5	DNBPG-io	76 × 5	1	168
	4.7	TBC	40 × 5	1	104
	4.2	CTA-I	100 × 5	1	34
	4	CTA-I	100 × 5	1	163
	4	CTA-I	60 × 5	1	25, 163
	4	Chiralcel OC	22 × 7	1	70
	4	DNBPG-io	76 × 5	1	164, 168
	4	TBC	70 × 5	1	104
γ-arylketo ester derivative	4	Chiralcel OD	10 × 5	9	154

(continued)

Table V. *Continued.* **Preparative Resolution of Racemic Compounds**

Structure	Sample Size per Run (g)	CSP	Column (cm)	Number of Cycle	Ref.
HO–CCl$_2$CF$_3$ (anthracene structure)	3.2	CTA-I	100 × 5	1	166
(phenyl lactone structure)	3	CTA-I	50 × 2.5	1	73, 163
O=S–C$_4$H$_9$ / CH$_2$Cl (anthracene structure)	3	DNBPG-io	76 × 5	1	164
O=, O, C(CH$_3$)$_3$, N, OCH$_2$Ph (oxazolidinone structure)	3	Chiralcel OD	50 × 5	1	104
COCH$_3$ COCH$_3$ (spiro structure)	2	CTA-I	66 × 6.3	4	151
Et (furanone structure)	2.47	CTA-I	60 × 5	1	168
(benzimidazole structure) C$_6$H$_5$	2.1	CTA-I	70 × 3.8	1	69
H$_3$C, N, O, NH$_2$, N, C$_6$H$_5$ (benzodiazepine structure)	2	Chiralcel OC	20 × 4	1	38
CH$_3$ COCH$_2$OCH$_3$, N, O, CH$_3$ (structure)	2	CTA-I	100 × 5	1	163
H (pyrrole fused structure)	2	TBC	45 × 5	2	104

(continued)

Table V. *Continued.* **Preparative Resolution of Racemic Compounds**

Structure	Sample Size per Run (g)	CSP	Column (cm)	Number of Cycle	Ref.
(triazole/dichlorophenyl/SR structure)	2	CTA-I	66 × 6.3	8	151
(pyridyl ketone / dichlorophenoxy / Et structure)	2	MMBC	60 × 2.5	1	163
(AcO, H₃C, OAc structure)	2	TBC	75 × 5	1	65
(isoindolinone / pyridine / Cl structure, X, O=C(CH₂)₃NHCOMe)	2	Chiralcel OC	22 × 7	1	70
	2	Chiralpak AD	40 × 10	3	177
(O₂N dinitrobenzamide, COOR, C(CH₃)₃, HN, Ph structure; R = (CH₂)₉-CH=CH₂)	1.92	NAP-Al	76 × 5	1	170
(phenyl, OH, CONH₂ structure)	1.87	CTA-I	40 × 10	1	110
(naphthalene hydantoin structure; R = (CH₂)₃-CH=CH₂)	1.6	DNBPG-io	76 × 5	1	164
(binaphthyl CH₂OH / HOH₂C structure)	1.6	CTA-I	66 × 6.3	5	151
(cyclophosphamide-type structure: CH₂CH₂Cl, N, P, O, NHCH₂CH₂Cl)	1.55	Poly-PEA	155 × 5	1	171, 172
Imidazole derivative	1.5	Hyd-β-CD	45 × 8	2	154

(continued)

Table V. *Continued.* **Preparative Resolution of Racemic Compounds**

Structure	Sample Size per Run (g)	CSP	Column (cm)	Number of Cycle	Ref.
	1.4	CTA-I	100 × 5	1	26
	1.35	ChiraSpher	25 × 2.5	4	154
	1.25	CTA-I	100 × 5	1	163
	1.13	DNBPG-io	76 × 5	1	164
	1	DNBPG-io	75 × 5	1	164
	1	TBC	75 × 5	1	104
	1	DNBPG-io	not mentioned	1	36
	1	ChiraSpher	40 × 10	1	110, 111
	1	DNBPG-io	25 × 1	1	173, 174

(continued)

Table V. *Continued.* **Preparative Resolution of Racemic Compounds**

Structure	Sample Size per Run (g)	CSP	Column (cm)	Number of Cycle	Ref.
	1	TBC	75 × 5	1	65
	1	PMBC	75 × 5	1	104
	1	DNBPG-co	22 × 8	1	78
	1	CTA-I	66 × 6.3	3	151
	1	CTA-I	66 × 6.3	1	151
	1	CTA-I	66 × 6.3	1	151
	1	DNBPG-io	76 × 5	1	164
	1	DNBPG-io	76 × 25	1	164
	1	CTA-I	100 × 5	3	163
	1	MMBC	60 × 2.5	1	163

(continued)

Table V. *Continued.* **Preparative Resolution of Racemic Compounds**

Structure	Sample Size per Run (g)	CSP	Column (cm)	Number of Cycle	Ref.
	1	ChyRoSine-A	26 × 4	1	77
	1	DNBPG-io	25 × 5	1	175

new CSPs meeting the requirements for preparative purposes will appear and that the costs of CSPs will fall.

Concerning the strategy to be adopted for preparative chiral resolutions, both options—selecting an appropriate CSP for a defined racemate and adapting the structure of the racemate to a defined CSP—should be considered, but the preparation of a new CSP specially designed for the resolution of a particular racemate could become a useful approach for compounds due to be produced in large amounts.

Acknowledgments

I am grateful to Mr. A. Kirkwood for linguistic assistance in preparing this manuscript.

References

1. Simonyi, M. *Med. Res. Rev.* **1984,** *4,* 359.
2. Williams, K.; Lee, E. *Drugs* **1985,** *30,* 333.
3. Stison, S. T. *Chem. Eng. News* **1992,** *28,* 46.
4. Wainer, I. W. In *Drug Stereochemistry, Analytical Methods and Pharmacology,* 2nd ed.; Wainer, I. W., Ed.; Marcel Dekker: New York, 1993; Chapter 6.
5. Ariens, E. J. *Eur. J. Pharmacol.* **1984,** *26,* 663.
6. Ramos Tombo, G. M.; Bellus, D. *Angew. Chem. Int. Ed.* **1991,** *30,* 1193.
7. De Camps, W. H. *Chirality* **1989,** *1,* 2.
8. Gross, M. *Ann. Rep. Med. Chem.* **1990,** *25,* 323.
9. Rauws, A. G.; Groen, K. *Chirality* **1994,** *6,* 72.
10. *Chiral Separations by Liquid Chromatography;* Ahuja, S., Ed.; ACS Symposium Series 471; American Chemical Society, Washington, DC, 1991.
11. *Chiral Separations by HPLC: Application to Pharmaceutical Compounds;* Krstulovic, A. M., Ed.; Ellis Horwood: Chichester, England; 1989, Chapter 11.

12. *Chromatographic Chiral Separations;* Zief, M.; Crane, L. J., Eds.; Chromatographic Science Series 40; Marcel Dekker: New York, 1988; p 91.
13. Henderson, G. M.; Rule, H. G. *J. Chem. Soc.* **1939**, 1568.
14. Lecoq, H. *Bull. Soc. R. Sci. Liege* **1943**, *12*, 316.
15. Prelog, V.; Wieland, P. *Helv. Chim. Acta* **1944**, *27*, 1127.
16. Mayer, W.; Merger, F. *Liebigs Ann. Chem.* **1961**, *644*, 65.
17. Legg, J. I; Douglas, B. E. *Inorganic Chem.* **1968**, *7*, 1452.
18. Lautsch, W.; Heinicke, D. *Kolloid Z.* **1957**, *154*, 1.
19. Steckelberg, W.; Bloch, M.; Musso, H. *Chem. Ber.* **1968**, *101*, 1519.
20. Hess, H.; Burger, G.; Musso, H. *Angew. Chem.* **1978**, *90*, 645.
21. Falk, H.; Schlögl, K. *Tetrahedron* **1966**, *22*, 3047.
22. Hesse, G.; Hagel, R. *Chromatographia* **1973**, *6*, 277.
23. Hesse, G.; Hagel, R. *Liebigs Ann. Chem.* **1976**, 996.
24. Schwanghart, A.-D.; Backmann, W.; Blaschke, G. *Chem. Ber.* **1977**, *110*, 778.
25. Francotte, E. *J. Chromatogr. A* **1994**, *666*, 565.
26. Francotte, E.; Wolf, R. M. *Chirality* **1990**, *2*, 16.
27. Pirkle, W. H.; McCune, J. E. *J. Liq. Chromatogr.* **1988**, *11*, 2165.
28. Dyas, A. M.; Robinson, M. L.; Fell, A. F. *Chromatographia* **1990**, *30*, 73.
29. Uzunov, D.; Stoev, G. *J. Chromatogr.* **1993**, *645*, 233.
30. Pawlowska, M.; Zukowski, J.; Armstrong, D. W. *J. Chromatogr. A* **1994**, *666*, 485.
31. Sato, K.; Nakano, H.; Hobo, T. *J. Chromatogr. A* **1994**, *666*, 463.
32. Francotte, E.; Stierlin, H.; Faigle, J. W. *J. Chromatogr.* **1985**, *346*, 321.
33. Kaneko, T.; Okamoto, Y.; Hatada, K. *J. Chem. Soc., Chem. Com.* **1987**, 1511.
34. Capraro, H.-G.; Francotte, E.; Kohler, B.: Rihs, G.; Schneider, P.; Scartazzini, R.; Zak, O.; Tosch, W. *J. Antibiotics* **1988**, *41*, 759.
35. Okamoto, Y.; Aburatani, R.; Kawashima, M.; Hatada, K.; Okamura, N. *Chem. Lett.* **1986**, 1767.
36. Howson, W.; Kitteringham, J.; Mistry, J.; Mitchell, M. B.; Novelli, R.; Slater, R. A.; Swayne, G. T. G. *J. Med. Chem.* **1988**, *31*, 352.
37. Miller, L.; Bush, H. *J. Chromatogr.* **1989**, *484*, 337.
38. Bouchaudon, J.; Dutruc-Rosset, G.; Alasia, M.: Beaudoin, F.; Bourzat, J.-D.; Chevé, M.; Cotrel, C.; James, C. Poster Presentation at the 2nd International Symposium on Chiral Discrimination, Rome, Italy, 1991.
39. Camacho-Torralba, P. L.; Beeson, M. D.; Vigh, G.; Thompson, D. H. *J. Chromatogr.* **1993**, *646*, 259.
40. Nicoud, R.-M.; Fuchs, G.; Adam, P.; Bailly, M.; Küsters, E.; Antia, F. D.; Reuille, R.; Schmid, E. *Chirality* **1993**, *5*, 267.
41. Miller, L.; Weyker, C. *J. Chromatogr. A* **1993**, *653*, 219.
42. Miller, L.; Honda, D.; Fronek, R.; Howe, K. *J. Chromatogr. A* **1994**, *658*, 429.
43. Pirkle, W. H.; Ho Hyun, M.; Bank, B. *J. Chromatogr.* **1984**, *316*, 585.
44. Pirkle, W. H.; Ho Hyun, M.; Tsipouras, A.; Hamper, B. C.; Bank, B. *J. Pharm. Biomed. Anal.* **1984**, *2*, 173.
45. Wulff, G.; Sarhan, A.; Zabrocki, K. *Tetrahedron Lett.* **1973**, *44*, 4329.
46. Wulff, G.; Kirstein, G. *Angew. Chem.* **1990**, *102*, 706.
47. O'Shannessy, D. J.; Andersson, L. I.; Mosbach, K. *J. Molec. Recogn.* **1989**, *2*, 1.
48. Fischer, L.; Müller, R.; Ekberg, B.; Mosbach, K. *J. Am. Chem. Soc.* **1991**, *113*, 9358.
49. Pirkle, W. H.; Burke III, J. A. *J. Chromatogr.* **1991**, *557*, 173.
50. Pirkle, W. H.; Welch, C. J. *J. Liq. Chromatogr.* **1992**, *15*, 1947.
51. Blaschke, G.; Markgraf, H. *Chem. Ber.* **1980**, *113*, 2031.
52. Blaschke, G.; Kraft, H.-P.; Fickentscher, K.; Köhler, F. *Arzneim. Forsch. Drug Res.* **1979**, *29*, 1640.
53. Welch, C. J. *J. Chromatogr.* **1994**, *666*, 3.

54. Francotte, E.; Junker-Buchheit, A. *J. Chromatogr.* **1992,** *576,* 1.
55. Haan, S. M.; Armstrong, D. W. In *Chiral Separations by HPLC;* Krstulovic, A. M., Ed.; Ellis Horwood: Chichester, England, **1989,** Chapter 10.
56. Blaschke, G.; Bröker, W.; Fraenkel, W. *Angew. Chem.* **1986,** *98,* 808.
57. Kinkel J. N.; Fraenkel, W.; Blaschke, G. *Kontakte Darmstadt* **1987,** 3.
58. Okamoto, Y.; Kaida, Y. *J. Chromatogr. A* **1994,** *666,* 403.
59. Allenmark, S. G. In *Chiral Separations by Liquid Chromatography;* Ahuja, S., Ed.; ACS Symposium Series 471; American Chemical Society: Washington, DC, 1991; pp. 114–125..
60. Isaksson, R.; Erlandsson, P.; Hansson, L.; Holmberg, A.; Berner, S. *J. Chromatogr.* **1990,** *498,* 257.
61. Rimböck, K.-H.; Kastner, F.; Mannschreck, A. *J. Chromatogr.* **1985,** *329,* 307.
62. Lohmann, D.; Auer, K.; Francotte, E. Presentation at the International Symposium on Chromatography, Weizmann Institute, Rehovot, Israel, 1988.
63. Brandt, K. H.; Hampe, Th. R.; Nagel, J.; Schmitt, E. Poster Presentation at the 3rd International Symposium on Chiral Discrimination, Tübingen, Germany, 1992.
64. Rimböck, K.-H.; Kastner, F.; Mannschreck, A. *J. Chromatogr.* **1986,** *351,* 346.
65. Francotte, E.; Wolf, R. M. *Chirality* **1991,** *3,* 43.
66. Francotte, E.; Wolf, R. M. *J. Chromatogr.* **1992,** *595,* 63.
67. Günther, K.; Carle, R.; Fleischhauer, I.; Merget, S. *Fresenius J. Anal. Chem.* **1993,** *345,* 787.
68. Küsters, E.; Loux, V.; Schmid, E.; Floersheim, P. *J. Chromatogr. A* **1994,** *666,* 421.
69. Blaschke, G. *J. Liquid Chromatogr.* **1986,** *9,* 341.
70. Mackenzie, R.; Dutruc-Rosset, G.; Beaudoin, F.; Bourzat, J.-D.; Chevé, M.; Manfré, F. Poster Presentation at the 3rd International Symposium on Chiral Discrimination, Tübingen, Germany, 1992.
71. Roussel, C.; Stein, J.-L.; Beauvais, F.; Chemlal, A. *J. Chromatogr.* **1989,** *462,* 95.
72. Krause, K.; Handke, G. *Tetrahedron Lett.* **1991,** *32,* 7225.
73. Wolf, R. M.; Francotte, E.; Hainmüller, J. *Chirality* **1993,** *5,* 538.
74. Gaffney, M. H.; Stiffin, R. M.; Wainer, I. W. *Chromatographia* **1989,** *27,* 15.
75. Rizzi, A. M. *J. Chromatogr.* **1989,** *478,* 101.
76. Pescher, P.; Caude, M.; Rosset, R.; Tambuté, A. *J. Chromatogr.* **1986,** *371,* 159.
77. Caude, M.; Tambuté, A.; Siret, L. *J. Chromatogr.* **1991,** *550,* 357.
78. Tambuté, A.; Gareil, P.; Caude, M.; Rosset, R. *J. Chromatogr.* **1986,** *363,* 81.
79. Dingenen, J.; Somers, I.; Mermans, R. Poster Presentation at the 2nd International Symposium on Chiral Discrimination, Rome, Italy, May 1991.
80. Balmer, K.; Persson, B.-A.; Lagerström, P.-O. *J. Chromatogr. A* **1994,** *660,* 269.
81. Nicholson, L. W.; Pfeiffer, C. D.; Goralski, C. T.; Singaram, B.; Fischer, G. B. *J. Chromatogr. A* **1994,** *687,* 241.
82. Levin, S.; Abu-Lafi, S.; Zahalka, J.; Mechoulam, R. *J. Chromatogr. A* **1993,** *654,* 53.
83. Barderas, A. V.; Duprat, F. *J. Liq. Chromatogr.* **1994,** *17,* 1709.
84. Miller, L.; Bergeron, R. *J. Chromatogr.* **1993,** *648,* 381.
85. Dyas, A. M.; Robinson, M. L.; Fell, A. F. *J. Chromatogr. A* **1994,** *660,* 249.
86. Pirkle, W. H.; Welch, C. J. *J. Liq. Chromatogr.* **1991,** *14,* 2027.
87. Beeson, M. D.; Vigh, G. *J. Chromatogr.* **1993,** *634,* 197.
88. Lee, S. H.; Berthod, A.; Armstrong, D. W. *J. Chromatogr.* **1992,** *603,* 83.
89. Li, S; Purdy, W. C. *J. Chromatogr.* **1992,** *625,* 109.
90. Balmer, K.; Lagerström, P.-O.; Persson, B.-A. *J. Chromatogr.* **1992,** *592,* 331.
91. Okamoto, M.; Nakazawa, H. *J. Chromatogr.* **1991,** *588,* 177.
92. Maris, F. A.; Vervoort, R. J. M.; Hindriks, H. *J. Chromatogr.* **1991,** *547,* 45.
93. Hampe, T. R.; Schlüter, J.; Brandt, K. H.; Nagel, J.; Lamparter, E.; Blaschke, G. *J. Chromatogr.* **1993,** *634,* 205.

94. Cabrera, K.; Lubda, D.; Jung, G. Poster Presentation at the 3rd International Symposium on Chiral Discrimination, Tübingen, Germany, October 1992.

95. Cabrera, K.; Lubda, D. *J. Chromatogr. A* **1994,** *666,* 433.

96. Fell, A. F.; Noctor, T. A. G.; Mama, J. E.; Clark, B. J. *J. Chromatogr.* **1988,** *434,* 377.

97. Tucker, R. P.; Fell, A. F.; Berridge, J. C.; Coleman, M. W. *Chirality* **1992,** *4,* 316.

98. Wainer, I. W. *Trends Anal. Chem.* **1987,** *6,* 125.

99. Wainer, I. W. In *A Practical Guide to the Selection and Use of HPLC Chiral Stationary Phase;* Baker, J. T., Ed.; J. T. Baker Inc.: Memphis, TN, 1988.

100. Roussel, C; Piras, P. *Pure Appl. Chem.* **1993,** *65,* 235.

101. Koppenhoefer, B.; Graf, R.; Holzschuh, H.; Nothdurft, A.; Trettin, U.; Roussel, C.; Piras, P. *J. Chromatogr. A* **1994,** *666,* 557.

102. Stauffer, S. T.; Dessy, R. E. *J. Chromatogr. Sci.* **1994,** *32,* 228.

103. Francotte, E., submitted to *Chirality.*

104. Francotte, E.; Richert, P., Ciba (Basel), unpublished results.

105. Mannschreck, A.; Koller. H; Wernicke, R. *Kontakte Darmstadt* **1985,** 40.

106. Blaschke, G.; Markgraf, H. *Arch. Pharm.* **1984,** *317,* 465.

107. Schlögl, K.; Widhalm, M. *Chem. Ber.* **1982,** *115,* 3042.

108. Isaksson, R.; Rochester, J. *J. Org. Chem.* **1985,** *50,* 2519.

109. Blaschke, G.; Donow, F. *Chem. Ber.* **1975,** *108,* 1188.

110. Kinkel J. N.; Reichert, K; Knöll, P. *GIT Fachz. Lab. Suppl.,* March 1989, p 104; *Chromatographie* **1989,** 104.

111. Seebach, D.; Müller, S. G.; Gysel, U.; Zimmermann, J. *Helv. Chim. Acta* **1988,** *71,* 1303.

112. Davankov, V. A. *J. Chromatogr. A* **1994,** *666,* 55.

113. Jeanneret-Gris, G.; Soerensen, C.; Su, H.; Porret, J. *Chromatographia* **1989,** *28,* 337.

114. Camacho-Torralba, P. L.; Beeson, M. D.; Vigh, G.; Thompson, D. H. *J. Chromatogr.* **1993,** *646,* 259.

115. Farkas, G.; Irgens, L. H.; Quintero, G.; Beeson, M. D.; Al-Saeed, A.; Vigh, G. *J. Chromatogr.* **1993,** *645,* 67.

116. Negawa, M.; Shoji, F. *J. Chromatogr.* **1992,** *590,* 113.

117. Nicoud, R. M. *LC-GC Int.* **1992,** *5*(5), 43.

118. Ching, C. B.; Lim, B. G.; Lee, E. J. D.; Ng, S. C. *J. Chromatogr.* **1993,** *634,* 215.

119. Broughtonand, D. B.; Gembicki, S. A. *AIChE. Symp. Ser.* **1962,** *80,* 233.

120. Barker, P. E.; Ganestos, G. *Sep. Sci. Technol.* **1987,** *22,* 2011.

121. Hashimoto, K.; Adachi, S.; Shirai, Y.; Morishita, M. *Chromatogr. Sci.* **1993,** *61,* 273.

122. Kubota, K.; Hayashi, S. *J. Chromatogr. A* **1994,** *658,* 259.

123. Gattuso, M. J.; Makino, S. Poster Presentation at Chiral 94, Reston, VA, May 1994.

124. Nicoud, R. M.; Bailly, M.; Kinkel J. N.; Devant, R. M.; Hampe, T. R.; Küsters, E. Presentation at the European Meeting on Simulated Moving Bed, Nancy, France, December 1993.

125. Fuchs, G.; Nicoud, R. M.; Bailly, M. *Proceedings of the 9th Symposium on Preparative and Industrial Chromatography ("Prep 92");* Société Française de Chimie: Paris, 1992; p 395.

126. Küsters, E.; Gerber, G.; Antia, F. D. *Chromatographia* **1995,** *40,* 387.

127. Macaudière, P.; Caude, M.; Rosset, R.; Tambuté, A. *J. Chromatogr. Sci.* **1989,** *27,* 383.

128. Fuchs, G.; Doguet, L.; Perrut, M.; Tambuté, A.; Le Goff, P. Presentation at the 2nd Symposium on Supercritical Fluids, Boston, MA, May 1991; personal communication.

129. Berger, C.; Perrut, M. *J. Chromatogr.* **1990,** *505,* 37.

130. Fuchs, G.; Doguet, L.; Barth, D.; Perrut, M. *J. Chromatogr.* **1992,** *623,* 329.

131. Blum, A. M.; Kumar, M. L. Poster Presentation at the 4th International Symposium on Chiral Discrimination, Montreal, Canada, September 1993.

132. Perrut, M. *J. Chromatogr. A* **1994,** *658,* 293.
133. Macaudière, P.; Caude, M.; Rosset, R.; Tambuté, A. *J. Chromatogr. Sci.* **1989,** *27,* 583.
134. Lynam, K. G; Nicolas, E. C. *J. Pharmacol. Biochem. Anal.* **1993,** *11,* 1197.
135. Anton, K.; Eppinger, J.; Francotte, E. Poster Presentation at the Supercritical Chemistry Symposium, Nottingham, England, September 1993.
136. König, W. A. *J. High Res. Chromatogr.* **1993,** 312, 338, 569.
137. Maas, B.; Dietrich, A.; Mosland, A. *J. High Res. Chromatogr.* **1994,** 169.
138. Fuchs, G.; Perrut, M. *J. Chromatogr. A* **1994,** *658,* 437.
139. Staerk, D. U.; Shitangkoon, A.; Vigh, G. *J. Chromatogr. A* **1995,** *702,* 251.
140. Schurig, V.; Grosenick, H. *J. Chromatogr.* **1994,** *666,* 617.
141. Schurig, V.; Grosenick, H.; Green, B. S. *Angew. Chem.* **1993,** *105,* 1690.
142. Staerk, D. U.; Shitangkoon, A.; Vigh, G. *J. Chromatogr. A* **1994,** *663,* 79.
143. Staerk, D. U.; Shitangkoon, A.; Vigh, G. *J. Chromatogr. A* **1994,** *677,* 133.
144. Schurig, V.; Leyer, U. *Tetrahedron Asym.* **1990,** *1,* 865.
145. Lindström, M.; Torbjörn, N.; Roeraade, J. *J. Chromatogr.* **1990,** *513,* 315.
146. Hardt, I; König, W. A. *J. Chromatogr. A* **1994,** *666,* 611.
147. Pirkle, W. H.; Murray, P. G. *J. High Res. Chromatogr.* **1993,** 285.
148. Schlögl, K.; Widhalm, M. *Monatsh. Chem.* **1984,** *115,* 1113.
149. Meyer, A.; Schlögl, K.; Schwager, H. *Monatsh. Chem.* **1987,** *118,* 535.
150. Bruche, G.; Mosland, A.; Kinkel J. N. *J. High Res. Chromatogr.* **1993,** *16,* 254.
151. Werner, A. *Kontakte Darmstadt* **1989,** 50.
152. Drabek, J.; Bourgeois, F.; Francotte, E.; Lohmann, D. Poster Presentation at the Symposium on Advances in the Chemistry of Insect Control, Oxford, England, July 1989.
153. Huxley-Tencer, A.; Francotte, E.; Bladocha-Moreau, M. *Pestic. Sci.* **1992,** *34,* 65.
154. Dingenen, J.; Kinkel J. N. *J. Chromatogr. A* **1994,** *666,* 627.
155. Jacobson, S. C.; Guiochon, G. *J. Chromatogr.* **1992,** *590,* 119.
156. Seidel-Morgenstern, A.; Guiochon, G. *AIChE J.* **1993,** *39,* 809.
157. Heuer, C.; Seidel-Morgenstern, A.; Hugo, P. *Chem. Eng. Sci.* **1995,** *50,* 1115.
158. Lohmann, D.; Francotte, E.; Auer, K., Ciba (Basel), unpublished results.
159. Francotte, E.; Auer, K.; Huxley, A.; Eckhardt, W. Poster Presentation at the 3rd International Symposium on Chiral Discrimination, Tübingen, Germany, October 1992.
160. Eckhardt, W.; Francotte, E.; Herzog, J.; Margot, P.; Rihs, G.; Kunz, W. *Pestic. Sci.* **1992,** *36,* 223.
161. Pirkle, W. H.; Pochapsky, T. C.; Mahler, G. S.; Corey, D. E.; Reno, D. S.; Alessi, D. M. *J. Org. Chem.* **1986,** *51,* 4991.
162. Arlt, D.; Börner, B.; Grosser, R.; Lange, W. Poster Presentation at the 2nd International Symposium on Chiral Discrimination, Rome, Italy, May 1991.
163. Francotte, E., Ciba (Basel), unpublished results.
164. Pirkle, W. H.; Finn, J. M. *J. Org. Chem.* **1982,** *47,* 4037.
165. Francotte, E.; Lohmann, D. *Helv. Chim. Acta* **1987,** *70,* 1569.
166. Francotte, E.; Lang, R. W.; Winkler, T. *Chirality* **1991,** *3,* 177.
167. Oehrlein, R.; Jeschke, R.; Ernst, B.; Bellus, D. *Tetrahedron Lett.* **1989,** *30,* 3517.
168. Pirkle, W. H.; Hamper, B. C. In *Preparative Liquid Chromatography;* Bidlingmeyer, B. A., Ed.; Elsevier: Amsterdam, Netherlands, 1987; pp 235-287.
169. Pirkle, W. H.; Sowin, T. J. *J. Chromatogr.* **1987,** *396,* 83.
170. Pirkle, W. H.; McCune, J. E. *J. Chromatogr.* **1988,** *441,* 311.
171. Blaschke, G.; Hilgard, P.; Maibaum, J.; Niemeyer, U. *J. Pohl, Arzneim. Forsch.* **1986,** *36,* 1493.
172. Blaschke, G.; Maibaum, J. *J. Chromatogr.* **1986,** *366,* 329.

173. Pirkle, W. H.; Tsipouras, A.; Sowin, T. J. *J. Chromatogr.* **1985,** *319,* 392.
174. Pirkle, W. H.; Tsipouras, A. *J. Chromatogr.* **1984,** *291,* 291.
175. Droux, S.; Gouraud, Y.; Gigliotti, G.; Petit, F. Presentation at the 4th International Symposium on Chiral Discrimination, Montreal, Canada, September 1993.
176. Davankov, V. A.; Zolotarev, Y. A.; Kurganov, A. A. *J. Liquid Chromatogr.* **1979,** *2,* 1191.
177. Dingenen, J. Presentation at the 19th International Symposium on Column Liquid Chromatography (HPLC 95), Innsbruck, Austria, June 1995.

11

Enantioselective Transport Through Liquid Membranes

L. Jonathan Brice and William H. Pirkle

This chapter is devoted to membrane systems used for enantioselective transport of chiral molecules. Basic treatments of membrane theory, morphology, and transport considerations are given. Following this, different types of enantioselective transport systems, grouped by structural class of organic carrier molecule, are discussed.

Membranes

A membrane, defined as any semipermeable barrier between two phases, may either restrict movement of solutes across the phase boundary or can prevent or regulate the movement of solutes between two separate phases (*1*). Although the best-known membranes (e.g., cell walls) are solid, a liquid can serve as a membrane between two other liquid phases with which it is immiscible. Liquid membranes may be either unsupported or supported.

Unsupported Liquid Membranes

Membranes are termed *unsupported* when they lack mechanical support. Such a system consists of two liquid phases separated by the membrane, which might be

3407–8/96/0309$16.50/0

a prefabricated emulsion (emulsion membrane) or an immiscible liquid phase simultaneously in contact with the two otherwise miscible phases (bulk-liquid membrane). In an emulsion membrane system, one of the two immiscible phases is stirred with a surfactant to generate an emulsion, which is then mixed with another liquid phase containing the material to be extracted. The phases are separated, and a deemulsification step is required to recover the emulsifying agents and any other previous material present in the original phase. The emulsification and deemulsification steps make these membrane systems more complex than bulk-liquid membrane systems, and they have not been used for enantioselective transport processes. Simple bulk-liquid membranes have been used in most published examples of enantioselective transport. Usually, but not exclusively, chloroform is in simultaneous contact with two physically separated aqueous phases.

Supported (Immobilized) Liquid Membranes

Planar porous solid supports that have been impregnated with liquid have a small surface area to volume ratio, leading to slow transport across the supported membrane. In contrast, hollow-fiber membranes may have ratios of surface area to volume of 10,000 m^2/m^3. Hollow-fiber membranes are just beginning to be adapted to enantioselective transport, but they clearly have considerable potential.

Enantioselective Transport

One attractive feature of liquid-membrane separations is that the diffusion coefficient of a solute in a liquid is orders of magnitude higher than it would be in a solid. The rate at which a solute is transported across a membrane, termed *flux*, depends linearly on the diffusion coefficient of the solute in the membrane, the concentration of the solute, and the inverse of the thickness of the membrane. Often, the flux of a solute can be increased by adding a carrier molecule that interacts with the solute so as to facilitate its passage through the membrane (2). A carrier can alter the relative fluxes of two competing solutes—enantiomers in the present context. If the carrier is present in the membrane in nonracemic form, one diastereomeric solute–carrier complex may form preferentially, possibly leading to enantioselective transport across the membrane (Figure 1). The search for suitable chiral carriers has dominated research in enantioselective transport across liquid membranes.

Binaphthyl-Based Carriers

The first enantioselective liquid-membrane transport was reported by Cram. Noting that Pedersen (3) had observed that crown ether complexes of metal

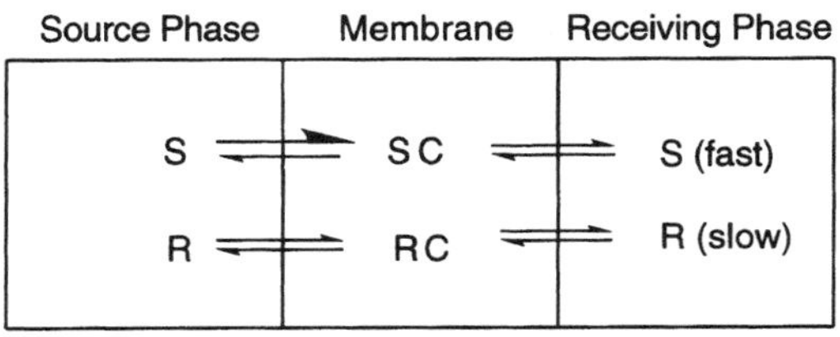

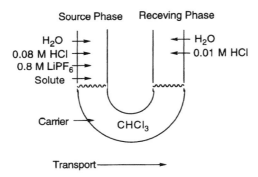

Figure 1. *Flux motif for enantioselective membrane transport.*

cations or ammonium ions are lipophilic, Cram and co-workers (*4*) were able to preferentially extract one enantiomer of protonated (±)-α-substituted benzylamines **4–6** from an aqueous solution into a chloroform phase containing chiral crown ethers **1–3** (Figure 2). The liquid-membrane apparatus used in the experiment was assembled by placing chloroform in the bottom of a U-tube and layering aqueous solutions above the organic phase in each arm of the tube (Figure 2). Data from the experiments, including the enantiomeric excess (ee) in the receiving phase, are presented in Table I.

Figure 2. *Cram's "single U-tube" experiment.*

Table I. Transport Data from Cram's "Single U-tube" Experiment

Run Time (h)	Carrier	Solute (Racemic)	Initial Source Phase Conc. (M)	Preferentially Transported Isomer	Receiving Phase ee (%)	k_A/k_B
133	none	4·HBr	0.05	none	0	1
45	(S,S)-1	5·HCl	0.21	(S)	35	2.2
19	(R,R)-2	5·HCl	0.28	(R)	78	9
12	(R,R)-3	5·HCl	0.28	(R)	82	12
32	(R,R)-3	5·HCl	0.28	(R)	77	8
66	none	5·HCl	0.28	none	0	1
182	(R,R)-2	6·HCl	0.28	(R)	85	14
182	none	6·HCl	0.28	none	0	1

NOTE: Carrier concentration was 0.027 M for all runs. Values for k_A (the rate constant for the preferentially transported isomer) and k_B (the rate constant for the other isomer) were reported after about 10% of the solute was transported.

Inspection of Table I reveals a number of features that pertain to all examples of enantioselective transport cited in this chapter. Ideally, transport should not occur in the absence of the chiral carrier. Maximum enantioenrichment in the receiving phase requires that the background rate of unassisted flux (equal for both enantiomers) be as small as possible. In Table I, the relative rates of transport of the enantiomers k_A/k_B have been calculated from the ee of the transported material after about 10% of the solute has been collected in the receiving phase. The entire system will ultimately come to equilibrium with an essentially racemic mix of solute in each aqueous phase. It is during the early stages of transport that one finds the greatest enantioenrichment. For most membrane resolutions, the enantioselectivities reported are those observed at extremely low transport levels (3% to 5% transport is typical).

To increase flux, Cram used the lipophilic anion from PF_6^- in the cationic ammonium–carrier complex. This anion has been used in almost all membrane resolutions of ammonium ions since reported. Different concentrations of hydrochloric acid on each side of the membrane were used to create a pH gradient to help sequester solute in the receiving phase and keep the entire transport system in the kinetic regime by minimizing back-transport from the receiving phase. In addition, the proton gradient helped drive the forward transport process by coupling it to the transport of protons from the receiving phase to the source phase, much as in active transport across cellular membranes.

From the data in Table I, the structure–selectivity relationship immediately becomes apparent. For the unfunctionalized carrier **1**, the degree of enantioenrichment is lower than that for either of the substituted carriers **2** and **3**. The highest ee is obtained with carrier **2** and solute **6**. Cram's explanation is that, in the crown ether–ammonium salt complex, substitution on the binaphthyl groups increases steric congestion in those areas and promotes a more selective solute–carrier recognition event (Figure 3). This model has also been used to ex-

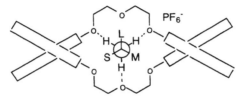

Unsubstituted Carrier

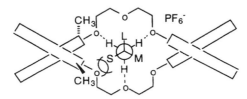

Substituted Carrier
(Me/S Steric Interaction)

Figure 3. Rationalization for the origin of enantioselectivity between crown ethers and solute ana-lytes. S, M, and L represent the (stereically) small-, medium-, and large sized groups appended to the stereogenic center.

plain the liquid–liquid chromatographic resolution of amino acid esters with crown ethers (*5*).

The logical conclusion of Cram's U-tube experiment is a machine that separates enantiomers by simultaneously transporting each from the source phase into two different receiving phases. Obviously, the machine requires an enantiopure sample of each enantiomer of the carrier. The architecture of such a "resolving machine" is displayed in Figure 4. The carrier and solute chosen were a pair showing a relatively high enantioselectivity in the U-tube experiments. The results are displayed in Table II (*6, 7*).

Conceptually, the resolving machine is simply two U-tube experiments run simultaneously, each using a different enantiomer of the carrier but feeding from the same source. Consequently, the source phase does not become enriched in the more slowly transported enantiomer, and the ratio of the fluxes of the solute enantiomers are the same. Owing to the design of the machine, the central aqueous phase must be stirred to prevent local depletion of the various components, which could result in slow transport or reduced enantioselectivity.

Yamaguchi (*8*) has reported the use of crown ether **7** (Figure 5) for the enantioselective transport of amino acids across a polymer-supported liquid membrane. Amino acids with bulky side chains evidence the greatest selectivity (*see* Table III). Changing the source-phase anion from PF_6^- to $HClO_4^-$ led to a higher

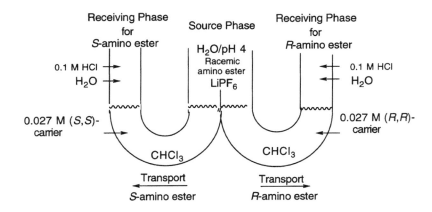

Figure 4. Cram's "double U-tube" experiment. Such a system could be developed into a systme that would continuously resolve a racemate.

Table II. Transport Data from Cram's "Double U-tube" Experiment

| | Carrier | | Amino Ester Delivered to Receiving Arms | | | | | |
| | | | Right | | | Left | | |
Time (h)	Right Arm	Left Arm	% Delivered	ee (%)	R/S	% Delivered	ee (%)	S/R
25	(R,R)	(S,S)	8.8	77	7.8	10	70	5.6
25	(R,R)	(S,S)	5.7	77	7.5	5.4	79	8.5
14.5	(R,R)	(S,S)	1.1	84	11	0.7	81	9.4
22	(R,R)	(S,S)	1.2	90	19	0.6	86	13

(R)-**7**   *Figure 5. A binaphthyl crown ether.*

selectivity but a much reduced flux across the membrane. In a final experiment, racemic phenylglycine was pumped against its concentration gradient for an extended time, and the receiving phase was found to be enriched in the D-enantiomer (the ratio was 3.3).

Active Transport and Antiport Ions

In most of the examples in this chapter, the racemic solute contains an ammonium ion and the receiving phase is at a slightly lower pH than the source phase. This pH gradient can be viewed simplistically as a way to minimize the back-transfer of solute from the receiving phase to the membrane, but other considerations linked to components in the receiving phase affect transport rates and selectivities. Lehn (9) has shown that the chemical potential formed by a KCl gradient provides enough driving force to transport tyrosine through a bulk toluene membrane against its concentration gradient, albeit at a reduced rate, in general, compared to membrane transport that is concentration driven (Figure 6). Although this is not an instance of enantioselective transport, such effects could influence enantioselectivity were this the goal.

In addition to Cl⁻, other counterions (antiport anions) may be used in these membrane systems and often affect both enantioselectivity and rate of trans-

Table III. Data for Enantioselective Transport of Solutes Using 7 as the Carrier

Amino Acid	Time (h)	k_A	k_B	k_A/k_B
Phenylglycine	12.7	5.97	0.26	22.7
Tryptophan	12.7	4.66	0.69	6.7
Phenylalanine	12.7	3.99	0.55	7.2
Leucine	25.5	3.87	0.29	13.5
Isoleucine	12.0	1.11	0.07	15.0
tert-Leucine	25.5	0.42	0.03	13.4
Methionine	25.3	3.40	0.32	11.2
Valine	25.5	0.75	0.10	7.7
Tyrosine	47.0	0.26	0.04	6.0

Nore: k_A and k_B values are in units of 10^{-7} mol cm^{-2} h^{-1}.

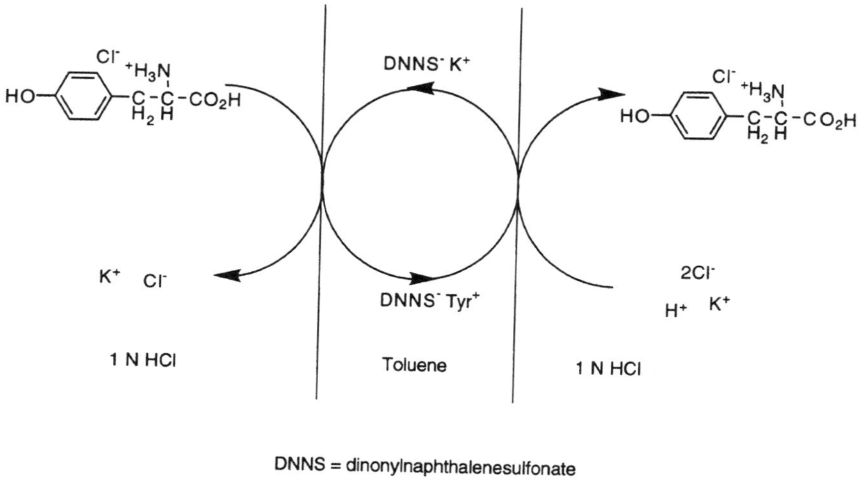

DNNS = dinonylnaphthalenesulfonate

Figure 6. Lehn's active transport system.

port. The anion of mandelic acid has been transported through a chloroform membrane, and a variety of acids, initially present in the receiving phase, have been back-transported as carboxylate anions. With (−)-N-(1-naphthyl)methyl-α-methylbenzylamine as a carrier, the rate of transport was found to be highest when antiport anions similar to mandelic acid in lipophilicity were used. Which enantiomer of mandelic acid is selectively transported depends on the antiport anion. In all cases, ee values of less than 10% were observed (*10*).

Biphenanthryl Crown Ethers

In an extension of the bisphenylbinaphthyl generation of membrane carriers, biphenanthryl carriers have been synthesized and their transport characteristics examined. Yamamoto (*11*) initially used biphenanthryl crown ethers **8** and **9** (Figure 7) to transport three optically active amino hydrochlorides across a bulk chloroform membrane (Table IV). Structurally analogous to the bisphenylbinaphthyl carriers (e.g., crown ether **7**), these ethers rely on additional aromatic rings to increase the steric constraints.

The 3-phenanthryl crown ethers **10** and **11** (Figure 8) were used to transport three different chiral ammonium species across a bulk chloroform membrane (Table V). These crown ethers do not have flanking substituents, as in the previous case, but the dihedral angle between the phenanthryl units may be increased because of the helicene-like backbone (*12*).

In a novel modification of carrier structure, the crown ether backbone of **8** was shortened by three atoms to give the biphenanthryl crown ether **12** (Figure

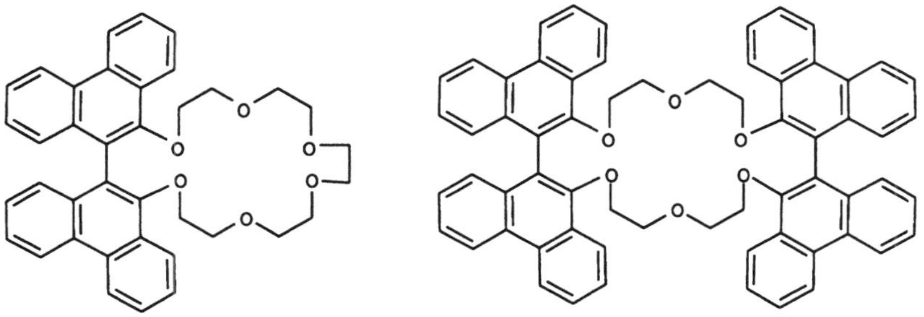

Figure 7. Biphenanthryl crown ethers.

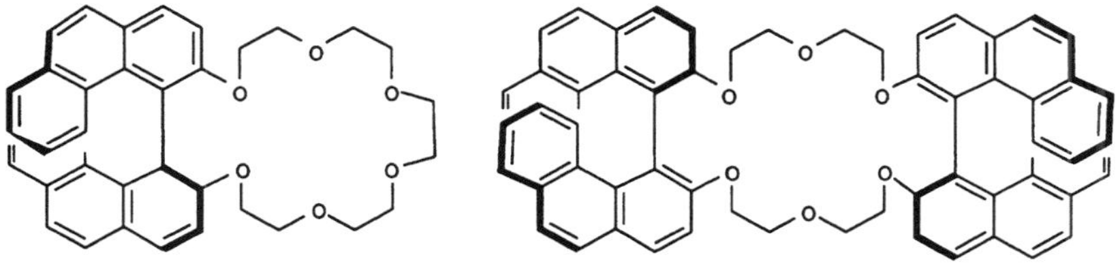

Figure 8.   3-Phenanthryl crown ether. Note helicene-like backbone.

Table IV. Data for Enantioselective Transport of Solutes Using 8 and 9 as the Carriers

Carrier	Solute[a]	Temp (°C)	Time (h)	Transport (%)	Dominant Isomer	ee (%)
(S)-8	MPG	19	1	2.8	(R)	24
(S)-8	PEA	19	0.5	3.7	(S)	31
(S)-8	PEA	−5	0.5	2.5	(S)	45
(S)-8	DPEA	19	0.5	2.5	(S)	32
(S)-8	DPEA	−5	0.5	3.8	(S)	78
(R,R)-9	DPEA	19	24	2.6	(S)	19
(R,R)-9	PEA	19	12	3.2	(R)	21
(R,R)-9	DPEA	19	12	3.0	(R)	20
(R,R)-9	DPEA	−5	12	3.2	(R)	23

[a]All solutes were racemic. DPEA is 1,2-diphenylethylamine hydrochloride; MPG is methyl phenylglycinate hydrochloride; PEA is 1-phenylethylamine hydrochloride.

Table V. Data for Enantioselective Transport of Solutes Using 10 and 11 as the Carriers

Carrier	Solute[a]	Time (h)	Transport (%)	Dominant Isomer	ee (%)
(S)-10	MPG	0.5	1.4	(S)	21
(S)-10	DPEA	0.5	1.6	(S)	66
(S)-10	AT	0.5	2.5	(R)	74
(R,R)-11	MPG	24	2.7	(R)	25
(R,R)-11	DPEA	0.5	1.4	(R)	35
(R,R)-11	AT	0.5	2.1	(S)	42

[a]All solutes were racemic. AT is 2-aminotetralin hydrochloride; DPEA is 1,2-diphenylethylamine hydrochloride; MPG is methyl phenylglycinate hydrochloride.

9), which shows increased enantioselectivity in the transport of a previously examined series of solutes across a chloroform membrane (*13*).

The most structurally complex biphenlanthryl crown ether yet studied is **13**, a bis crown ether with a naphthyl spacer between the phenanthryl subunits (Figure 10). This carrier was shown to transport (±)-1,6-diphenyl-1,6-hexanediamine with 82% ee, versus 26% for **8** (*14*).

Helicene-Based Crown Ethers

Crown ethers that incorporate a helicene moiety into the ether backbone have been successfully employed as carriers in liquid-membrane systems. Crown ethers **14** and **15** (Figure 11) have been shown to transport a series of amine hydrochlorides across a chloroform membrane (Table VI) (*15*).

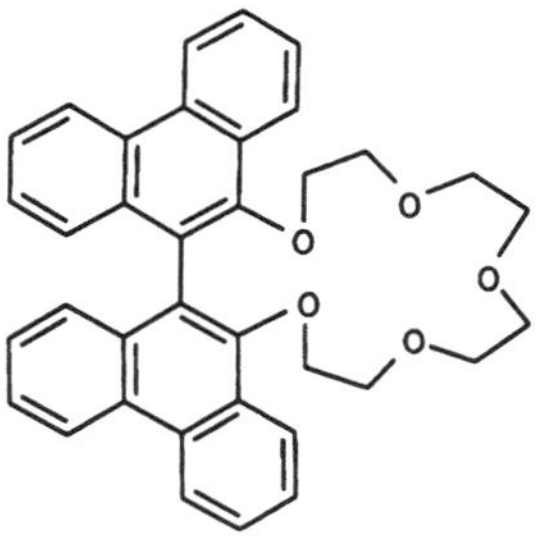

12

Figure 9. A biphenanthryl crown ether with a shortened tether.

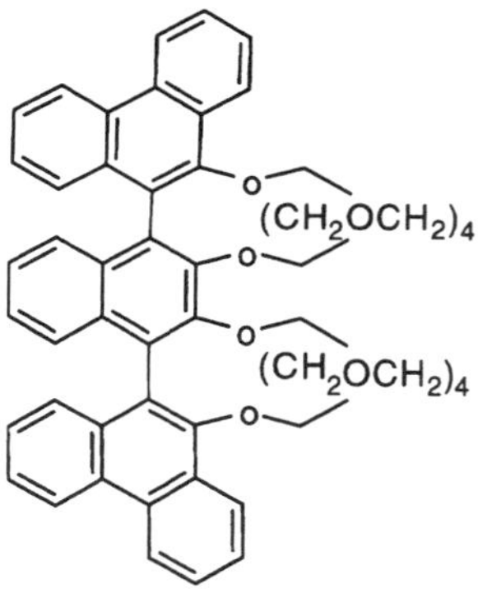

13

Figure 10. A bis-biphenyl crown ether with a shortened tether. In some cases, this carrier shows superior enantioselectivity compared to other biphenanthryl crown ethers.

Table VI. Data for Enantiomeric Transport of Solutes Using 14 and 15 as the Carriers

Carrier	Solute[a]	Time (h)	Transport (%)	Dominant Isomer	ee (%)
(−)-14	MPG	6	6	(S)	75
(+)-14	MPG	6	6	(R)	77
(−)-14	PEA	5	12	(R)	26
(+)-14	PEA	5	11	(S)	29
(−)-15	MPG	6	2	(R)	26
(+)-15	MPG	6	2	(S)	28
(−)-15	PEA	4	5	(S)	20
(+)-15	PEA	4	5	(R)	18

[a]All solutes were racemic. MPG is methyl phenylglycinate hydrochloride; PEA is 1-phenylethylamine hydrochloride.

Moderate enantioselectivities have been observed for the circulohelicene **16** (Figure 12 and Table VII) (*16*). However, the absence of groups flanking the biphenyl portion of the carrier is presumably the reason it is less effective than either **14** or **15**.

Crown Ethers Based on the 2,3,6,7-Dibenzobicyclo[3.3.1]nona-2,6-diene Subunit

Numerous enantioselective membrane transport systems have been reported that incorporate chiral moieties based on a 2,3,6,7-dibenzobicyclo[3.3.1]nona-2,6-diene subunit. Both mono and bis structures have been investigated, with crown ethers **17** and **21** (Figure 13) showing the greatest selectivity (Table VIII). Inter-

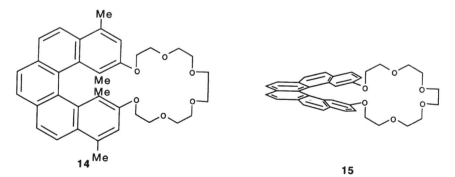

Figure 11. Helicene derived crown ethers.

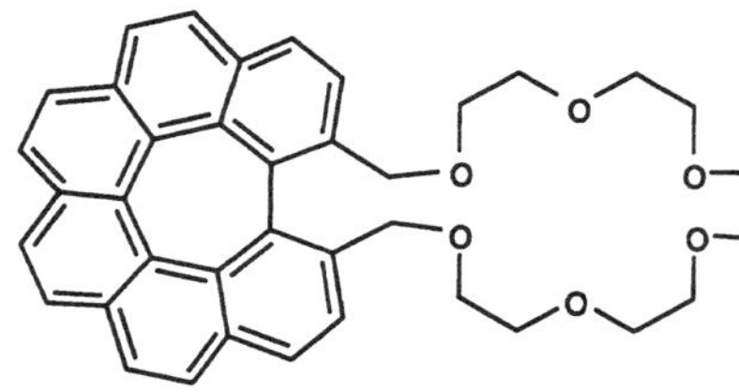

Figure 12. Circulohelicene derived crown ethers.   **16**

Table VII. Data for Enantioselective Transport of Solutes Using 16 as the Carrier

Carrier	Solute[a]	Time (h)	Transport (%)	Dominant Isomer	ee (%)
(−)-16	MPG	10.0	5.8	(S)	28
(−)-16	PEA	4.5	6.0	(R)	23
(−)-16	DPEA	8.0	5.9	(R)	41
(+)-16	MPG	10.0	5.9	(R)	30
(+)-16	PEA	4.5	6.1	(S)	22
(+)-16	DPEA	8.0	6.0	(S)	42

[a]All solutes were racemic. DPEA is 1,2-diphenylethylamine hydrochloride; MPG is methyl phenylglycinate hydrochloride; PEA is 1-phenylethylamine hydrochloride.

Table VIII. Data for Enantioselective Transport of Solutes Using 17–22 as the Carrier

Carrier	Solute[a]	Time (h)	Transport (%)	Dominant Isomer	ee (%)
(−)-17	DPEA	3.5	14	(S)	74
(+)-18	DPEA	4.0	17	(S)	38
(−)-19	DPEA	1.5	11	(S)	84
(+)-22	DPEA	1.5	12	(S)	53
(+)-20	DPEA	3	14	(S)	32
(+)-21	DPEA	2.5	15	(S)	80
(−)-17	MPG	48	14	—	0
(+)-18	MPG	22	26	(R)	23
(−)-19	MPG	72	15	(S)	8
(+)-22	MPG	42	11	(R)	5
(+)-20	MPG	36	12	—	0

[a]All solutes were racemic. DPEA is 1,2-diphenylethylamine hydrochloride; MPG is methyl phenylglycinate hydrochloride.

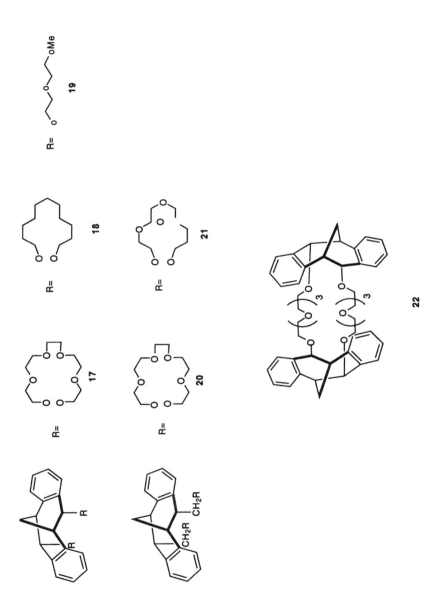

Figure 13.   Carriers based on the 2,3,6,7-dibenzo [3.3.1] nonane subunit.

estingly, the open-chain analogue **19** is also capable of enantioselective transport, unlike the open-chain analogues of the biphenyl-based crown ethers (*17*). Crown ethers **23** and **24** (Figure 14), both lacking flanking phenyl rings, are not as selective as their dibenzo analogue **20** in transport processes (Table IX) (*18*). Since these bicyclo[3.3.1]nonane skeletons afford enantioselective transport across membranes even when they are incorporated into linear polyethers instead of crown ethers, researchers have attempted to optimize these acyclic structures. Their enantioselectivity has been shown to depend on the size of the polyether side chains. An example of this dependence, using the homologous series of carriers **25–31** for the transport of (±)-1,2-diphenylethylamine hydrochloride across a chloroform membrane, is shown in Figure 15 (*19*).

Tetrahydrofuran-Based Ethers

As an alternative to ethers in which the chirality originates from an aromatic subunit, tetrahydrofuran-containing ethers **32–39** have been synthesized and their transport characteristics examined (Figure 16). As in the previous section, both cyclic and acyclic ethers were found to provide enantioselective transport, with extended acyclic ether **34** giving the highest selectivity. Crown ethers analogous

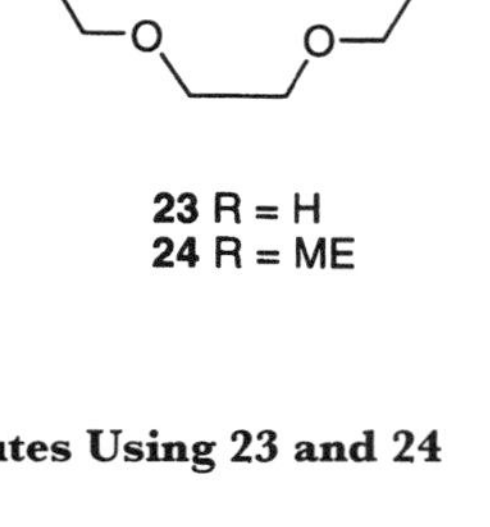

23 R = H
24 R = ME

Figure 14. Carriers based on the [3.3.1] nonane subunit.

Table IX. Data for Enantioselective Transport of Solutes Using 23 and 24 as the Carriers

Carrier	Solute[a]	Time (h)	Transport (%)	Dominant Isomer	ee (%)
(–)-23	DPEA	2.5	10.7	(S)	21
(–)-23	MPG	25.0	9.9	(R)	20
(+)-24	DPEA	3.0	10.8	(S)	24
(+)-24	MPG	24.0	9.3	(R)	8

[a]All solutes were racemic. DPEA is 1,2-diphenylethylamine hydrochloride; MPG is methyl phenyglycinate hydrochloride.

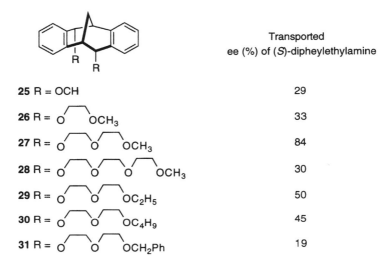

	Transported ee (%) of (S)-dipheylethylamine
25 R = OCH	29
26 R = O⌒OCH₃	33
27 R = O⌒O⌒OCH₃	84
28 R = O⌒O⌒O⌒OCH₃	30
29 R = O⌒O⌒OC₂H₅	50
30 R = O⌒O⌒OC₄H₉	45
31 R = O⌒O⌒OCH₂Ph	19

Figure 15. 2,3,6,7-dibenzo [3.3.3] nonane carriers with acyclic ether moieties.

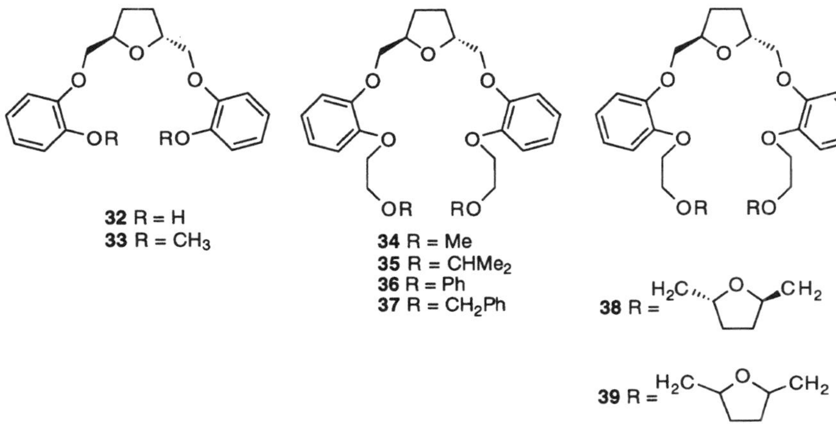

32 R = H
33 R = CH₃

34 R = Me
35 R = CHMe₂
36 R = Ph
37 R = CH₂Ph

38 R =

39 R =

Figure 16. Some tetrahydrofuran derived carriers.

to **34** but lacking the catechol moiety were investigated and found to provide no enantiodiscrimination in transport. A sample of the results reported is presented in Table X *(20)*.

The tetrahydrofuran subunit has been incorporated into a tricyclic kryptand-like structure to give carrier **40**, and its membrane transport properties have been investigated (Figure 17). The enantioselectivities for the transport of amine hydrochlorides across a bulk-liquid membrane are reported after 10% to 11% of the solute has been transported (Table XI). Most workers report enan-

Table X. Data for Enantioselective Transport of Solutes Using 33, 34, 38, and 39 as the Carriers

Carrier	Solute[a]	Time (h)	Transport (%)	Dominant Isomer	ee (%)
(+)-33	MPG	24	2.8	(S)	18
(+)-33	DPEA	2	9.4	(S)	30
(+)-33	AT	5	4.5	(R)	35
(+)-34	MPG	5	2.7	(S)	32
(+)-34	DPEA	2	10.0	(S)	51
(+)-34	AT	3	5.5	(R)	18
(+)-38	MPG	3	3.1	(R)	18
(+)-38	DPEA	3	10.5	(R)	28
(+)-38	AT	4	8.2	(S)	25
(+)-39	MPG	3	9.0	none	0
(+)-39	DPEA	2	8.5	(R)	5
(+)-39	AT	4	8.8	none	0

[a]All solutes were racemic. AT is 2-aminotetralin hydrochloride; DPEA is 1,2-diphenylethylamine hy-

tiomeric excesses determined earlier in the transport process, so the values may understate the kinetic preference (21).

Cyclohexanediol-Based Crown Ethers

The most recent class of ethers to be used as carriers in membrane transport systems relies on the *trans*-1,2-cyclohexanediol subunit for chirality (Figure 18). For crown ethers with one diol subunit, the *trans*-isomer appears to afford greater se-

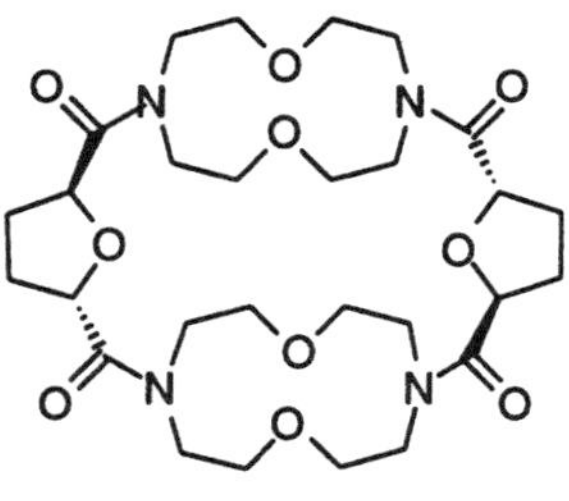

Figure 17. A Kryptand-like carrier.

Table XI. Data for Enantioselective Transport of Solutes Using 40 as the Carrier

Solute[a]	Time (h)	Transport (%)	Dominant Isomer	ee (%)
DPEA	3	10.1	(S)	46
AT	5	10.3	(R)	25
MPG	12	11.0	(R)	16

[a]All solutes were racemic. AT is 2-aminotetralin hydrochloride; DPEA is 1,2-diphenylethylamine hydrochloride; MPG is methyl phenylglycinate hydrochloride.

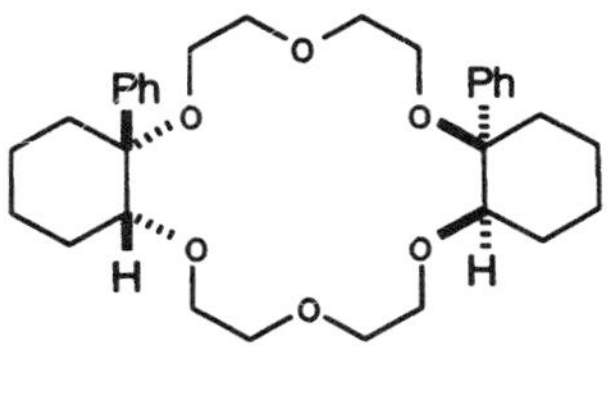

Figure 18.   1,2 cyclohexanediol derived carriers.

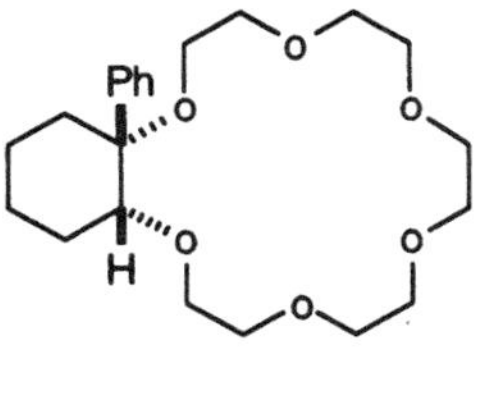

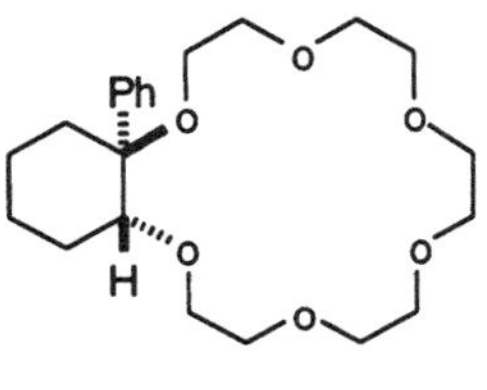

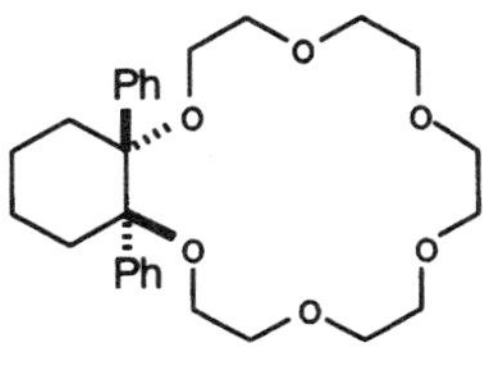

lectivity than the *cis*-isomer, but crown ether **41**, with two *cis*-diol subunits, performs as well as ethers with a *trans*-configuration (*22*) (Table XII).

Other Types of Ethers

Throughout the literature, there are isolated reports of structures that do not fall into any of the previous classes of chiral membrane carriers. Crown ethers **45** and **46** (Figure 19) have shown enantioselective membrane transport properties with amine hydrochlorides (Table XIII) (*23*).

Table XII.   Data for Enantioselective Transport of Solutes Using 41–44 as the Carriers

Carrier	Solute[a]	Time (h)	Transport (%)	Dominant Isomer	ee (%)
(–)-41	DPEA	3.0	9.3	(S)	38
(–)-41	MPG	8.5	10.2	(S)	3
(–)-42	DPEA	7.3	9.8	(S)	70
(–)-42	MPG	8.0	10.0	(S)	16
(–)-43	DPEA	4.8	9.6	(S)	81
(–)-43	MPG	6.0	9.9	(S)	14
(–)-44	DPEA	2.0	9.6	(S)	30
(–)-44	MPG	8.0	10.6	(S)	8

[a]All solutes were racemic. DPEA is 1,2-diphenylethylamine hydrochloride; MPG is methyl phenylglycinate hydrochloride.

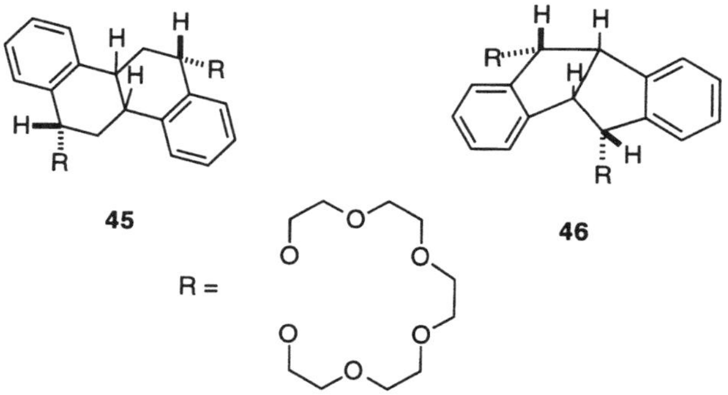

45   **46**

R =

Figure 19. Dibenzodecalin-derived carriers.

Table XIII. Data for Enantioselective Transport of Solutes Using 45 and 46 as the Carriers

Carrier	Solute[a]	Time (h)	Transport (%)	Dominant Isomer	ee (%)
(−)-45	DPEA	0.5	11.2	(S)	23
(−)-45	AT	2.0	9.5	(R)	10
(−)-45	MPG	17.0	10.0	(R)	7
(−)-46	DPEA	4.0	10.5	(S)	40
(−)-46	AT	9.0	9.8	(R)	22
(−)-46	MPG	3.4	10.0	(R)	15

[a]All solutes were racemic. AT is 2-aminotetralin hydrochloride; DPEA is 1,2-diphenylethylamine hydrochloride; MPG is methyl phenylglycinate hydrochloride.

Table XIV. Data for Enantioselective Transport of Solutes Using 47–51 as the Carriers

Carrier	Time (h)	Transport (%)	Dominant Isomer	ee (%)
(+)-47	2	5.9	(S)	5.9
(+)-48	22	5.5	(R)	5.4
(+)-49	4	5.5	(R)	13.0
(−)-50	70	4.7	(R)	4.0
(+)-51	90	4.9	(S)	8.0

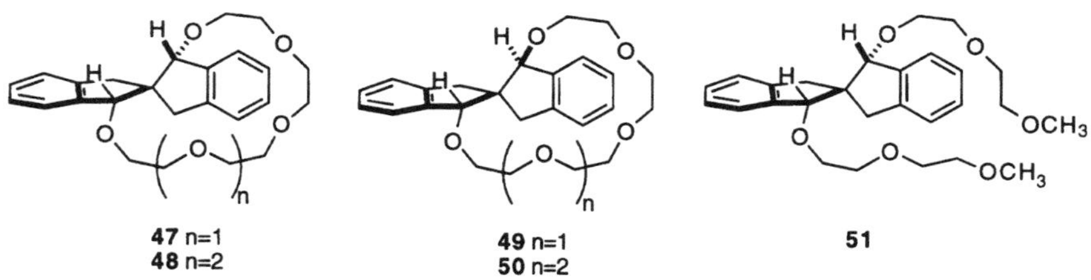

47 n=1
48 n=2

49 n=1
50 n=2

51

Spiro compounds **47–51** have been investigated in membrane transport of methyl phenylglycinate hydrochloride (Table XIV) (*24*). Like the tetrahydrofuran-based carriers, these spiro compounds demonstrate transport enantioselectively as either cyclic or acyclic polyethers.

The 1,2,4-triazole skeleton has been incorporated into cyclic ethers **52** and **53**, both of which have been used successfully as chiral centers in bulk-liquid membrane systems (*25*). Interestingly, changing the lipophilicity by modifying the group at the triazole 1-position drastically affects the transport selectivity for 1-(α-naphthyl)ethylamine but not for methyl phenylglycinate (Table XV).

For the acyclic triazole-containing polyethers, both triazole structure and

Table XV. Data for Enantioselective Transport of Solutes Using 52 and 53 as the Carriers

Carrier	Solute[a]	Time (h)	Dominant Isomer	ee (%)
52	NEA	5	(R)	25
52	MPG	26	(S)	24
53	NEA	3	(R)	9
53	MPG	3.3	(S)	23

[a]All solutes were racemic. MPG is methyl phenylglycinate hydrochloride; NEA is 1-(α-naphthyl)eth-

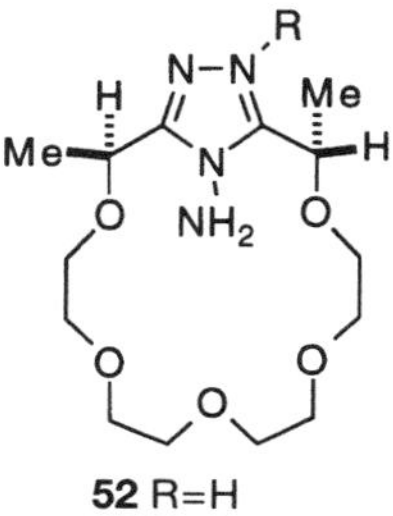

52 R=H
53 R=C$_{12}$H$_{25}$

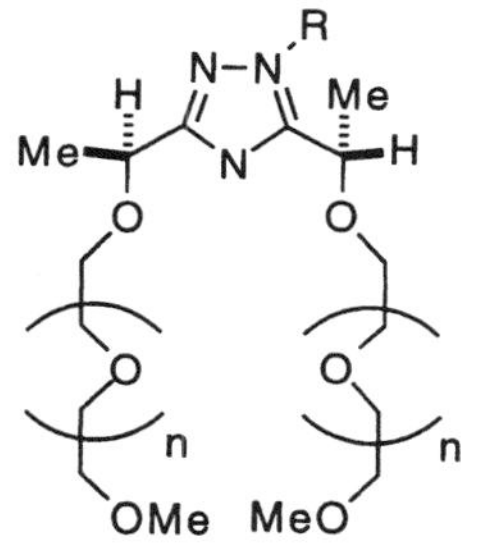

54

n = 1, 2

R = H, C$_{12}$H$_{25}$, CH$_2$CO$_2$cholesteryl

Figure 20. A 1,2,4-triazole carrier with acyclic ether moieties.

polyether chain length were varied (Figure 20). Neither variation changed the enantioselectivity much, but in all cases, the selectivities are inferior to those of the cyclic counterparts.

Other Carriers

Lasalocid A, a naturally occurring ionophore, has been evaluated as a chiral membrane carrier for transport of a series of chiral cobalt complexes through a bulk-liquid membrane (Table XVI) (*26*).

Using enantiodiscriminating interactions elucidated in a research program devoted to the development of chiral stationary phases for high-performance liquid chromatography, Pirkle and Doherty (*27*) have demonstrated that the octadecyl ester of *N*-(1-naphthyl)leucine (Figure 21) can serve as a chiral carrier in the selective membrane transport of a series of *N*-(3,5-dinitrobenzoyl) amino acid esters (Table XVII). Instead of a simple bulk-liquid membrane, the apparatus employed was a 3-m length of silicone rubber tubing coiled around two spools, one immersed in a source phase and one immersed in a receiving phase. A decane solution of the carrier was then circulated through the tubing. One solute enantiomer preferentially entered the tubing in the source phase, flowed downstream, and left the tubing in the receiving phase. The carrier solution was continuously recirculated.

Table XVI. Data for Enantioselective Transport of Solutes Using Lasalocid A as the Carriers

Solute	Source Phase Conc. (mM)	Flux (10^{-7} mol h^{1} cm^{-2})	ee (%)	Transport (%)
[Co(en)$_3$]Cl$_3$	20.0	5.9	10	4.3
μ-*cis*-[Co(dien)$_2$](ClO$_4$)$_3$	20.0	7.1	27	5.2
[Co(sep)]Cl$_3$	20.0	8.4	16	6.2

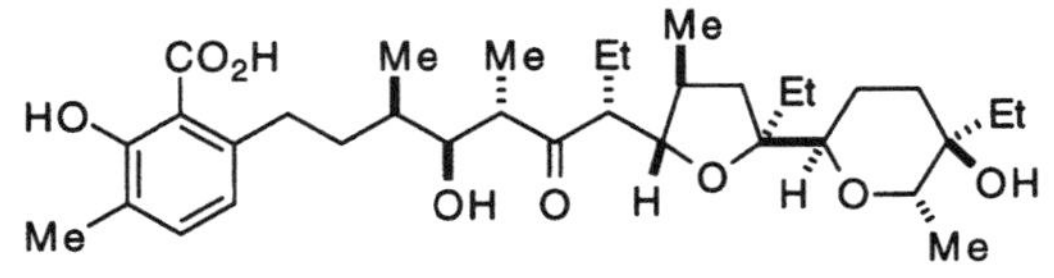

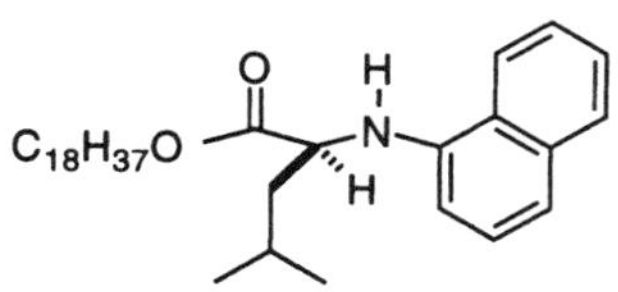

Figure 21. The octadecyl ester of N-(1-Naphthyl) leucine.

Table XVII. Data for Enantioselective Transport of Solutes Using the Octadecyl Ester of *N*-(1-Naphthyl) Leucine as the Carrier

Analyte (as Dinitrobenzoyl Derivative)	Rate ($\mu g/min$)		k_A/k_B	ee (%)
	(S)-Isomer	(R)-Isomer		
Alanine butyl ester	3.8	0.78	4.9	66
Valine butyl ester	6.1	1.9	3.3	53
α-Methylvaline butyl ester	4.4	1.7	2.6	44
Leucine methyl ester	5.0	0.71	7.2	76
Leucine butyl ester	6.4	1.6	3.9	59
Leucine octyl ester	21	12	1.7	26
Leucine butyl amide	2.1	0.28	7.6	77
Leucine butyl ester	10.6	1.5	6.9	75

Hollow-Fiber Membranes

In addition to the bulk-liquid membrane separations, related selectors have been employed in a conceptually similar apparatus that uses hollow-fiber membranes. Two membrane modules, each containing bundled polyacrylonitrile hollow fibers, are used (Figure 22). An aqueous phase containing the (initially) racemic solute passes over the fibers in the first module while a water-immiscible organic phase (in this case, hexane) containing the chiral selector is passed through the bores of the fibers. Solute transported into the hexane is circulated to and through the second module, where it is transported into a second aqueous receiving phase, thus extracting it from the hexane and freeing it from the selector.

With either ester or amide derivatives of *N*-(1-naphthyl)leucine as the chiral selector, the dinitrobenzoyl derivative of racemic leucine was used as a solute for

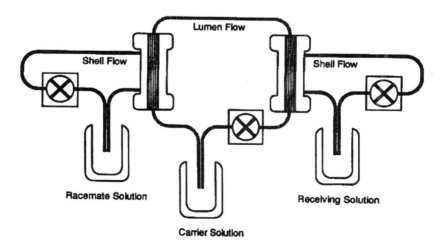

Figure 22. Architecture of a hollow fiber membrane apparatus.

this hollow-fiber liquid-membrane system. The ee of the transported material, initially 95%, decayed over time as thermodynamic equilibrium was approached. The flux across the membrane depended on the ion-pairing reagent used; the best results reported were obtained with tetrahexylammonium chloride (28). Using a second unit containing the other enantiomer of the selector so that each analyte enantiomer is transported from the source phase into a separate receiving phase (6, 7) should maintain the initial high enantioselectivity as long as the driving force for transport remains adequate. As described, this is basically a single-plate system and relies on a highly enantiodifferentiating selector for its success. Solute flux rates can be relatively high in such a system because high flow rates can be used. As in previous examples of membrane transport, the chiral carrier is contained in the organic phase, although there is no inherent reason that it must be.

Another report concerning hollow-fiber membrane systems deals with the transport of the underivatized amino acid leucine, with N-dodecylhydroxyproline as the carrier in an octanol phase (29). The architecture consists of a single hollow-fiber module, with organic and aqueous phases being run countercurrently. Measurements of ee in both source and receiving phases, made after a short equilibration period, indicate in all cases a near-complete separation of leucine enantiomers (Table XVIII), even though the enantioselectivity of the chiral selector is not high. The ability to use chiral selectors of relatively low enantioselectivity and still attain high levels of solute enrichment is a great advantage of the countercurrent approach. The drawback is the relatively low throughput owing to the need to use low countercurrent velocities for the fluids involved.

An alternative geometry of hollow-fiber modules for countercurrent separation of enantiomer has been described. Conceptually, the system seems to combine the ideas of earlier workers. It is reported to afford separations of enantiomers (e.g., the commercial pharmaceuticals shown in Table XIX) even though the enantioselectivities afforded by the chiral selectors (both enantiopure enantiomers are used, as in references 6 and 7) are quite modest, as indicated in Table XIX (30). The use of countercurrent systems to offset low enantioselectivities is analogous to the use of longer chromatography columns to afford a greater number of theoretical plates to compensate for low selectivity. As in chromatography, the disadvantages are increased time and decreased throughput.

Table XVIII. Enantioselective Transport in a Hollow Fiber Membrane System Using N-n-Dodecyl Hydroxyproline as the Carrier

Aqueous Flow (mL/min)	Aqueous Velocity (cm/s)	(+)-Leucine in Extract	(−)-Leucine in Extract
0.14	0.05	1.00	0.00
0.21	0.08	1.00	0.00
0.32	0.12	0.98	0.00
0.39	0.15	1.00	0.00
0.63	0.24	0.98	0.03
0.81	0.31	0.94	0.03

Table XIX. Hollow Fiber Membrane Separations Using Tartaric and Lactic Acid Derived Carriers

Analyte	Carrier[a]	Solvent	Transport Counterion	α
Phenylpropanolamine	0.025 M DHT	heptane	—	1.19
Phenylpropanolamine	0.025 M DHT	heptane	—	1.50
Phenylpropanolamine	10% wt PLA	chloroform	—	1.07
Phenylpropanolamine	10% wt PLA	chloroform	—	1.10
Ephedrine	0.10 M DHT	heptane	0.06 M PF_6^-	1.30
Mirtazapine	0.10 M DHT	heptane	—	1.05
Mirtazapine	0.25 M DHT	heptane	—	1.06
Mirtazapine	0.50 M DHT	heptane	—	1.08
Mirtazapine	0.25 M DBT	decanol	—	1.06
Mirtazapine	10% wt PLA	chloroform	—	1.04
Mirtazapine	0.10 M DHT	heptane	0.10 M PF_6^-	1.07
Mirtazapine	0.50 M DHT	heptane	0.03 M PF_6^-	1.16
Phenylglycine	0.25 M DHT	heptane	—	1.06
Albuterol	0.25 M DHT	heptane	0.018 M BPh_4^-	1.06
Albuterol	0.25 M DHT	cyclohexane	0.0035 M BPh_4^-	1.04
Terbutaline	0.43 M DHT	dichloroethane	0.05 M PF_6^-	1.14
Terbutaline	2.15 M DHT	heptane	0.0045 M BPh_4^-	1.05
Propranolol	0.25 M DHT	heptane	—	1.03
Ibuprofen	0.10 M DHT	heptane	—	1.10

NOTE: — means none.
[a]DBT is dibenzoyltartrate; DHT is dihexyltartrate; PLA is poly(lactic acid).

Cyclodextrins

Another class of carriers whose potential has been demonstrated is the cyclodextrins (*31*). An aqueous cyclodextrin solution is adsorbed onto a piece of filter paper placed between two organic phases to form the supported membrane. Alternatively, an aqueous cyclodextrin solution has been interposed between two organic phases (one the source phase and one the receiving phase) in a capillary tube for smaller scale separations. As Table XX shows, multiple membranes can be used in series to improve the enantiomeric ratio in the final receiving phase (membranes A and B).

Poly(amino acids)

Poly(amino acid) membranes have been used to separate amino acids (*32*). A membrane consisting of poly(L-glutamate) appended with amphiphilic nonylphenoxyoligo(oxyethylene) side chains (Figure 23) has shown good selectivity in the enantioselective transport of both tryptophan and tyrosine. In addition to the chirality inherent in the poly(amino acid), the backbone itself is thought to adopt an ordered helical structure that is important to enantiodiscrimination. Because

Table XX. Membrane Separations Using Cyclodextrins as Carriers

Compound (Racemic)	Membrane[a]	Dominant Isomer	k_A/k_B
rac-S-(1-Ferrocenylethyl)thiophenol	A	(+)	7.0
rac-1'-Benzylnornicotine	B	(+)	3.1
rac-2,2-Binaphthyldiyl-14-crown-4	C	(+)	2.0
rac-2,2-Binaphthyldiyl-20-crown-6	A	(S)	3.0
rac-Mephenytoin	D	(−)	2.0
rac-Disopyramide	D	(−)	1.4

[a]A is three β-cyclodextrin capillary membranes in series; B is two β-cyclodextrin capillary membranes in series; C is one α-cyclodextrin–maltosyl polymer capillary membrane; D is β-cyclodextrin capillary membrane.

Average $m = 6$
Backbone MW = 10000

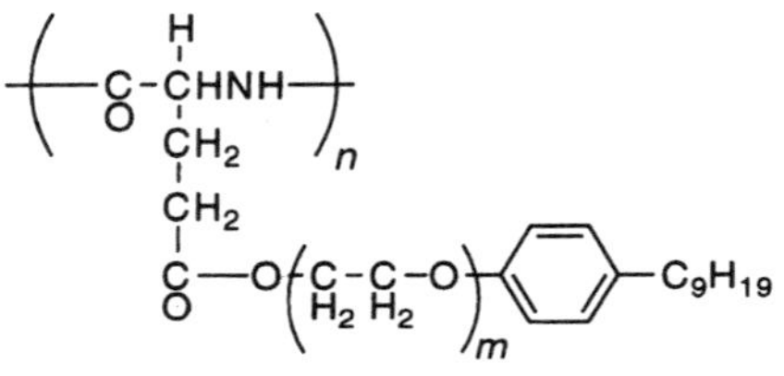

Figure 23. A polymeric membrane.

this helicity is temperature-dependent, varying the temperature of the membrane system allows fine-tuning of a separation.

Conclusion

With increasing understanding of the relationship between stereochemistry and bioactivity, the field of chiral separations takes on an ever more important role. While classical enantioseparation methods such as chemical resolution and chromatography have addressed some of these needs, the quest for novel and more efficient means of enantioseparation continues. Liquid-membrane separations are among the most attractive of these emerging technologies because they can be adapted into continuous-flow operations to increase throughput while reducing cost. While simple liquid membranes have been invaluable for their contributions to the early studies of the underlying principles of membrane dynamics and carrier structure requirements, the future of this field probably lies more toward supported membrane systems. These have much higher surface area and, because of shorter path lengths, better transport kinetics. Undoubtedly, there is a need for chiral selectors that afford high levels of selectivity for chiral compounds of economic importance. Commercial considerations aside, this field of separation science is still relatively unexplored and undoubtedly represents a fecund area for research.

References

1. *Liquid Membranes: Theory and Application;* Noble, D. L.; Way, J. D., Eds.; ACS Symposium Series 347; American Chemical Society: Washington, DC, 1986; Chapter 1, p 1.
2. *Liquid Membranes: Theory and Application;* Noble, D. L.; Way, J. D., Eds.; ACS Symposium Series 347; American Chemical Society: Washington, DC, 1986; Chapter 1, p 4.
3. Pedersen, C. D. *J. Am. Chem. Soc.* **1967,** *89*, 7017.
4. Newcomb, M.; Helgeson, R. C.; Cram, D. J. *J. Am. Chem. Soc.* **1974,** *96*, 7367.
5. Sousa, L. R.; Hoffman, D. H.; Kaplan, L.; Cram, D. J. *J. Am. Chem. Soc.* **1974,** *96*, 7100.
6. Newcomb, M.; Toner, J. L.; Helgeson, R. C.; Cram, D. J. *J. Am. Chem. Soc.* **1979,** *101*, 4941.
7. An earlier report of enantioselective liquid-liquid chromatography using crown ethers presages the formal liquid membrane work. *See* Cram, D. J.; Cram, J. M. *Science (Washington, D.C.)* **1974,** *183*, 803.
8. Yamaguchi, T.; Nishimura, K.; Shinbo, T.; Sugiura, M. *Chem. Lett.* **1985,** 1549.
9. Lehn, J. M. *Pure Appl. Chem.* **1979,** *51*, 979.
10. Lehn, J. M.; Moradpour, A.; Behr, J. P. *J. Am. Chem. Soc.* **1975,** *97*, 2532.
11. Yamamoto, K.; Fukushima, H.; Okamoto, Y.; Hatada, K.; Nakazaki, M. *J. Chem. Soc. Chem. Comm.* **1984,** 1111.
12. Yamamoto, K.; Noda, K.; Okamoto, Y. *J. Chem. Soc. Chem. Comm.* **1985,** 1065.
13. Yamamoto, K.; Kitsuki, T.; Okamoto, Y.; *Bull. Chem. Soc. Jpn.* **1986,** *59*, 1269.
14. Yamamoto, K.; Yumioka, H.; Okamoto, Y.; Chikamatsu, H. *J. Chem. Soc. Chem. Comm.* **1987,** 168.
15. Nakazaki, M.; Yamamoto, K.; Ikeda, T.; Kitsuki, T.; Okamoto, Y. *J. Chem. Soc. Chem. Comm.* **1983,** 787.
16. Yamamoto, K.; Ikeda, T.; Kitsuki, T.; Okamoto, Y.; Chikamatsu, H.; Nakazaki, M. *J. Chem. Soc. Chem. Comm.* **1990,** 271.
17. Naemura, K.; Fukunaga, R.; Yamanaka, M.; *J. Chem. Soc. Chem. Comm.* **1985,** 1560.
18. Naemura, K.; Matsumura, T.; Masanori, K. *J. Chem. Soc. Chem. Comm.* **1988,** 239.
19. Naemura, K.; Fukunaga, R.; Komatsu, M.; Yamanaka, M.; Chikamatsu, H. *Bull. Chem. Soc. Jpn.* **1989,** *62*, 83.
20. Naemura, K.; Ebashi, I.; Matsuda, A.; Chikamatsu, H. *J. Chem. Soc. Chem. Comm.* **1986,** 666.
21. Naemura, K.; Kanda, Y.; Iwasaka, H.; Chikamatsu, H. *Bull. Chem. Soc. Jpn.* **1987,** *60*, 1789.
22. Naemura, K.; Miyabe, H.; Shingai, Y. *J. Chem. Soc. Perkins Trans. I* **1991,** 957; Naemura, K.; Miyabe, H.; Shingai, Y.; Tobe, Y. *J. Chem. Soc. Perkins Trans. I* **1991,** 1073.
23. Naemura, K.; Komatsu, M.; Adachi, K.; Chikamatsu, H. *J. Chem. Soc. Chem. Comm.* **1986,** 1675.
24. Naemura, K.; Ebashi, I.; Nakazaki, M. *Bull. Chem. Soc. Jpn.* **1985,** *58*, 767.
25. Echegoyen, L.; Martinez-Diaz, M.; de Mendoza, J.; Torres, T.; Vicente-Arana, M. *Tetrahedron* **1992,** *43*, 9545.
26. Chia, P.; Lindoy, L.; Walker, G.; Everett, G. *J. Am. Chem. Soc.* **1991,** *113*, 2533.
27. Pirkle, W.; Doherty, E. *J. Am. Chem. Soc.* **1989,** *111*, 4113.

28. Pirkle, W.; Bowen, W. *Tetrahedron Asym.* **1994,** *5,* 773.
29. Ding, H.; Carr, P.; Cussler, E. *AIChE J.* **1992,** *38,* 1493.
30. Keurentjes, J. T.; Nabuurs, L. J.; Vegter, E. A.; Aerts, J. *Chemical Specialties USA 1994 Symposium;* p 41.
31. Armstrong, D.; Jin, H. *Anal. Chem.* **1987,** *59,* 2237.
32. Maruyama, A.; Adachi, N.; Takatsuki, T.; Torii, M.; Sanui, K.; Ogata, N. *Macromolecules* **1990,** *23,* 2748.

Index

Highlights from ACS Books

Good Laboratory Practice Standards: Applications for Field and Laboratory Studies
Edited by Willa Y. Garner, Maureen S. Barge, and James P. Ussary
ACS Professional Reference Book; 572 pp; clothbound ISBN 0–8412–2192–8

Silent Spring Revisited
Edited by Gino J. Marco, Robert M. Hollingworth, and William Durham
214 pp; clothbound ISBN 0–8412–0980–4; paperback ISBN 0–8412–0981–2

The Microkinetics of Heterogeneous Catalysis
By James A. Dumesic, Dale F. Rudd, Luis M. Aparicio, James E. Rekoske,
and Andrés A. Treviño
ACS Professional Reference Book; 316 pp; clothbound ISBN 0–8412–2214–2

Helping Your Child Learn Science
By Nancy Paulu with Margery Martin; Illustrated by Margaret Scott
58 pp; paperback ISBN 0–8412–2626–1

Handbook of Chemical Property Estimation Methods
By Warren J. Lyman, William F. Reehl, and David H. Rosenblatt
960 pp; clothbound ISBN 0–8412–1761–0

Understanding Chemical Patents: A Guide for the Inventor
By John T. Maynard and Howard M. Peters
184 pp; clothbound ISBN 0–8412–1997–4; paperback ISBN 0–8412–1998–2

Spectroscopy of Polymers
By Jack L. Koenig
ACS Professional Reference Book; 328 pp;
clothbound ISBN 0–8412–1904–4; paperback ISBN 0–8412–1924–9

Harnessing Biotechnology for the 21st Century
Edited by Michael R. Ladisch and Arindam Bose
Conference Proceedings Series; 612 pp;
clothbound ISBN 0–8412–2477–3

From Caveman to Chemist: Circumstances and Achievements
By Hugh W. Salzberg
300 pp; clothbound ISBN 0–8412–1786–6; paperback ISBN 0–8412–1787–4

The Green Flame: Surviving Government Secrecy
By Andrew Dequasie
300 pp; clothbound ISBN 0–8412–1857–9

For further information and a free catalog of ACS books, contact:
American Chemical Society
Distribution Office, Department 225
1155 16th Street, NW, Washington, DC 20036
Telephone 800–227–5558